AF361813

Prospects from Diagnosis to Treatment in Cancer Using Magnetic Methods

Prospects from Diagnosis to Treatment in Cancer Using Magnetic Methods

Editors

Moriaki Kusakabe
Akihiro Kuwahata

MDPI • Basel • Beijing • Wuhan • Barcelona • Belgrade • Manchester • Tokyo • Cluj • Tianjin

Editors
Moriaki Kusakabe
University of Tokyo
Japan

Akihiro Kuwahata
University of Tokyo
Japan

Editorial Office
MDPI
St. Alban-Anlage 66
4052 Basel, Switzerland

This is a reprint of articles from the Special Issue published online in the open access journal *Cancers* (ISSN 2072-6694) (available at: https://www.mdpi.com/journal/cancers/special_issues/magnetic_diagnosis_methods_cancer).

For citation purposes, cite each article independently as indicated on the article page online and as indicated below:

LastName, A.A.; LastName, B.B.; LastName, C.C. Article Title. *Journal Name* **Year**, *Volume Number*, Page Range.

ISBN 978-3-0365-6884-3 (Hbk)
ISBN 978-3-0365-6885-0 (PDF)

Cover image courtesy of Moriaki Kusakabe.

Contents

Review

Understanding MNPs Behaviour in Response to AMF in Biological Milieus and the Effects at the Cellular Level: Implications for a Rational Design That Drives Magnetic Hyperthermia Therapy toward Clinical Implementation

David Egea-Benavente [1,†], Jesús G. Ovejero [2,3,†], María del Puerto Morales [2] and Domingo F. Barber [1,*]

1   Department of Immunology and Oncology, and NanoBiomedicine Initiative, Centro Nacional de Biotecnologia (CNB)-CSIC, Darwin 3, 28049 Madrid, Spain; degea@cnb.csic.es
2   Department of Energy, Environment and Health, Instituto de Ciencia de Materiales de Madrid (ICMM)-CSIC, Sor Juana Inés de la Cruz 3, 328049 Madrid, Spain; jesus.g.ovejero@csic.es (J.G.O.); puerto@icmm.csic.es (M.d.P.M.)
3   Department of Dosimetry and Radioprotection, Hospital General Universitario Gregorio Marañón, Doctor Esquerdo 46, 28007 Madrid, Spain
*   Correspondence: dfbarber@cnb.csic.es
†   David Egea-Benavente and Jesús G. Ovejero contributed equally to this work.

Citation: Egea-Benavente, D.; Ovejero, J.G.; Morales, M.d.P.; Barber, D.F. Understanding MNPs Behaviour in Response to AMF in Biological Milieus and the Effects at the Cellular Level: Implications for a Rational Design That Drives Magnetic Hyperthermia Therapy toward Clinical Implementation. *Cancers* **2021**, *13*, 4583. https://doi.org/10.3390/cancers13184583

Academic Editors: Moriaki Kusakabe and Akihiro Kuwahata

Received: 28 July 2021
Accepted: 7 September 2021
Published: 12 September 2021

Publisher's Note: MDPI stays neutral with regard to jurisdictional claims in published maps and institutional affiliations.

Simple Summary: Magnetic hyperthermia therapy is an alternative treatment for cancer that complements traditional therapies and that has shown great promise in recent years. In this review, we assess the current applications of this therapy in order to understand why its translation from the laboratory to the clinic has been less smooth than was anticipated, identifying the possible bottlenecks and proposing solutions to the problems encountered.

Abstract: Hyperthermia has emerged as a promising alternative to conventional cancer therapies and in fact, traditional hyperthermia is now commonly used in combination with chemotherapy or surgery during cancer treatment. Nevertheless, non-specific application of hyperthermia generates various undesirable side-effects, such that nano-magnetic hyperthermia has arisen a possible solution to this problem. This technique to induce hyperthermia is based on the intrinsic capacity of magnetic nanoparticles to accumulate in a given target area and to respond to alternating magnetic fields (AMFs) by releasing heat, based on different principles of physics. Unfortunately, the clinical implementation of nano-magnetic hyperthermia has not been fluid and few clinical trials have been carried out. In this review, we want to demonstrate the need for more systematic and basic research in this area, as many of the sub-cellular and molecular mechanisms associated with this approach remain unclear. As such, we shall consider here the biological effects that occur and why this theoretically well-designed nano-system fails in physiological conditions. Moreover, we will offer some guidelines that may help establish successful strategies through the rational design of magnetic nanoparticles for magnetic hyperthermia.

Keywords: hyperthermia; magnetic hyperthermia; magnetic nanoparticles; magnetic nanoparticle-induced biological effects; clinical trial; new therapies

1. Introduction: From Cancer to Magnetic Hyperthermia Therapy via Nanomedicine

1.1. Cancer

Cancer is a multifactorial disease in which a variety of parameters influence its development, progression or outcome, such as the type of cancer, tissue localization, genetic predisposition, immune status of the patient, etc. For this reason, it is one of the most challenging diseases to treat and develop new and effective therapies, which in turn requires the cooperation of multidisciplinary teams. Many types of therapies have been approved to treat cancer and the specific therapy or a combination of these that patients receive depends on factors like the type or stage of development of the cancer. Traditionally, the most common treatments for cancer involve surgery, radiotherapy and chemotherapy. However, more recently, strategies like immunotherapy have been developed and implemented in combination with these established approaches due to their capacity to improve these treatments. Complementary therapies like stem cell transplant also help restore blood-forming stem cells in patients after particularly harsh treatments. In addition, biomolecular advances have helped us better understand the causes of certain types of cancer, guiding the use of more specific and precise treatments, for example, using biomarkers or genetic studies [1].

1.2. Hyperthermia

Hyperthermia (HT) is a cancer treatment strategy first shown to produce benefits in the 1940s when it was contemplated that rising the temperature of a tissue might combat fibrosis and cancer [2]. To treat cancer, HT involves exposing malignant tissues to supraphysiological temperatures [3]. Damage to tumour cells or their death are the main desirable effects of such heating but also, HT may improve tumour antigen presentation, the activation of dendritic and NK cells, and leukocyte trafficking through the endothelium [4]. These are phenomena that enhance the anti-tumour immune response, and that make cancer cells more sensitive to the effects of radiotherapy and chemotherapy. HT is generally considered to be defined as a rise in temperature of the tumour region to between 39 and 43 °C (known also as mild hyperthermia) [3,5], although an increase up to 45 °C may also be considered [6]. However, when the temperature rises above 45 °C the situation is usually referred to as thermal ablation, which may have dramatic side-effects due to the damage caused to normal tissue and the death of healthy cells. For this reason, careful temperature control is necessary during HT treatment. Usually, MHT is administered such a multivalent oncological strategy in combination with other anti-cancer approaches, specifically when HT has been demonstrated to produce an improved synergic effect [7].

HT therapies are mainly classified according to the area of the body treated, which is usually closely related with the method used to increase the temperature (energy source) [6,7].

1.2.1. Whole-Body Hyperthermia

Whole-body HT is the systemic heating of the body in an attempt to obtain benefits treating widely disseminated metastatic cancer. The recommended upper limits are 42 °C maintained for 1 h, or 40 °C in combination with cytotoxic drugs. Technically, this method has important drawbacks, one of which is the need to sedate or submit the patient to general anaesthesia. Moreover, rising the body temperature from 37 °C to 42 °C is a lengthy process (90–180 min). In many cases the side effects of whole-body HT are unacceptable and much effort has been made to develop equipment that can resolve these disadvantages, e.g., IRATHERM-2000, currently in phase I/II clinical trials [8]. However, the success of this approach is limited by the poor balance between risk and benefit, and by the increasing interest in applying the fundamentals of HT in a safer and more specific manner [9].

1.2.2. Regional Hyperthermia

Regional HT is the application of the HT to a whole organ, limb or region but not to a specific tumour area. The most commonly used strategy to raise the temperature in selected regions is perfusion HT. Perfusion HT consists of inducing a heated fluid flow, normally of the patient's own blood, through a specific area, and it is usually directed at tumours localized to a limb. Nevertheless, there are variants where stomach cancer or other tumours in the abdominal cavity are treated in this way, such as continuous hyperthermia peritoneal perfusion (CHPP). This approach is not technically difficult and it is safe to employ by heating the perfusion fluid up to 43 °C for 2 h. However, this technique alone does not produce outstanding results and only in synergism with cytotoxic drugs are the desired results produced. This combined therapy needs a precise adjustment of certain parameters, such as the flow rate (30–40 mL/min), the cells, pH and O_2 of the blood perfused, or the amount of drug administered (which will be higher than in more specific therapies) in order to avoid undesirable toxicity [6,10]

1.2.3. Local Hyperthermia

Local HT is based on the controlled heating of a specific tumour zone in an attempt to avoid side-effects in the healthy surrounding area. This approach is based on the application of electromagnetic waves or ultrasound through a physical stimulator, enabling local heat to be applied in an external or an invasive way. The external application operates through micro- or radio-waves directed at a superficial or slightly deeper solid tumour (only a few centimetres below the skin) by an external device (e.g., microwave antennas, radiofrequency electrodes, laser fibres, electromagnetic coils or ultrasound transducers). Alternatively, invasive local HT (or interstitial HT) requires a thin needle or probe to be inserted into the tumour and serve as in situ energy applier. In this case, invasive local HT can be used in a deep tumour but it is restricted to small tumours (less than 5 cm in diameter) located in an accessible organ or tissue, such as the head, neck, bladder or prostate. In general terms, the success of local HT is determined by the tissue characteristics and blood flow, factors on which the energy and heat distribution is strongly dependent. Indeed, the heat distribution is often not as homogenous as would be desired and to resolve this problem, segmented radio-frequency electrodes that allow a three-dimensional control of heating have been evaluated for clinical implementation [6,7,10].

1.2.4. The Drawbacks of Conventional Hyperthermia

Without any doubt, HT is a very promising approach for cancer treatment and various clinical trials support its application for a wide range of different cancers: head, neck, melanoma, sarcoma, breast, glioblastoma multiforme (GBM), bladder, cervix, rectum, oesophagus, lung, mesothelioma and paediatric germ cell tumours [6]. However, since its first clinical application in the 1980s, its implementation has not been as widespread as might have been expected, probably reflecting several of the problems that traditional HT presents.

Firstly, HT requires specific equipment and technically, it is more complex than other more standard therapies like chemotherapy. Currently, there is considerable effort being directed at developing new, improved equipment. Secondly, the limited effectiveness of HT means it cannot compete with the standard protocols of cancer therapy. For an acceptable result, HT must be applied in combination with other therapies, mainly chemotherapy and radiotherapy, producing notable synergic effects. Consequently, HT is often considered as a sensitizing adjuvant therapy more than a cancer therapy by itself [11]. Traditionally, this reduced effectiveness can be explained by the uncontrolled dispersion of the heat, which is in turn caused by the lack of powerful devices to control and monitor the local temperature. In addition, this situation is exacerbated by the physical and physiological handicaps, such as the non-homogeneity of the tissues, the physical distance between the tumour cells to be treated and the heat source, or the thermal dissipation produced by the circulatory system. Thirdly, HT also produces side-effects and the rise in temperature may provoke non-desirable toxicity in healthy tissues or cells, even after local application. This is again due to the heat source not being exactly adjacent to the cancer cells, producing local side-effects. Finally, more studies into thermal biology and how HT affects individual cells molecularly are needed to better understand the process. Above all, it is necessary to assess the thermotolerance that has often been proposed [5,7,10] and to analyse the sensitivity of different types of cancer to temperature.

A general roadmap to address the aforementioned limitations involves improving the real-time temperature control of the tumour region and effectively localizing the induction of heat using a contactless stimulus. Regarding the former, technical developments have been quite successful, for example with Magnetic Resonance Thermometry (MRT). MRT is a MRI (magnetic resonance imaging) based technique that involves non-invasive

3D measurement of temperature distributions that could substitute the currently used invasive thermal probes. Since combining thermometry with MRI was first proposed four decades ago [12], MRT techniques have improved in accuracy and robustness for in vivo applications [13], now reaching pre-clinical stages of development [13,14]. Moreover, new approaches based on nanomedicine have been explored to obtain non-invasive heat sources that can be precisely targeted to the tumour cell in order to avoid side-effects and enhance the effectiveness of HT. Nevertheless, it would be desirable to have a single equipment where in vivo location and MHT could be performed simultaneously. However, limitations intrinsic to the technique (as different magnetic field requirements between MRI and MHT) do not allow a real-time guidance [15].

1.3. Nanomedicine: A Trip through the Hyperthermia Based Nanotherapies to Treat Cancer

Medicine is evolving towards more specific and personalised therapies, and HT must also move in this direction. Faced by this challenge, nanomedicine emerges as a very promising alternative to convert HT into a well-implanted and important cancer treatment. In recent decades, the application of nanotechnology and nanoparticles (NPs) to medicine and cancer has revolutionized techniques for diagnosis and treatment, now offering a broad range of alternatives. In terms of diagnosis, magnetic nanoparticles (MNPs) are contemplated as MRI contrast agents [16,17] or organic and inorganic NPs as nano-biosensors [18,19]. In addition, NPs also play an important role in cancer treatment, improving chemotherapy delivery [20–22], and they are being used in the development of innovative techniques such as gene [23,24] and HT therapies. Another interesting approach is the use of NP-loaded cells instead of individual NPs, a good example being the use of MNP-loaded immune cells for magnetic targeting in adoptive cell transfer therapies [25]. Both the possibilities for diagnosis and treatment make these nanomaterials, and specifically MNPs, very powerful candidates in the fight against cancer, know referred to as theranostic agents [16,26]. The success of NPs can be largely explained by the possibility of using them as more specific guided medicines and from the growing number of clinical trials, the use of NPs in cancer is becoming a reality [27–29].

First of all, NPs have the intrinsic ability to accumulate passively in tumours due to an Enhanced Permeability and Retention (EPR) effect, a concept coined in 1986 [30]. This EPR effect is based on the disruption and consequently, the loss of impermeability in the tumour vasculature, allowing the extravasation of proteins and macromolecules, and also of NPs into the interstitial space of tumours [31]. Moreover, the absence of functional lymphatic vessels contributes to the retention and non-clearance of the NPs [32]. The use of NPs (or macromolecular drugs, polymeric drugs or liposomes) is crucial to obtain an adequate EPR effect, since the use of these nanocarriers allows the desired drug to reach and accumulate in the tumour avoiding renal clearance caused by its small size (renal clearance threshold = 40 kDa) [30]. The success of the EPR effect is not exclusively dependent on the size of the nanoparticle or nanomedicine, but is much more conditioned by the tumour environment: size and concentrations of endothelial cells fenestrations, grade of fibrinolysis or thrombocytopenia in tumour hypoxic area, pericyte coverage of tumour microvessels, amount of collagen IV in basement membrane or density of extracellular matrix, and it, in turn, it is closely dependent of the type of tumour and organism. Nevertheless, an adequate diffusion extravasation and retention of NPs with sizes larger than 100 nm would be compromised [32]. Another way to direct NPs to cancer cells is through their active accumulation. This strategy involves the functionalization of the NP surface with antibodies or other specific ligands that recognize and promote NP uptake into target cells [33,34]. Even local NP injection into the tumour is a less aggressive and more specific minimally invasive option than conventional therapies not based on NPs [32].

Returning to HT, the possibility of a rationally targeting NPs to cancer cells offers two significant improvements over non-specific therapies. On the one hand, treatment effectiveness is notably enhanced due to the fact that each loaded NP can act as an individual heat source, only increasing the temperature in the areas where NPs accumulate and not affecting the surrounding tissues. Consequently, the HT produced with NPs is associated with fewer side-effects. Moreover, research into NPs is still constantly refining their synthesis and preparation, which is becoming cleaner, faster and cheaper, as witnessed by their single-step synthesis by microwaves [35,36].

Consequently, nanomedicine is particularly relevant to the future of HT therapies, which are generally based on the accumulation of NPs at target sites and the application of an external stimulus that induces NP heating. There are four different ways to achieve nanomaterial-induced MH, being the nature of the stimulus and the type of nanomaterials used (with their intrinsic properties) crucial parameters to achieve hyperthermia phenomena (Figure 1; Table 1).

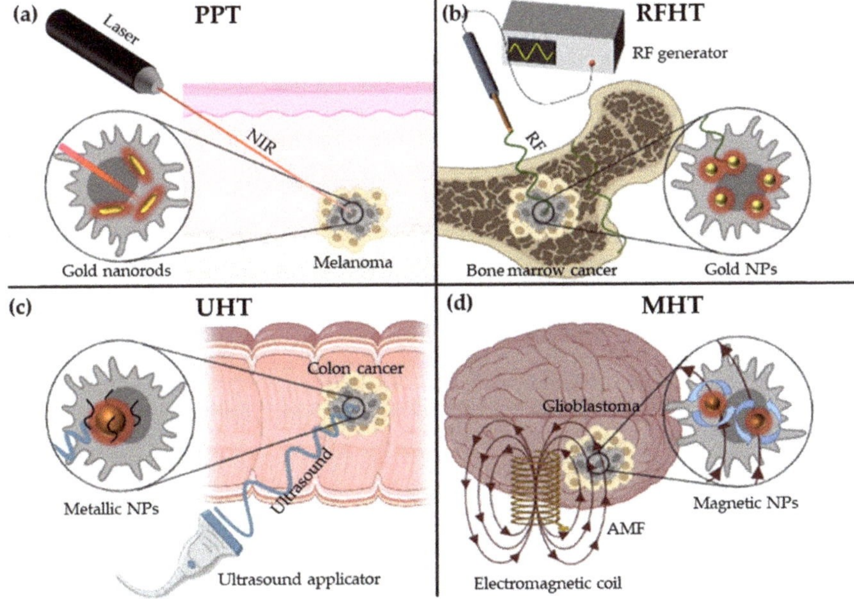

Figure 1. Therapies based on the accumulation of nanoparticles in the tumour and on the application of an external stimulus to induce nanoparticle heating: (**a**) Photothermal nano therapy (PTT), (**b**) Radiofrequency nano-hyperthermia (RFHT), (c) Ultrasound nano-hyperthermia (UHT) and (**d**) Magnetic nano-hyperthermia therapy (MHT).

Table 1. Main features of the different nanotherapies.

	Nanotherapy			
	PTT	**RFHT**	**UHT**	**MHT**
Equipment	Laser	Radiofrequency generator	Ultrasound applicator	Electromagnetic coil
External stimulus	NIR	RF-EF	Ultrasound	AMF
Physical principles	Surface Plasmon Resonance (SPR) or Optical absorption	Joule heating and Fermi electrons	Thermal interactions and mechanical interactions	Néel relaxation and Brown relaxation
Underlying effects	-	-	Cavitation	NPs Rotation
NPs employed	Mainly: Au nanorods Others: CuS NPs or carbon nanomaterials	Mainly: Au NPs Others: Pt, Si NPs or carbon nanotubes	Si, Au or iron-oxide NPs	MNPs (SPIONs)
Type of cancer treated	Non-deep tumours: Melanoma	Bone marrow, liver pancreas, colon or lung cancer	All types of cancer (breast, melanoma, colon, etc …)	Mainly glioblastoma and Pancreatic cancer
Main advantage	Feasible and cheap	Great penetration	No depth restriction	High specificity
Main disadvantage	Limited penentration	Invasive (needle insertion)	Lower specificity Expensive equipment	Lack of response in biological milieus (see point 5)

1.3.1. Photothermal Nano Therapy (PTT)

Photothermal nano therapy, usually shortened as photothermal therapy (PTT), is based on the capacity of agents to convert optical energy applied through a laser into heat. The laser beam can be tightly controlled, adjusting parameters such as the power density, duration or wavelength. Near-infrared (NIR) wavelengths (700–1400 nm) are normally used due to their good tissue penetration without producing undesirable damage or burns. Different nanomaterials can be used as photothermal agents but possibly, the best studied are gold NPs. When gold NPs are illuminated with NIR light, the energy produced by the excitation of their surface plasmon resonance (SPR) is transformed into heat and released to the local environment. The SPR effect is due to the oscillation of the free surface conduction electrons when excited by light, and they oscillate more with a resonance wavelength determined by the geometrical features of the gold nanomaterial [37,38]. For this reason, the size, surface and shape of gold-nanorods, and hence, their anisometry means they produce the best response to NIR excitation [39]. The PPT is a phenomenon shared by other nanomaterials, such as carbon nanomaterials (nanotubes, fullerene, graphene) [40,41] and CuS NPs [42] related to optical absorption. The effectiveness of PTT therapy is largely dependent on the penetration capacity of NIR, and it is more effective in melanoma and non-deep tumours. However, NIR-I (1000–1150 nm) penetrates well into tissues and organs like kidney, spleen and liver, whereas NIR-II (700–1000 nm) is the best option for muscles, and for stomach, heart and brain tissue in rats [43].

1.3.2. Radiofrequency Nano-Hyperthermia Therapy (RFHT)

Radiofrequency nano-hyperthermia therapy (RFHT) is based on the capacity of different nanomaterials to absorb non-invasive radiofrequency electric fields (RF-EFs) and release heat in response [44]. The physical mechanisms explaining RFHT remain controversial and although Joule heating is considered the main mechanism, others will also contribute [45]. The longitudinal acoustic vibrational mode is another theoretical mechanism proposed to explain RFHT, for example involving the absorption of RF-EF energy by Fermi electrons [46]. Different nanomaterials release heat after exposure to RF-EFs, such as Pt or Si NPs and carbon nanotubes. However, the best studied are again Au NPs, for which high concentrations and small sizes have been demonstrated to be important parameters to enhance RFHT. The main advantage of this strategy is the greater penetration of RF-EF, which enables deep tumours to be treated. Indeed, RF therapies have been tested against bone marrow, liver, pancreas, colon or lung cancer [44].

1.3.3. Ultrasound Nano-Hyperthermia Therapy (UHT)

Ultrasound nano-hyperthermia therapies (UHT) are based on the capacity of some nanomaterials to enhance the effects produced by exposure to ultrasound. Ultrasound therapy produces a non-specific response, such that extensive ultrasound exposure could provoke thermal damage to healthy tissues. However, high intensity focused ultrasound (HIFU, 0.1–1 kW/cm^2) is a less harmful alternative [47]. The selection of nanomaterials used in UHT is crucial due to their importance to enhance the effects of ultrasound therapy where NPs are located. Ultrasound affects tissues in two ways: thermally and mechanically. In UHT the thermal interactions depend on the attenuation coefficient and thermal conductivity of the NPs. The first of these contributes to the absorption and scattering of the ultrasound waves so that large NPs provoke a major attenuation and therefore, greater thermal dissipation. In terms of conductivity, small metallic NPs are the best candidates to improve the thermal conductivity when cell are loaded with these NPs. On the other hand, NPs maximize the mechanical interactions, consequently the cavitation nucleation threshold decrease and cavitation phenomenon is induced, which causes mechanical cell-membrane damage and cell lysis. Theoretically, this therapy could be applied to all types of tumours without any restriction in terms of depth. However, in practice it has been little studied due to the lower specificity and the expensive equipment required, and it has mainly been carried out with silica, gold and iron-oxide NPs against breast, melanoma or colon tumours [44].

1.3.4. Magnetic Nano-Hyperthermia Therapy (MHT)

Magnetic nano hyperthermia therapy (MHT) is essentially based on the intrinsic ability of MNPs to respond to alternating magnetic fields (AMFs) by converting magnetic energy into heat [37,44]. This approach is a valuable alternative for some kind of cancers located close to vital organs and in particular, for those difficult to remove surgically like some brain tumours [48]. This therapeutic approach allows the main clinical limitation of conventional radiotherapy to be overcome, that is the lack of selectivity of ionizing radiation which damages healthy and tumour tissues alike [49], thereby limiting the treatment to certain tumours. In MHT, the tumour cells (in vitro)/tumour (in vivo) are first treated with functionalized MNPs that are specifically internalized by these cells and then, the cells/tumour are subjected to an AMF to increase the local temperature of the tumour cells and induce controlled apoptosis. The main disadvantage for this therapy is having pre-clinical implementation problems and it is necessary a well-understanding actuation mechanism in order to exploit its full potential. Concretely, these handicaps are discussed in the point 5 of this manuscript: nanoparticles aggregation and consequently loss of ability to releasing heat.

The purpose of this review has not been to carry out an exhaustive search of all the articles published in recent years in medical databases that include the term "magnetic hyperthermia therapy", but we have preferred to include those articles that we believe have contributed the most to understanding MNPs behavior in response to AMF in biological milieus, which is a critical step to drive magnetic hyperthermia therapy toward clinical implementation. The focus of this review is placed on the specific issues related to the physical basis of this therapy, the type of MNPs used and the tumours that are best suited to this therapy. The concepts underlying MHT will be addressed to understand what type of NPs are the best candidates to use. The biological effects of MHT at the cellular and molecular level will be explained and ultimately, the most promising and novel rationally improved strategies that currently produce the best results will be described. Finally, the current ongoing clinical trials will be reviewed and prospects for the clinical implementation of MHT therapies based on MNPs will be reviewed.

2. Physical Concepts of Magnetic Hyperthermia (MH)

The clinical application of magnetic hyperthermia is referred to MHT. This therapy is based on certain physical principles that we consider important to explain below. MNPs present unique magnetic properties that can be taken advantage of to achieve selective contactless heating mediated by AMF. The ferromagnetic materials used in the production of magnets are characterized by the remanent magnetization (M_R) they present even in the absence of a magnetic field. This magnetization is a consequence of the alignment of atomic moments in a specific direction defined by the anisotropy of the material (easy magnetization axis). When ferromagnetic materials are reduced to the nanoscale, thermal fluctuations become more and more important until the thermal energy surpasses the anisotropy energy of the NPs, making the magnetic moment flip between two bistable positions of the easy axis and leading to what is called superparamagnetic behaviour. As a consequence of these spontaneous fluctuations, the M_R of the MNPs disappears and the magnetization (M) presents a S-like reversible response to low frequency magnetic fields (H) like that indicated in Figure 2a. The time taken for the magnetic material to lose its magnetization is called the relaxation time [50].

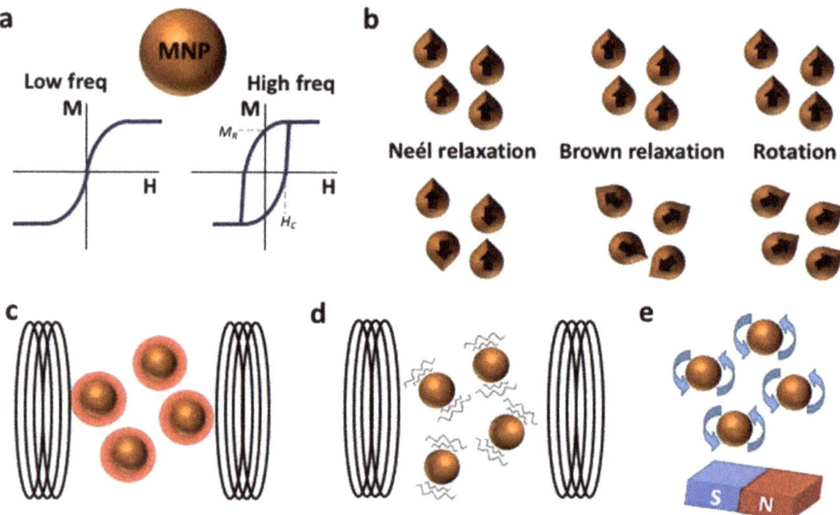

Figure 2. Physical properties of magnetic nanoparticles (MNPs), and the magnetic response to low and high frequency magnetic fields. (**a**) At low frequencies the MNPs present a S-like magnetic response with a high susceptibility (superparamagnetic) and they transit towards a hysteretic magnetic response when exposed to high frequency AMFs. (**b**) The MNPs may lose spontaneously their magnetization by Néel and Brownian relaxation when the applied field is subtracted or rotate coherently in response to a rotatory magnetic field. Through these mechanisms the MNPs can convert the energy of magnetic fields into (**c**) heat, (**d**) mechanical vibration or (**e**) local torque in the cell environment depending on the features of the field applied.

MNPs recover a ferromagnetic response when the oscillation of the applied field is faster than the relaxation time of the magnetic moments (high frequencies) and the magnetization processes take place through dissipative loops. This dual response is of interest for their biomedical application as they become contactless nanoheaters. Thanks to the lack of the M_R of superparamagnetic MNPs in the low frequency regime, they can be prepared as a colloidal suspension, avoiding aggregation. Consequently, they can be injected intravenously without any fear of them obstructing capillaries [51,52]. In addition, once they are situated in the target tissue they can be remotely activated as nanoheaters by applying a high frequency field, producing minimal effects in the surrounding biological tissues where only a weak magnetic response is produced. In contrast to other contactless mechanisms of activation, such as that required in photothermal or photodynamic therapies, AMFs can penetrate the body with minimal attenuation, ensuring homogeneous field conditions in the whole tumour without any shadow effect [53].

There are two dissipative mechanisms by which MNPs may lose their magnetization when the magnetic field is removed: Néel relaxation and Brownian relaxation (Figure 2b). The former relaxation is associated with the inversion of the magnetic moments between the two directions of the easy axis magnetization, and it depends on the magnetic features of the MNPs. The latter is produced by the physical rotation of the MNP within the liquid media. In both cases, relaxation leads to the misalignment of the easy magnetization axes and consequently, a cancelation the global magnetization of the system. Brownian rotation is a consequence of the random interaction with the surrounding media and therefore, it is controlled by the hydrodynamic size, the temperature and the viscosity of the medium [54]. In both cases, relaxation leads to the misalignment of the easy magnetization axes and consequently, a cancelation the global magnetization of the system. MNPs will adopt a faster relaxation mechanism depending on their intrinsic properties and those of the surrounding medium [55], although combinations of both mechanisms may exist. It has been postulated that in certain AMF conditions MNPs may physically align their easy magnetization axis with the applied field before undergoing Néel relaxation [56].

It has generally been claimed that for translational MHT, MNPs design must focus on optimizing the mechanisms of Néel relaxation [57] since the natural MNPs-aggregation induced by contact with biological milieus (for instance, during lysosome encapsulation) and the high viscosity of these media blocks the Brownian effect [58]. However, the heat produced by MNPs with pure Néel relaxation requires tumour cells to have a high concentration of Fe in order to increase the temperature, which is difficult to achieve by intravenous injection and that has been resolved by intratumour injection [59].

An additional response of MNPs to an AMF is mechanical damage due to the magnetic torque generated in the presence of the field, which is in turn due to the misalignment between the field applied and the easy magnetization axis. This is a different concept to the Brownian relaxation indicated above. In this case, the magnetic moments of the different MNPs maintain their relative alignment but they are rotated collectively to reduce the angle between their magnetic moment and the field applied. Equation (1) shows how the magnetic torque (T) applied by a magnetic field (B) to a magnetic moment (μ) grows with the misalignment (θ) between them [60].

$$T = \mu \cdot B \cdot \sin(\theta) - 6\eta V_H \frac{d\theta}{dt} \tag{1}$$

This equation also takes into account the resistance to rotation produced by the medium depending on its viscosity coefficient (η), the hydrodynamic volume of the MNP (VH) and the angular speed of rotation ($d\theta/dt$). The order of magnitude of the mechanical torque in normal conditions for a single MNP is 10^{-21} Nm [61]. However, the torque generated by the magnetic field is enhanced for an aggregate of MNPs [62], like those observed inside the cell's lysosomes.

The most relevant parameters (η, V_H, AMF and frequency, etc.) have been revised in detail to determine the transition between heat dissipative mechanisms and magneto-mechanical actuation. As a result, in viscous media the mechanical rotation induced by AMFs is mainly relevant for large MNPs (15–50 nm) and low frequency AMFs (<10 kHz) [63,64], although it can be exploited to exert effective mechanical torque on biological components [65]. This transition with AMF frequency becomes even clearer in the case of large anisometric NPs [63] and some empirical studies reported that AMFs induce a torque [66] or a physical movement of the MNPs [67] that may produce mechanical damage of the lysosome and cell membrane [68,69].

The use of rotatory fields is more convenient when inducing mechanical damage to cells. These fields maintain the MNPs magnetized throughout rotation and it is technologically simpler to generate homogenous fields of high intensity (>0.1 T) using permanent magnets. According to theoretical simulations, the torque induced by rotatory fields is 30-fold that created by AMFs [61]. Hence, many of the studies that seek to use MNPs to induce mechanical damage now focus on this kind of magnetic stimulus [66,70].

2.1. Determining the Heating Power of MNPs

One of the main limitations for the critical analysis of MHT measurements is how to determine the heating power of MNPs [71]. The Specific Absorption Rate (SAR) is an empirical parameter frequently used by radiological protection departments to regulate the amount of radiation absorbed by patients when exposed to radiofrequency fields and it is normalized to the mass of biological tissue irradiated in terms of W/g [72]. This parameter was taken by the MHT community to quantify the amount of magnetic energy transformed into heat by a suspension of MNPs. However, it is important to note that in the latter case, the temperature increase ($\Delta T/\Delta t$) is generally normalized to the Fe concentration in the magnetic colloid (m_{Fe}) according to Equation (2), where C_V is the specific heat of the colloid [73,74].

$$SAR(W/g_{Fe}) = \frac{C_V}{m_{Fe}} \frac{\Delta T}{\Delta t} \tag{2}$$

It is also important to note that this is a system-dependent parameter that varies with the field intensity (H) and frequency (f) of the AMF. With the aim of standardizing such parameters and comparing the heating power of MNPs studied under different AMF conditions, an alternative parameter has been introduced, the Intrinsic Loss Power (ILP). This parameter divides the SAR by the frequency of the field and the square of the field intensity (Equation (3)).

$$ILP(W/g_{Fe}) = \frac{SAR}{f\,H^2} \tag{3}$$

This definition is based on the theoretical model proposed that considers a linear response of the magnetic moment to the AMF applied [75]. This model predicts a linear dependence of the SAR on the frequency and a quadratic dependence on H, although this is only valid for highly anisotropic MNPs and a small H, and thus, it cannot be assumed as a universal parameter. Besides, recent double-blind experiments showed that the specific features of the experimental set-ups and the thermal curves analysed may produce inconsistent SAR values between laboratories [71]. A promising solution to achieve a global and consistent parameter to determine heating power is to measure the high frequency magnetic loop of the MNPs [76]. As indicated previously, the amount of heat dissipated by a collection of MNPs is strictly related to the area of the high frequency hysteresis loops (A), see Figure 2a right). Therefore, the theoretical SAR can be derived from the product of this area to the number of magnetic cycles per second, i.e., the frequency of the AMF (Equation (4)).

$$SAR = A \cdot f \tag{4}$$

Hysteresis loops are intrinsic to magnetic systems and they provide information about the magnetic response to AMFs of different frequencies and intensities. Therefore, they provide information about the dissipative properties of the MNPs that are independent of the thermal diffusion of the medium and do not depend on the thermal loses of the calorimetric system. However, the AC magnetometers required to study these are still scare and generally homemade [77].

The ideal field and frequency AMF conditions for MHT are also still to be defined. The former limit was established for the maximum field-frequency product as $H \times f \leq 4.85 \times 10^8$ $A\,m^{-1}s^{-1}$ based on the feeling of discomfort in irradiated subjects [78], a subjective test of patient comfort. New AMF application systems can concentrate the radiation in a restricted volume, reducing the radiation dose received by the patient and enabling more flexible limits to be proposed, ranging between $1.8 \times 10^9\ A\,m^{-1}s^{-1}$ [79] to $8.3 \times 10^9\ A\,m^{-1}s^{-1}$ [80], and up to $18.7 \times 10^9\ A\,m^{-1}s^{-1}$ [81].

2.2. Other Advantages of Using MNPs

An additional functionality of MNPs that enhances their applicability in MHT is the possibility to concentrate them in a certain region using magnets. When MNPs are magnetized with a magnetic gradient they minimize their energy by shifting towards the region of the maximum field. This principle has been exploited to concentrate magnetic nanoagents to a target superficial tumour by locating a set of permanent magnets in the proximal skin area. Using the same principle, MNPs have been used to label circulating tumour cells so that they can be concentrated and detected for an early diagnosis of metastasis risk [82,83]. More recently, the magnetic guiding of immune cells loaded with MNPs has been proposed as an advanced solution to reduced vascular accessibility [84], and as a means to activate mechanosensitive membrane receptors that inhibit cancer proliferation [85].

MNPs also present interesting properties for clinical imaging techniques, such as MRI. The strong permanent field used to align the magnetic moment of water protons magnetizes MNPs, creating local regions of enhanced magnetic fields in the tissues where these MNPs lie. The local field created by MNPs modifies the relaxation time of the surrounding water molecules [86]. This changes the MRI contrast of the tissue loaded with MNPs and offers an interesting pathway for personalized therapy. In addition, Magnetic

Particle Imaging (MPI) has emerged as a promising technique to solve the incompatibility of a simultaneous MHT and MRI (mentioned in 1.2.4) and ideally achieve theragnostic NPs which are useful as heat generators for MHT, and at the same time, for diagnosis through real-time in vivo image during the therapy. MPI is an emergent image modality which works by detecting the nonlinear magnetization of the flipping MNP. MPI present several advantages: ideal penetration and signal-noise ratio, no view limitations, highly sensitive, linear and quantitative signal, high contrast, zero ionizing radiation, and safer and persistent. All this makes this technique an excellent non-invasive 3D tomographic imaging method to be combined with MHT for a real-time therapy image guided [15,87].

3. MNPs for MHT

Maximizing the heating power of MNPs is an interesting approach to minimize the dose required for effective HT therapy. In this section we will summarize the most important strategies to maximize the amount of heat dissipated by MNPs. Considering equation 4, it is easy to identify that the SAR can be maximized by increasing the frequency of the AMF, the area (A) of the high frequency hysteresis loops, or both parameters at the same time. The simplest strategy to increase A is to increase the magnetization of the sample. Taking magnetite (Fe_3O_4) as a reference for biocompatible magnetic materials, saturation magnetization can be increased by doping its crystalline structure with other transition metals. The magneto-crystalline structure of magnetite is an inverse spinel made of two sub-lattices of magnetic moments aligned in opposite directions that occupy octahedral and tetrahedral positions [88]. Due to the higher number of octahedral positions and the Fe^{2+}/Fe^{3+} occupancies, the magnetic moment of these two sub-lattices is not compensated, which classifies magnetite as a ferrimagnetic material. Thus, in the presence of a magnetic field the octahedral and tetrahedral moments lie in parallel and antiparallel, respectively. As such, the global magnetization of magnetite can be enhanced by substituting Fe ions with other transition metals with a higher atomic moment, like Mn that occupies octahedral positions [89,90], or with transition metals like Zn that have no atomic moment and occupy tetrahedral positions. Both effects can be combined in ternary $Zn_xMn_{(1-x)}Fe_2O_3$ ferrites to maximize the saturation magnetization [91]. However, it must be borne in mind that the magnetic coupling between the lattices might be affected when high concentrations of dopants are used to reduce the magnetic order in the MNP, averting the enhanced magnetization. Besides, such complex formulations compromise the homogeneity of the stoichiometry in the sample.

An alternative strategy involves widening the magnetic cycle by increasing the coercivity field (H_C) of the MNP. The Hc is related to the field required to cancel the remanent magnetization and thus, to the anisotropy of the magnetic system. MNPs with a high magnetic anisotropy (K) are harder to magnetize but they dissipate more energy as their magnetization is reversed. If the AMF is not sufficiently intense, most of the magnetic moments of the system remain fixed in their easy magnetization axis without dissipating any thermal energy. Thus, it is necessary to reach an AMF threshold to partially or completely overcome the anisotropy barrier of the MNP. The anisotropy field (H_K) is an interesting parameter to define such a threshold in theoretical simulations (Equation (5)).

$$H_K = 2K/M_S \qquad (5)$$

According to the numerical simulations based on the Stoner-Wolfarth theory, MNPs requires an AMF higher than approximately $0.4\,H_K$ to begin heat dissipation, which reaches it maximum value when the AMF approximates to the H_K [92]. At higher AMFs, magnetic moments reach saturation and heat dissipation does not further increases. This theory also establishes a direct relationship between the H_K and H_C ($H_K = 0.48\,H_C$) for a set of MNPs with randomly oriented easy magnetization axes, making this parameter an interesting link between theoretical simulations and empirical data [93]. In summary, increasing the magnetic anisotropy of the system can help increase the SAR as long as the intensity of the AMF applied is similar to the H_K of the system.

The magnetic anisotropy constant (K) is a composite of two components: the magnetocrystaline anisotropy and the shape anisotropy. The former is associated with the crystalline structure of the MNPs and the coupling between their atomic magnetic moments. The magnetocrystalline anisotropy of magnetite can be enhanced by doping the structure with other transition metals, such as Co [94]. These cobalt ferrite MNPs have a K value that is 2 orders of magnitude higher than magnetite [95], creating MNPs with magnetic moments strongly fixed in the easy magnetization axes. Heat dissipation for highly anisotropic MNPs like cobalt ferrites is generally produced by Brownian mechanisms, which might be not ideal for in vivo applications.

The second contribution to anisotropy is defined by the geometry of the MNP. Magnetic materials tend to minimize their magnetic poles by locating their magnetization along their longest axis, establishing the geometrical easy magnetization axis. The geometric anisotropy can be enhanced by creating anisometric nanostructures like nanocubes, nanorods and nanodiscs [63,96–98]. These nanostructures are of general interest as they are good mechano-transducers that may apply large torques on biological components although they must be prepared with reduced dimensions to preserve their superparamagnetic response [99,100]. An alternative to superparamagnetism is to create nanomaterials with an exotic magnetic order known as magnetic-vortex that can be mechanically manipulated and also presents a lack of remanence [69].

As a rule of thumb, increasing the saturation magnetisation (M_S) of the sample is a good approach for AMF inducers that operate at low intensity and high frequency, whereas increasing the K of the system is more convenient for AMFs that operate in high-intensity and low frequency conditions [101]. Both strategies could be combined by creating magnetically coupled composites with two magnetic components [102]. The area of the high frequency hysteresis loops can be maximized by coupling a "soft" magnetic phase with the high M_S and a "hard" magnetic phase with large coercivity through an exchange interaction [103]. Indeed, $CoFe_2O_4@MnFe_2O_4$ core-shell MNPs represent a paradigmatic example of this kind of system [104]. The synergetic effect of such magnetic coupling generates an increase in the SAR value of one order of magnitude with respect to similar MNPs with a single ferrite phase (3.03 kW/g). Even more impressive results were obtained with $Zn_{0.4}Fe_{2.6}O4@CoFe_2O_4$ core-shell MNPs prepared with a cubic shape in which the SAR was above 10 kW/g [105]. Although these are outstanding SAR values compared to other MNPs [106], the synthesis of homogeneous core-shell structures with a controlled shell thickness remains a challenge.

A more feasible approach to exchange coupling to develop advanced magnetic nanoheaters is the preparation of multicore MNPs, also known as nanoflowers. These structures are made of aggregates of magnetic nanocrystals with common epitaxial interfaces that couple the magnetic responses of the nanocrystals. Exchange coupling creates a cooperative magnetization process between the nanocrystals formed that increases the susceptibility of the aggregated nanostructures [107]. This increase in susceptibility implies a rise in the SAR values up to c.a. 2 kW/g at very high frequencies (700 kHz), even at moderate AMF intensities (25 kA/m) [108]. The amount of heat dissipated strongly depends on the primary crystal size, and on the composition and extent of the interfacial surface [109], creating an energetic balance of the dipolar interactions and exchange coupling between the primary nanocrystals that may favour or hamper heat dissipation [110].

Dipolar interactions are also a crucial parameter when using MNPs as nanoheating agents. When MNPs are magnetized they create their own dipolar field that affects the magnetic response of the surrounding MNPs. This effect is especially relevant for highly concentrated colloids and in aggregated systems in which the MNPs are in close proximity, such as endosomes. For small MNPs, the dipolar field may increase the anisotropy barriers and enhance their heating performance [111]. But in most cases the effect is the opposite, the dipolar field cancelling the effect of the applied field, reducing the susceptibility of the system and consequently diminishing their SAR [112,113]. This has been postulated as a possible cause for the discrepancies observed in the heating performance of MNPs between ex vivo and in vivo studies [114]. Only in the case of elongated aggregates, such as chains, does the dipolar interaction result in cooperative magnetization that favours susceptibility and increases the coercivity of the system [115].

Although these parameters are interesting from the point of view of the magnetic properties, superparamagnetic iron oxide nanoparticles (SPIONs) are clearly the most advanced candidate in terms of commercial availability, regulation and clinical trials. In terms of magnetic properties, SPIONs present a relatively high susceptibility and low residual magnetization in the absence of an external magnetic field. Moreover, SPIONs are well-tolerated and they have low toxicity profiles, even in long-term studies [116]. In addition, they can be biotransformed from SPIONs to other iron compounds, facilitating their clearance [117–120]. SPIONs have been in clinical use for years, and several types have been approved for use in humans by the European Medicines Agency (EMA) and the Food and Drug Administration (FDA) in the USA, especially as anti-anaemic drugs and contrast agents for MRI [121].

Preparation of Candidate MNPs for MHT

In designing MNPs for HT, a balance must be reached between the size of the magnetic core to maximize the heat released (>10 nm) and the colloidal stability in biological media required for intravenous injection (<50 nm). Above these sizes, magnetic interactions between NPs are very strong and it is difficult to keep them apart despite their coating, such that they tend to aggregate and precipitate. One NP formulation already approved for use in cancer therapy is NanoTherm®, approved in 2010 by the EMA to treat recurrent GBM and in 2018 by the FDA for human prostate cancer (https://www.magforce.com/en/home/about_magforce/#highlights; accessed on 27 July 2021). NanoTherm® is a colloidal suspension of aminosilane coated 15 nm iron oxide NPs (with an iron concentration of 112 mg/mL) that can be delivered percutaneously into the tumour tissue. However, the challenge remains to develop NPs with enhanced specific loss of power and efficient delivery, within clinical AMF design constraints [122].

Iron oxide MNPs commercially available for HT are produced by precipitation of iron salts (Fe(II) and Fe(III)) in alkaline aqueous solutions. The size of the particles does not exceed 20 nm and it can be ensured by thermal treatment for long periods of time or by controlling the pressure in autoclaves. Thus, high pressure homogenisation processes allow the formation of individual crystals with mean diameters of 15–20 nm [123]. In a similar way, core–shell NPs and mixed ferrites are obtained by co-precipitation of stoichiometric mixtures of solutions containing divalent (Mn(II), Co(II), Zn(II)) and trivalent metals (Fe(III)) in alkaline medium [124]. Although core-shell MNPs can be finely tuned in organic media synthesis, the synthetic methods in aqueous media are preferred because they do not need additional coatings for water transference and are fully scalable to mass production, although controlling the size distribution and crystal order is limited due to the use of temperatures below 100 °C.

Controlling the shape of NPs and producing larger MNPs (>20 nm) can be achieved by thermal decomposition in organic media at temperatures as high as 300 °C, while cubes or rods are prepared using shape directing agents, either carboxylic acids or amines, respectively [96,97]. Autoclaves have recently been used to produce up to gram quantities of cubes, using iron pentacarbonyl as a precursor (Patent: WO2020222133A1). Other ferrites, such as manganese ferrite, zinc ferrite or a mixture of them in the form of core@shell or alloys prepared by this method, have showed excellent properties for HT [101].

Moreover, the assembly of magnetic cores into regular structures has been shown to significantly influence the HT behaviour of the particles, requiring the control of some key synthetic parameters that drive the self-assembly and growth process, such as surfactants and the viscosity of the medium. Thus, flower-like iron oxide assemblies between 25 and 250 nm can be obtained by heating a mixture of Fe(II) and Fe(III), or of an Fe(III) salt alone, to 200 °C in a heating mantle or in an autoclave [108,125]. In this sense, polyols are very interesting polar solvents that work as reducer, surfactant and high temperature synthesis media, allowing the use of inorganic salts as precursors [126]. Moreover, it is possible to combine the polyol procedure with more efficient heating technologies, such as microwave heating, leading to higher production yields over shorter reaction times [127].

The assembly of magnetic cores can be exploited even in the absence of an interface between them. The chains of magnetic nanocubes naturally produced by magnetotactic bacteria have for long been the best candidates for MHT, with a SAR of 2.38 kW/g at 310 kHz and 30 kA/m [128]. For this reason, several mimetic systems have been produced using silica as a template or other anisometric nanomaterials [129]. Finally, further coating and functionalisation is possible for NPs obtained in either aqueous, organic and polyol media. Coating with aminosilanes and/or polysaccharides is mainly used for targeted HT applications [130]. These coatings provide an excellent first layer for the bioconjugation of biomolecules, such as antibodies or peptides, drugs or biomarkers.

4. The Biological Effects of the Application of AFM to Cells Loaded with MNPs: Is It Always Hyperthermia?

Having established the physical principles and the materials used in the development of MHT, we can focus on the biological effects that they produce. However, rather than adopting a general or macroscopic view of this issue, we shall focus on the cellular and subcellular effects of this treatment. There are many in vitro and in vivo studies that have correlated MNPs and AMF therapies with extended life expectancy or tumour regression [131], although the biochemical mechanisms responsible for such improvements remain unclear. Depending on the magnetic features of the MNPs, and on the AMF amplitude and frequency, the MNPs can transform the energy of the magnetic field into heat or mechanical effects (Figure 2c,d,e). The application of the AMF may cause the magnetic moments of the internalized MNPs to rotate in the direction of the field (Néel's relaxation) and the actual MNPs to physically rotate (Brown's relaxation). These two responses to the magnetic field are manifested to a greater or lesser extent depending on the intrinsic characteristics of the NMPs (size, shape, anisotropy, crystallinity) and the magnitude of the AMF. When the AMF is applied and it alternates at high frequency, the continuous re-orientation of the MNPs with the magnetic field alters the release of heat by the MNPs [132] or the physical movement of the MNPs, provoking mechanical damage.

Consequently, there is a controversy around if it could be termed hyperthermia for all effects provoked by MNPs + AMF treatment. So that, we propose to classify the biological effects into four groups, depending on the main cause of the biological effects observed. First, we talk about the most intuitive mechanism, the rise of temperature, that is, hyperthermia. Then, we mention several studies where the authors believe that the resulting effects are not related to temperature, but due to the physical and mechanical MNPs movements. Later, we introduce a section to propose an explanation for these apparently non-temperature-related effects, offering the possibility of a macroscopically undetectable hyperthermia phenomenon. Finally, we mention other indirect treatment processes that help us to tumour regression.

4.1. Biological Effects of Heating

It is well known that a rise in temperature triggers cell death, yet not all cell types are equally sensitive. Cancer cells are considered to be more susceptible to HT than healthy cells due to their higher metabolic rates [133], the hyperthermic inhibition of DNA repair [134] and the poorer heat dissipation through the blood flow [135]. More specifically, the biological effects of HT include (Figure 3): an increase in oxidative stress (Figure 3.①) [136,137]; inactivation of membrane receptors and increase in ion permeability that affects cell transport (inhibition of amino acid transport and increased Na^+, K^+, and especially Ca^{2+}: Figure 3.②); a lack of stability and an increase in membrane fluidity (Figure 3.③); changes in cytoskeletal organization, involving microtubule, microfilament and intermediate filament depolymerization (Figure 3.④); increased protein denaturation and insoluble protein aggregation in the nucleus (Figure 3.⑤), which promotes heat shock protein expression (Figure 3.⑥) and centrosome damage, as well as mitotic dysfunction (Figure 3.⑦); and eventually, DNA damage or denaturation can occur (Figure 3.⑧) [138].

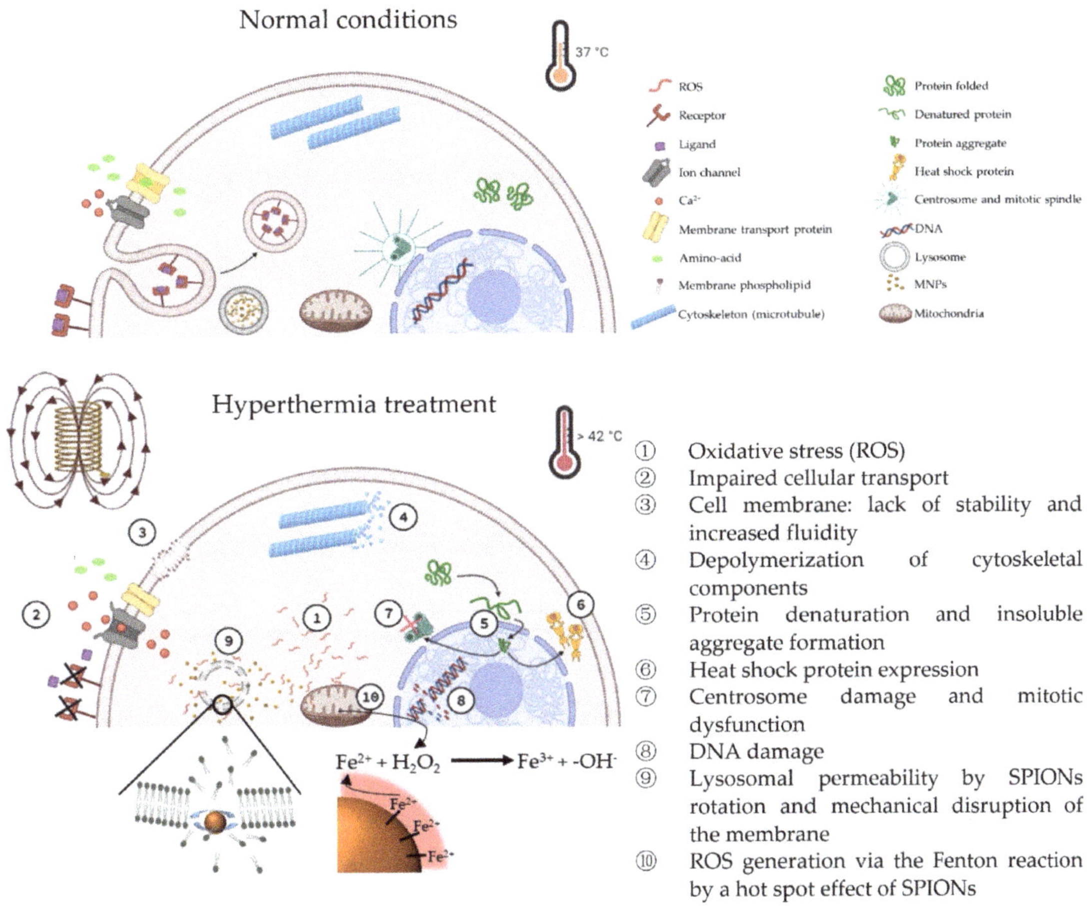

Figure 3. Cellular and sub-cellular biological effects derived from hyperthermia. **Upper:** normal conditions (37 °C). **Bottom:** hyperthermia treatment (>42 °C).

4.2. Biological Effects of Mechanical Rotation or Vibration

It is interesting that there have recently been several reports of cell death after MHT without any perceptible rise in temperature [139–142]. The mechanisms responsible for these effects have not yet been elucidated, although there is data indicating that they are related to an increase in lysosomal permeability, which triggers an enhanced reactive oxygen species (ROS) production and enhanced activity of the lysosomal protease cathepsin D in the cytoplasm diminishing tumour cell viability [47,143,144]. This lysosomal permeabilization might be also caused by mechanical rotation or vibration of SPIONs altering lipid membrane stability (Figure 3.⑨). Indeed, dynamic magnetic fields induce a slow rotation of lysosome targeted SPIONs, tearing the lysosomal membrane and activating apoptosis [145]. Moreover, according to theoretical simulations, the rotation of MNPs in a liquid media can be induced by either rotatory fields or AMFs. However, at moderate field intensities the torque induced by rotatory fields is 30 times higher than those created by

AMFs [61]. Evidence of the mechanical damage produced by MNPs under an AMF has come from reports of lysosomes rupture inside the cells [144].

4.3. Biological Effects Derived from Non-Perceptible Heating: The "Hot Spot" Effect

Another possible explanation for the biological effects of AMFs that are apparently unrelated to temperature changes might be very local intracellular heat release from SPIONs (not detected macroscopically), known as a "hot spot" effect. This local heat-release enhances biological effects, such as the generation of ROS by the iron oxide surface of the NPs through the Fenton reaction (Figure 3.⑩), which is known to be accelerated directly by temperature [146]. Hence, we think that markers of sub-cellular temperature rises (e.g., Hsp70 [147] or thermal nanoprobes) should be implemented routinely in these studies to determine if tumour cell death can be always attributed to HT (even if these occur on a subcellular scale), or whether the biological effects that occur are unrelated to temperature.

4.4. Biological Effects Derived from Other Indirect Process

It seems clear that independently of whether they are due to MNP heating or their mechanical rotation and vibration, biological effects not directly or not exclusively related to temperature play a crucial role in tumour regression. ROS formation through Fenton reactions is probably the best studied of these, given that ROS can severely damage cell elements due to oxidation, such as DNA, proteins, lipids and enzyme cofactors, thereby inducing apoptosis [148]. ROS formation can occur by lysosome degradation or through the breakdown of other subcellular structures but also, by MNP interactions and Fenton reactions at the MNP's surface. Interestingly, and concomitant with ROS production, an increase in fluidity and a loss of cytoplasmic membrane integrity also activates cell death, either necrosis or apoptosis [5]. Other effects not directly related with the MNP-tumour cell interaction have an enormous importance in the fight against cancer. MHT has also been associated with activation of DCs and NK cells [149].

5. MNP Behaviour in Response to AMF in the Biological Milieu

In the previous section, we have analyzed the biological effects driven directly or indirectly by MNP exposure to AMF in MHT settings, that may or may not be trigger by MNPs heat release in response to AMF exposure. In recent years, many studies have suggested that the ability of MNPs to produce heat in response to AMF exposure when MNPs are in biological milieus is severely reduced, even sometimes undetectable, a circumstance that would not be desirable for MHT therapies. A possible explanation for this undesired behaviour could be that the magnetic response of MNPs to AMF was modified as a consequence of MNPs-cell interaction [150,151], being the causes of these changes in the magnetic properties a reduction in MNPs mobility, dipolar interactions, milieu viscosity, and MNPs clustering or aggregation [58,114,150–153]. This alteration of MNPs magnetic properties implies a dramatic reduction of SAR values, that could be observed when MNPs are aggregated by contact with cells, but also by contact with physiological milieus or viscous media emulating cellular environment, and depending on the intrinsic properties of the MNPs, these SAR decreasing values could be more than 60% [114,152], even in MNPs that, after being tested in aqueous medium, showed a promising heating capacity [153]. Most of the observed alterations in the MNPs magnetic properties could be explained by the restriction of the Brownian relaxation, since MNPs cannot respond to AMF with rotation because MNPs are physically immobilized and blocked. Therefore, when Brownian relaxation component is suppressed in biological conditions, and it was demonstrated that Néel relaxation was unaffected by changes to their biological microenvironment, emphasizing the importance of MNP intrinsic magnetic properties for MHT when particle mobility cannot be kept. So that, Néel relaxation component becomes the only possible heat induction mechanism [58]. Likewise, doping MNPs with Zn allows a strong Néel relaxation that was preserved after MNPs-cell interaction, which is suitable for heat releasing in MHT [151].

Although Néel relaxation contributes for heating generation during MHT the most important component for heating during an ideal MHT is the Brownian relaxation. Therefore, it is a key point to understand why the MNPs are immobilized in and if it is a reversible process. It has been shown that MNPs that were blocked as a consequence of cell internalization, can recover their original magnetic properties, including the Brownian relaxation, when the cells that contained them were lysed, due to the integrity of the magnetic core is preserved during this process [57] suggesting that MNPs immobilization or aggregation is the final cause of loss of Brownian relaxation. In another study, three systems with MNPs different spatial distributions and grades of aggregation were analyzed in order to compare their magnetic properties: isolated MNPs, MNPs-liposomes system, and MNPs-cell interaction (using, in turn, Jurkat cell line that attached MNPs to the outer membrane and Pan02 cell line that internalized the MNPs). Results showed that the biological environment played a crucial role in the dynamic magnetic response of the MNPs, being more altered for MNPs-cell system, and concluding that the simple fact of being in contact with the cells triggers MNPs aggregation [154].

Other studies tried to determine if this aggregation derived from MNPs-cell contact was a process dependent on the intrinsic properties of the MNPs or the host cell line. For that, co-precipitated maghemite nanoparticles, assembly of the same maghemite nanoparticles in liposomes, cobalt ferrite nanoparticles, iron oxide/gold dimers, iron oxide nanocubes and iron oxide nanoflowers were tested in three different biological environments: MNPs in water, MNPs attached to adenocarcinoma SKOV-3 cells membranes, or MNPs internalized in SKOV-3 cells. As result, a rapid fall in the heating capacity of all the nanomaterials tested (regardless of its different composition, shape or size) has been observed when MNPs were associated with the cell membrane or were internalized [153]. Likewise, different core size MNPs (6, 8 and 14 nm), coating (APS: 3-aminopropyltriethoxysilane, and DMSA: dimercaptosuccinic acid), cell line (Jurkat and Pan02) and subcellular localization (membrane or internalized in endosomes/lysosomes) were tested in biological milieus being demonstrated that the aggregation process was independent of MNPs core size, coating, cellular environment, host cell line and MNPs subcellular localization [150].

Currently, these problems of aggregation or blockade of MNPs are the main bottlenecks in MHT therapies, so that more efforts will be required to develop strategies directed to avoid them to achieve a satisfactory MHT.

6. Rational Design of Strategies Based on MNPs for MHT and Their Applications to Tumours *In Vitro* and *In Vivo*

Having described the physical and biological behaviour of MNPs in relation to MHT, we shall assess how this knowledge has been directed towards extracting the full potential of this therapy through the rational design of MHT strategies based on MNPs. Hence, it is important to consider the different intrinsic (size [155], shape [156], doping, etc . . .) and extrinsic (magnetic field intensity and frequency, subcellular location, intracellular aggregation, etc . . .) parameters that govern the success of such treatments (Scheme 1). As such, the different studies where MHT has been successfully achieved based on the rational design of *in vitro* and *in vivo* experiments will be considered.

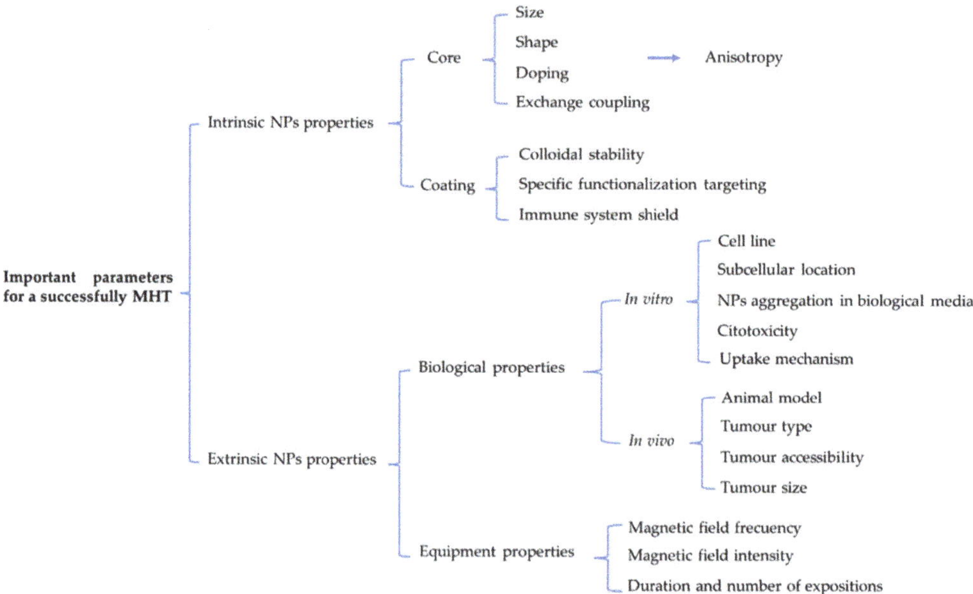

Scheme 1. Main parameters to be considered to achieve a successfully magnetic hyperthermia treatment.

6.1. Fine-Tuning the Intrinsic and Extrinsic MNP Properties for In Vitro Magnetic Hyperthermia

We can find many examples of rationalized MNP designs that have been considered for in vitro MHT. For instance, size-optimized MNPs enhance cell death in Jurkat cells after MHT [157]. Likewise, phospholipid-PEG coating has been used to concurrently deliver Doxorubicin and to generate heat for an enhanced multimodal cancer treatment in HeLa cells [158]. Likewise, MNP functionalization with the folate-receptor (a tumour marker) has been employed for smart delivery to the MCF7 and G1 cell lines, with no uptake by a control L929 cell line [159]. Furthermore, it has been described that MNPs pegylation coating is capable of counteracting the interactions between dipolar particles while maintaining a low level of nanoparticle aggregation in environments of different ionic strengths and viscosities [160]. Each of these serves as a good example of the rationalization of MNP size, coating and targeting for in vitro MHT.

We carried out studies that combined different strategies for the rational design of MNPs [161], synthesizing them by thermal decomposition to obtain 18 nm flower-like Mn-doped SPIONs covered with DMSA and functionalized with cRGD (an $\alpha v \beta 3$-Integrin-Ligand) peptide (from now on named NF-DMSA-PEP) that targets the U87MG glioblastoma cell line. These NF-DMSA-PEP had higher SAR values than 12 nm spherical MNPs covered with DMSA (NP-REF) and 20 nm flower-like MNPs covered with citric acid (NF-REF), demonstrating the notable role of rationalized intrinsic MNPs properties (size, shape, doping and coating MNPs). Furthermore, extrinsic properties related to the biological features of the target cells must also be considered. MNPs uptake was tested by comparing our NF-DMSA-PEP system with the same MNPs without the cRGD peptide (NF-DMSA). NF-DMSA-PEP uptake by U87MG cells was enhanced 5–6 fold, while endocytosis-exocytosis cycles avoided compact aggregation inside lysosomes and the resulting decrease in NP-induced HT. Consequently, peptide effectiveness was demonstrated, establishing 2 h as the optimal time to then apply the magnetic field. Finally, a 2 h NF-DMSA-PEP incubation followed by a 1 h AMF application (25 kA/m, 250 kHz) efficiently induced intracellular cell heating (Hsp70 over-expression), ROS production and cell death (but without inducing apoptosis). The biological effects observed were always stronger with NF-DMSA-PEP than with control MNPs.

6.2. Tuning Intrinsic and Extrinsic MNP Properties for In Vivo Magnetic Hyperthermia

One further step in the study of these approaches is the application of rational design to treat tumours in animal models, mainly glioblastoma, pancreas, breast and prostate. The most common way of administering nanoparticles for antitumor hyperthermia treatments is by intratumoral injection of MNPs, with or without the aid of the use of advanced imaging techniques to deliver MNPs into the tumour [162,163]. Systemic administration of MNP through intravenous injection followed by the biological or physical targeting of those MNPs to the tumour is another possibility, especially in the case of hard-to-reach tumors. However, MNPs doses needed is greater than for intratumoral administration, because the amount of MNPs that reach the tumour depend on several factors such as biodistribution [51], EPR effect [31], active targeting [33,34] and renal clearance [31]. These factors must to be taken account during the rational design process of the MNPs for MHT. For instance, the pegylation of MNPs allows long blood circulation times avoiding the rapid uptake by mononuclear phagocytic system and renal clearance [164]. Moreover, thinking in translational therapy, several studies conclude that EPR effect work properly in rodents but not in humans [32] making essential the use of active targeting strategies that could make the process more complex. Nevertheless, satisfactory MHT through rational designed MNPs intravenously administered has been done, as exemplified a study where a rational design based on pegylation of MNPs and functionalization with c(RGDyK) peptide solved the renal clearance and active targeting issues respectively, and finally glioblastoma regression in mice was achieved [165].

For intratumoral administration rational MNPs design is also needed to improve the regression of tumours. For instance, MNPs covered with DMSA and functionalized with doxorubicin as chemotherapeutic agents were administered intratumorally achieving breast cancer xenograft regression through a synergic effect [166]. Likewise, the development of a biocompatible magnetic lipid nanocomposite vehicle for encapsulate MNPs and doxorubicin was demonstrated to provoke a synergic effect sensitizing the tumour cells to cancer chemotherapy in a subcutaneous melanoma mice model [167]. Another example is the design of Janus MNPs charged with doxorubicin achieving the decrease of tumour weight in subcutaneous breast solid tumour models [168]. Another example, cubic-shaped MNPs, since cubic-shaped MNPs are better heaters than spherical MNPs [97], coated individually with a polymer shell to avoid MNPs aggregation, have been shown to be effective heat mediators for MH and heat-mediated chemotherapy on an in vivo xenograft tumour model using A431 epidermoid carcinoma cells [169].

In addition to the intrinsic properties of MNPs, the extrinsic properties must be considered. It was shown above that controlling and optimizing the biological parameters related to the MNP-cell interactions is important for satisfactory MHT. However, since intra-tumour injection is normally chosen in vivo, it is more interesting to consider other parameters like the animal model or the type, size and location of the tumour. In addition, optimizing the HT equipment conditions will play an even more crucial role. Modifying the magnetic field frequency and intensity will allow the MNP induced heating capacity be fine-tuned, with a higher frequency and a stronger magnetic field intensity translated into more heating, which is crucial for certain nanomaterials. Normally, MNPs require a threshold field to open their hysteresis loop and then achieve heat release. Nevertheless, higher magnetic field frequencies and intensities are not permitted for in vivo or translational therapies and thus, a compromise between these parameters must be found in order to remain within the safety limits [79–81]. The duration and the repetition of applications are other parameters to be considered. A full comparison of the in vitro and in vivo conditions for MHT has recently been prepared by Vilas-Boas et al. [170].

7. From the Laboratory to the Clinic

Despite the improvements implemented and the exponential growth of studies into MHT to treat cancer, translational investigation in this area has not progressed as desired and its clinical implementation has not yet occurred [79]. Various factors have influenced this delay, technological challenges being the most important. As mentioned previously, only one NP formulation has currently been approved for HT: NanoTherm® (MagForce AG, Berlin, Germany). This sole alternative rules out the possibility employing a rational design of NPs with better physical characteristics and a potentially improved heating capacity at the same concentrations or doses [171]. In addition, NanoTherm® is a ferrofluid that agglomerates in the tissue, and entrapment by macrophages as opposed to glioblastoma cells might result in cancer cells receiving insufficient doses [172]. Moreover, intratumour NPs injection must be employed to avoid aggregation, restricting their use to solid and accessible tumours like glioblastoma or prostate cancer. Consequently, a homogeneous NP distribution and therefore, constant heat distribution across the tumour is difficult to achieve [171]. Moreover, the inability to use active rational design also translates into a loss of effectiveness. On the other hand, the AMF applicator MFH 300F® (later implemented as Nanoactivator® F100: MagForce s AG, Berlin, Germany) is the only apparatus used in the clinical trials carried out to date, and always operating at a fixed frequency of 100 kHz and with a field strength of 0–18 kA/s [79]. However, changes to these parameters might produce better results while still respecting the safety limits.

Several clinical trials have been perfromed by Charité—Universitätsmedizin Berlin (Germany) and the spin-off company MagForce AG, Berlin (Germany). The pipeline that has driven the initial idea towards the realization of clinical trials can been easily traced (Figure 4). In 1993, the potential use of SPIONs for HT therapy was first noted and how NP application localised to a tumour might be less invasive than other techniques. With clinical implementation in mind, a moderate concentration of ferrite 5 mg/g tumour was considered, coupled to clinically acceptable magnetic fields that were comparable to radiofrequency heating by local application and superior to regional RF heating [173]. Nearly a decade later, in 2001 this idea was developed further and a new magnetic field therapy system was introduced for the treatment of human solid tumours with magnetic fluid HT. In this study, two of the three pillars of HT therapy used in the Charité-MagForce clinical trials were well defined. Firstly, the aminosilane magnetite NPs used were subsequently manufactured by MagForce AG and called NanoTherm® and they were seen to be more significantly taken-up by malignant cells than normal cells. Secondly, the first prototype of a magnetic fluid hyperthermia (MFH) therapy system was designed (Applicator MFH 300F) and which was later developed as Nanoactivator® F100 by MagForce AG [174,175]. Finally, and just one year later the third pillar appeared, the software initially called HyperPlan and now NanoPlan® (MagForce AG) that enables treatments to be planned through a thin-sliced CT or MRI scan. The software developed, in combination with the AMIRA® visualization package (Mercury Computer Systems, Berlin, Germany), allows us to obtain 3D reconstructions of the NP distribution in the tumour and the localization of the thermometry catheter. Moreover, the physician can modify the parameters to simulate different scenarios and to determine the optimal magnetic field strength for the treatment, estimating the possible temperature distribution during the treatment [176].

Once the basic concepts have been fixed and a well-defined route obtained, a pilot clinical trial was carried out. The aim of this pilot study was to evaluate whether the MFH technique can be used for minimally invasive treatment of prostate cancer. The results indicated that HT using MNPs injected transperineally into the prostate was feasible and well-tolerated. Moreover, NPs were retained for at least 6 weeks in the prostate, making sequential HT treatment possible without the need for new NP application. This study formed the basis on which future clinical trials could be designed [177].

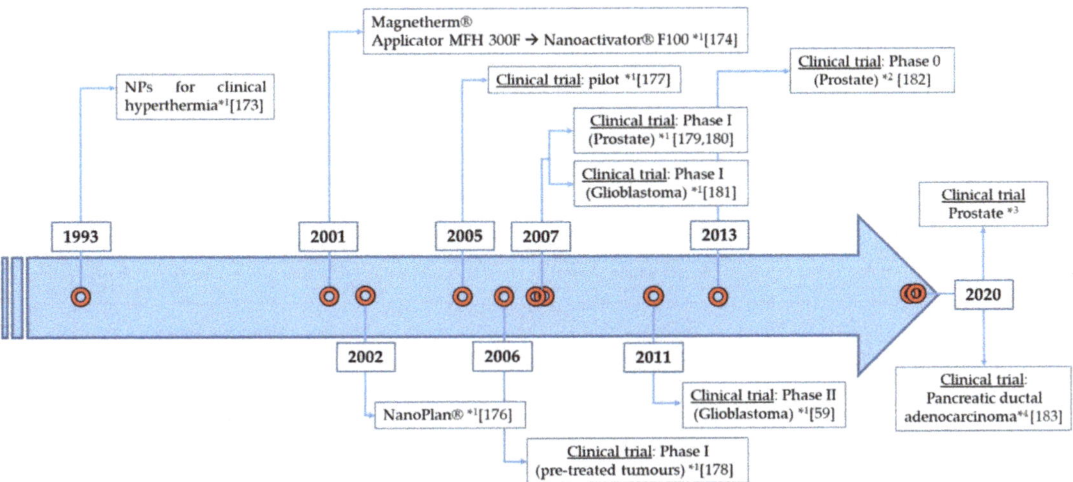

Figure 4. Timeline of the major milestones and clinical trials carried out on humans where Magnetic Hyperthermia Treatment was employed.

In the following years different clinical trials were performed, the details of which are shown in Table 2. These clinical trials focused on understanding, optimizing and improving particular aspects of the technique to enhance the results. First, in 2006 a phase I trial was performed to evaluate the feasibility and tolerability of thermotherapy using MNPs in different pre-treated tumours, as well as testing three different NP injection methods. The results showed that magnetic fluid and thermotherapy treatment was well-tolerated, with no or only moderate side-effects, respectively. Moreover, there was a clear need to further improve the temperature distribution by refining the implantation techniques, or simply by increasing the amount of NPs or the magnetic field strength [178].

A year later, a phase I trial was carried out in 2007 to investigate the feasibility of using thermotherapy with biocompatible SPIONs in patients with locally recurrent prostate cancer, evaluating an image-based approach for the non-invasive calculation of the 3D temperature distribution. It was concluded that heating using MNPs was feasible. Hyperthermic to thermoablative temperatures were achieved in the prostate at 25% of the available magnetic field strength, indicating the potential to reach higher temperatures and that a specific non-invasive thermometry method could be developed that may be used for thermal dosimetry [179]. In the same clinical trial, the treatment-related morbidity and quality of life (QoL) during thermotherapy was studied, and it appeared that interstitial heating using MNPs was feasible and well-tolerated by patients with locally recurrent prostate cancer. Furthermore, deposition of NPs in the prostate was evident 1 year later, even though a homogeneous distribution was not achieved. Finally, a refinement of the technique was needed to enable higher magnetic field strengths to be applied [180].

In parallel, a phase I trial was presented in 2007 to evaluate the feasibility and tolerability of the newly developed thermotherapy, using MNPs to treat recurrent GBM and guiding the intratumour NP injection by 3D imaging. The study demonstrated that thermotherapy using MNPs is safe to use in the treatment of brain tumours and that therapeutic temperatures ranging from HT to thermoablation can be achieved. These promising results opened the door to further studies [181] and consequently, a few years later a phase II clinical trial was carried out on 59 GBM patients. The objective of the study was to determine the efficacy of intratumour thermotherapy in conjunction with fractionated stereotactic radiotherapy for glioblastoma. This clinical trial confirmed the aforementioned advantages, demonstrating MHT to be a safe and well-tolerated cancer therapy. Moreover, the clinical trial concluded that the combination of HT and radiotherapy was clinically effective, augmenting the overall survival of patients. Finally, it was proposed that the combination of HT and chemotherapy (particularly temozolomide), in conjunction with intratumour NP targeting using convection-enhanced delivery (CED) is a promising approach for the treatment of other solid tumours that should be evaluated in future clinical trials [59].

Table 2. Summary of the different MHT clinical trials carried out by Charité –MagForce.

Trial Phase	N° of Patients	Cancer Type	NPs Used and Route of Administration	MFA and Conditions	Hyperthermia Sessions, Duration and Tª Reached	Year / Reference
I (Pilot)	1	Prostate	NanoTherm® injected transperineally into the prostate (NPs retained for at least 6 weeks).	Applicator MFH 300F: (f = 100 kHz) H = 4.0–5.0 kA/m	Hyperthermia, six 1-h sessions: - 1st session: 40.0–48.5 °C - 6th session: 39.4–42.5 °C	2005 [177]
I	22	Different pre-treated recurrences tumours: chondrosarcoma, rectal carcinoma, cervical carcinoma, prostate cancer, sarcoma, rhabdomyosarcoma, ovarian carcinoma	NanoTherm® injected by 3 implantation methods: - Infiltration under CT fluoroscopy (group A) - TRUS (transrectal ultrasound)— implantation guided by X-fluoroscopy (group B) - Intra-operative infiltration under visual control (group C)	Applicator MFH 300F: (f = 100 kHz) - H in pelvis = 3.0–6.0 kA/m - H in thoracic and neck up to 7.5 kA/m - H in head > 10 kA/m	Hyperthermia (six 1-h sessions) + radiotherapy/chemotherapy: - Tª ≥ 40.0 °C: 86% of tumour volume - Tª ≥ 43.0 °C: 30% of tumour volume (Group A) and 0.2% (Group B)	2006 [178]
I	10	Prostate	NanoTherm® injected transperineally into the prostate guided by TRUS/fluoroscopy (89.5% were retained for at least 6 weeks and NPs still observed 1 year later).	Applicator MFH 300F (f = 100 kHz) H = 2.5–18.0 kA/m	Hyperthermia, six 1-h sessions: - Median prostate Tª: 40.1 °C (7.8 min at 43 °C in 90% of cases) - Maximum prostate Tª: 55.0 °C - Median urethral Tª: 40.5 °C - Median rectal Tª: 39.8 °C	2007 [179,180]
I	14	Glioblastoma multiforme	Intratumoural injection guided by 3D image of 0.1–0.7 mL of NanoTherm®	Applicator MFH 300F (f = 100 kHz) H = 2.5–18.0 kA/m	Hyperthermia (six 1-h sessions) + radiotherapy/chemotherapy: - Intratumoural Tª: 42.4–49.5 °C	2007 [181]
II	59	Glioblastoma multiforme	Intratumoural injection guided by 3D image of 0.5–11.6 mL of NanoTherm®	Nanoactivator® F100 (f = 100 kHz) H = 2.0–15.0 kA/m	Hyperthermia (six 1-h sessions) + radiotherapy/chemotherapy: - Median Tª = 51.2 °C	2011 [59]

Finally, the "Magnetic Nanoparticle Thermoablation-Retention and Maintenance in the Prostate: A Phase 0 Study in Men (MAGNABLATE I)" was the first clinical trial performed outside of the Charité –MagForce collaboration, promoted by University College London Hospitals (NCT02033447 [182].

8. Conclusions

From this review we can draw some important conclusions regarding the clinical implementation of MHT. In the case of MNPs, there are many studies about how their intrinsic properties (size, shape, composition, coating, etc) can affect their ability to generate heat in response to AMFs in non-biological milieus. This knowledge can be used to improve their rational design for MHT therapies to obtain the better heat release. However, the behaviour of MNPs in response to AMF in biological media, and the underlying cellular mechanisms that are triggered, are not yet fully understood, representing a major bottleneck in the application of MHT under physiological conditions. Therefore, it is important that more studies focus on the behaviour of MNPs inside cells. For example, it is important to study MNP biodegradation in different cell types and tissues, since MNP degradation in cells could affect their magnetic properties during treatment, and issue that could be critical in therapies that require the repeated application of the magnetic pulses long after the administration of NPs. In terms of the development of HT equipment, it is important to emphasize the importance of selecting the appropriate field intensity and frequency for each case, so that treatments produce specific effects. For clinical applications, it should be remembered that safety limits exist above which we cannot operate.

In general terms, the clinical implementation of MHT has not progressed as might have been anticipated, despite the success of some clinical trials. The global conclusions that we can extract from the clinical trials carried out to date is that the first steps taken have yielded promising results. However, in recent years no more clinical trials have been carried out. Fortunately, this tendency seems to be changing and a few months ago, a new and very ambitious pivotal single-arm clinical study for the focal ablation of intermediate-risk prostate cancer using NanoTherm® was presented by MagForce USA, Inc. following its FDA approval. Along similar lines, in Europe both the Vall d'Hebron University Hospital and the Fuenlabrada University Hospital are involved in a new feasibility study on treating locally advanced pancreatic ductal adenocarcinoma (PDAC) as part of the NoCanTher project [183].

In summary, we strongly believe that increasing knowledge in the key biological aspects above-mentioned is highly necessary to achieve fine control of the process that could trigger the desired clinical implementation of MNPs-based magnetic hyperthermia.

Author Contributions: D.E.-B., J.G.O., M.d.P.M. and D.F.B. conceived and designed the review. D.E.-B. and J.G.O. wrote sections of the manuscript, and prepared the figures and tables. M.d.P.M. and D.F.B. coordinated, critically revised, modified and completed the manuscript. All authors have read and agreed to the published version of the manuscript.

Funding: DEB is a pre-doctoral scholar working under a FPI Contract (PRE2018-084189) from the Spanish Ministry of Science and Innovation. This work was supported in part by grants from the Spanish Ministry of Science and Innovation (SAF-2017-82223-R and PID-2020-112685RB-100 to DFB, and MAT2017-88148-R, to MPM) and by the European Commission through the HOTZYMES project (H2020-FETOPEN-RIA 829162 to MPM). The groups of DFB and MPM are part of the Network "Nanotechnology in Translational Hyperthermia" (HIPERNANO, RED2018-102626-T) supported by the Spanish Ministry of Science and Innovation.

Acknowledgments: We thank members of the laboratories led by Domingo F Barber (CNB-CSIC, Madrid) and María del Puerto Morales (ICMM-CSIC, Madrid) for their helpful comments and discussions. The authors are also grateful to Mark Sefton (BiomedRed SL) for English language editing of the manuscript.

Conflicts of Interest: The authors have no conflicts of interest to declare.

References

1. NIH. Cancer Treatment. Available online: https://www.Cancer.Gov/about-Cancer/Treatment (accessed on 27 July 2021).
2. Corbus, B.C.; Corbus, B.C., Jr. The Utilization of Heat in the Treatment of Tumors of the Urinary Bladder: A Presentation of Technique. *J. Urol.* **1947**, *57*, 730–737. [CrossRef]
3. Scutigliani, E.; Liang, Y.; Crezee, H.; Kanaar, R.; Krawczyk, P. Modulating the Heat Stress Response to Improve Hyperthermia-Based Anticancer Treatments. *Cancers* **2021**, *13*, 1243. [CrossRef]
4. Rao, W.; Deng, Z.-S.; Liu, J. A Review of Hyperthermia Combined with Radiotherapy/Chemotherapy on Malignant Tumors. *Crit. Rev. Biomed. Eng.* **2010**, *38*, 101–116. [CrossRef]
5. Van den Tempel, N.; Horsman, M.; Kanaar, R. Improving efficacy of hyperthermia in oncology by exploiting biological mechanisms. *Int. J. Hyperth.* **2016**, *32*, 446–454. [CrossRef]
6. Van der Zee, J.; Vujaskovic, Z.; Kondo, M.; Sugahara, T. The Kadota Fund International Forum 2004—Clinical group consensus. *Int. J. Hyperth.* **2008**, *24*, 111–122. [CrossRef] [PubMed]
7. Wust, P.; Hildebrandt, B.; Sreenivasa, G.; Rau, B.; Gellermann, J.; Riess, H.; Felix, R.; Schlag, P. Hyperthermia in combined treatment of cancer. *Lancet Oncol.* **2002**, *3*, 487–497. [CrossRef]
8. Wust, P.; Riess, H.; Hildebrandt, B.; Logfel, J.; Deja, M.; Ahlers, O.; Kerner, T.; Felix, R.; Von Ardenne, A.; Wust, H.R.P. Feasibility and analysis of thermal parameters for the whole-body hyperthermia system IRATHERM-2000. *Int. J. Hyperth.* **2000**, *16*, 325–339. [CrossRef] [PubMed]
9. Hildebrandt, B.; Dräger, J.; Kerner, T.; Deja, M.; Löffel, J.; Stroszczynski, C.; Ahlers, O.; Felix, R.; Riess, H.; Wust, P. Whole-body hyperthermia in the scope of von Ardenne's Systemic Cancer Multistep Therapy (sCMT) combined with chemotherapy in patients with metastatic colorectal cancer: A phase I/II study. *Int. J. Hyperth.* **2004**, *20*, 317–333. [CrossRef]
10. Bettaieb, A.; Wrzal, P.K.; Averil-Bates, D.A. Hyperthermia: Cancer Treatment and Beyond. In *Cancer Treatment—Conventional and Innovative Approaches*; IntechOpen: London, UK, 2013; pp. 257–283. [CrossRef]

11. Mortezaee, K.; Narmani, A.; Salehi, M.; Bagheri, H.; Farhood, B.; Haghi-Aminjan, H.; Najafi, M. Synergic effects of nanoparticles-mediated hyperthermia in radiotherapy/chemotherapy of cancer. *Life Sci.* **2021**, *269*, 119020. [CrossRef] [PubMed]
12. Tanaka, H.; Eno, K.; Kato, H.; Ishida, T. Possible application of non-invasive thermometry for hyperthermia using NMR. *Nippon. Acta Radiol.* **1981**, *41*, 897–898. [PubMed]
13. Feddersen, T.V.; Hernandez-Tamames, J.A.; Franckena, M.; Van Rhoon, G.C.; Paulides, M.M. Clinical Performance and Future Potential of Magnetic Resonance Thermometry in Hyperthermia. *Cancers* **2021**, *13*, 31. [CrossRef]
14. Winter, L.; Oberacker, E.; Paul, K.; Ji, Y.; Oezerdem, C.; Ghadjar, P.; Thieme, A.; Budach, V.; Wust, P.; Niendorf, T. Magnetic resonance thermometry: Methodology, pitfalls and practical solutions. *Int. J. Hyperth.* **2015**, *32*, 63–75. [CrossRef]
15. Lu, Y.; Rivera-Rodriguez, A.; Tay, Z.W.; Hensley, D.; Fung, K.B.; Colson, C.; Saayujya, C.; Huynh, Q.; Kabuli, L.; Fellows, B.; et al. Combining magnetic particle imaging and magnetic fluid hyperthermia for localized and image-guided treatment. *Int. J. Hyperth.* **2020**, *37*, 141–154. [CrossRef]
16. Dadfar, S.M.; Roemhild, K.; Drude, N.; von Stillfried, S.; Knüchel, R.; Kiessling, F.; Lammers, T. Iron oxide nanoparticles: Diagnostic, therapeutic and theranostic applications. *Adv. Drug Deliv. Rev.* **2019**, *138*, 302–325. [CrossRef]
17. Cai, X.; Zhu, Q.; Zeng, Y.; Zeng, Q.; Chen, X.; Zhan, Y. Manganese Oxide Nanoparticles as MRI Contrast Agents in Tumor Multimodal Imaging and Therapy. *Int. J. Nanomed.* **2019**, *14*, 8321–8344. [CrossRef] [PubMed]
18. Lin, D.; Tseng, C.-Y.; Lim, Q.F.; Tan, M.J.; Kong, K.V.; Chinh-Yu, T. A rapid and highly sensitive strain-effect graphene-based bio-sensor for the detection of stroke and cancer bio-markers. *J. Mater. Chem. B* **2018**, *6*, 2536–2540. [CrossRef]
19. Omer, W.E.; El-Kemary, M.A.; Elsaady, M.; Abou-Omar, M.N.; Youssef, A.O.; Sayqal, A.A.; Gouda, A.A.; Attia, M.S. Highly Efficient Gold Nano-Flower Optical Biosensor Doped in a Sol–Gel/PEG Matrix for the Determination of a Calcitonin Biomarker in Different Serum Samples. *ACS Omega* **2020**, *5*, 5629–5637. [CrossRef] [PubMed]
20. De Jong, W.H.; Borm, P.J.A. Drug delivery and nanoparticles: Applications and hazards. *Int. J. Nanomed.* **2008**, *3*, 133–149. [CrossRef]
21. Sun, T.; Zhang, Y.S.; Pang, B.; Hyun, D.C.; Yang, M.; Xia, Y. Engineered Nanoparticles for Drug Delivery in Cancer Therapy. *Angew. Chem. Int. Ed.* **2014**, *53*, 12320–12364. [CrossRef]
22. Pérez-Herrero, E.; Fernández-Medarde, A. Advanced targeted therapies in cancer: Drug nanocarriers, the future of chemotherapy. *Eur. J. Pharm. Biopharm.* **2015**, *93*, 52–79. [CrossRef] [PubMed]
23. Soto-Sánchez, C.; Martínez-Navarrete, G.; Humphreys, L.; Puras, G.; Zarate, J.; Pedraz, J.L.; Fernández, E. Enduring high-efficiency in vivo transfection of neurons with non-viral magnetoparticles in the rat visual cortex for optogenetic applications. *Nanomed. Nanotechnol. Biol. Med.* **2015**, *11*, 835–843. [CrossRef]
24. Zaimy, M.A.; Saffarzadeh, N.; Mohammadi, A.; Pourghadamyari, H.; Izadi, P.; Sarli, A.; Moghaddam, L.K.; Paschepari, S.R.; Azizi, H.; Torkamandi, S.; et al. New methods in the diagnosis of cancer and gene therapy of cancer based on nanoparticles. *Cancer Gene Ther.* **2017**, *24*, 233–243. [CrossRef]
25. Sanz-Ortega, L.; Rojas, J.; Barber, D. Improving Tumor Retention of Effector Cells in Adoptive Cell Transfer Therapies by Magnetic Targeting. *Pharmaceutics* **2020**, *12*, 12. [CrossRef]
26. Zhu, L.; Zhou, Z.; Mao, H.; Yang, L. Magnetic nanoparticles for precision oncology: Theranostic magnetic iron oxide nanoparticles for image-guided and targeted cancer therapy. *Nanomedicine* **2017**, *12*, 73–87. [CrossRef]
27. Paz-Ares, L.; Luft, A.; Vicente, D.; Tafreshi, A.; Gümüş, M.; Mazières, J.; Hermes, B.; Çay Şenler, F.; Csőszi, T.; Fülöp, A.; et al. Pembrolizumab plus Chemotherapy for Squamous Non–Small-Cell Lung Cancer. *N. Engl. J. Med.* **2018**, *379*, 2040–2051. [CrossRef]
28. Taruno, K.; Kurita, T.; Kuwahata, A.; Yanagihara, K.; Enokido, K.; Katayose, Y.; Nakamura, S.; Takei, H.; Sekino, M.; Kusakabe, M. Multicenter clinical trial on sentinel lymph node biopsy using superparamagnetic iron oxide nanoparticles and a novel handheld magnetic probe. *J. Surg. Oncol.* **2019**, *120*, 1391–1396. [CrossRef] [PubMed]
29. Makita, M.; Manabe, E.; Kurita, T.; Takei, H.; Nakamura, S.; Kuwahata, A.; Sekino, M.; Kusakabe, M.; Ohashi, Y. Moving a neodymium magnet promotes the migration of a magnetic tracer and increases the monitoring counts on the skin surface of sentinel lymph nodes in breast cancer. *BMC Med. Imaging* **2020**, *20*, 1–11. [CrossRef] [PubMed]
30. Matsumura, Y.; Maeda, H. A new concept for macromolecular therapeutics in cancer chemotherapy: Mechanism of tumoritropic accumulation of proteins and the antitumor agent smancs. *Cancer Res.* **1986**, *46*, 6387–6392. [PubMed]
31. Maeda, H.; Tsukigawa, K.; Fang, J. A Retrospective 30 Years After Discovery of the Enhanced Permeability and Retention Effect of Solid Tumors: Next-Generation Chemotherapeutics and Photodynamic Therapy-Problems, Solutions, and Prospects. *Microcirculation* **2016**, *23*, 173–182. [CrossRef] [PubMed]
32. Danhier, F. To exploit the tumor microenvironment: Since the EPR effect fails in the clinic, what is the future of nanomedicine? *J. Control. Release* **2016**, *244*, 108–121. [CrossRef]
33. Bazak, R.; Houri, M.; El Achy, S.; Kamel, S.; Refaat, T. Cancer active targeting by nanoparticles: A comprehensive review of literature. *J. Cancer Res. Clin. Oncol.* **2015**, *141*, 784. [CrossRef]
34. Swain, S.; Sahu, P.K.; Beg, S.; Babu, S. Nanoparticles for Cancer Targeting: Current and Future Directions. *Curr. Drug Deliv.* **2016**, *13*, 1290–1302. [CrossRef] [PubMed]
35. Pellico, J.; Ruiz-Cabello, J.; Fernández-Barahona, I.; Gutiérrez, L.; Lechuga-Vieco, A.V.; Enríquez, J.A.; Morales, M.P.; Herranz, F. One-Step Fast Synthesis of Nanoparticles for MRI: Coating Chemistry as the Key Variable Determining Positive or Negative Contrast. *Langmuir* **2017**, *33*, 10239–10247. [CrossRef]

36. Pellico, J.; Ruiz-Cabello, J.; Saiz-Alía, M.; Del Rosario, G.; Caja, S.; Montoya, M.; Fernández de Manuel, L.; Morales, M.P.; Gutiérrez, L.; Galiana, B.; et al. Fast synthesis and bioconjugation of68Ga core-doped extremely small iron oxide nanoparticles for PET/MR imaging. *Contrast Media Mol. Imaging* **2016**, *11*, 203–210. [CrossRef] [PubMed]

37. Kang, J.K.; Kim, J.C.; Shin, Y.; Han, S.M.; Won, W.R.; Her, J.; Park, J.Y.; Oh, K.T. Principles and applications of nanomaterial-based hyperthermia in cancer therapy. *Arch. Pharmacal Res.* **2020**, *43*, 46–57. [CrossRef]

38. Chang, M.; Hou, Z.; Wang, M.; Li, C.; Lin, J. Recent Advances in Hyperthermia Therapy-Based Synergistic Immunotherapy. *Adv. Mater.* **2020**, *33*, e2004788. [CrossRef] [PubMed]

39. Onaciu, A.; Braicu, C.; Zimta, A.-A.; Moldovan, A.; Stiufiuc, R.; Buse, M.; Ciocan, C.; Buduru, S.; Berindan-Neagoe, I. Gold nanorods: From anisotropy to opportunity. An evolution update. *Nanomedicine* **2019**, *14*, 1203–1226. [CrossRef]

40. Singh, R.; Torti, S.V. Carbon nanotubes in hyperthermia therapy. *Adv. Drug Deliv. Rev.* **2013**, *65*, 2045–2060. [CrossRef] [PubMed]

41. Gupta, B.D.; Pathak, A.; Semwal, V. Carbon-Based Nanomaterials for Plasmonic Sensors: A Review. *Sensors* **2019**, *19*, 3536. [CrossRef]

42. Tian, Q.; Jiang, F.; Zou, R.; Liu, Q.; Chen, Z.; Zhu, M.; Yang, S.; Wang, J.; Wang, J.; Hu, J. Hydrophilic Cu9S5 Nanocrystals: A Photothermal Agent with a 25.7% Heat Conversion Efficiency for Photothermal Ablation of Cancer Cells in Vivo. *ACS Nano* **2011**, *5*, 9761–9771. [CrossRef] [PubMed]

43. Sun, A.; Guo, H.; Gan, Q.; Yang, L.; Liu, Q.; Xi, L. Evaluation of visible NIR-I and NIR-II light penetration for photoacoustic imaging in rat organs. *Opt. Express* **2020**, *28*, 9002–9013. [CrossRef]

44. Beik, J.; Abed, Z.; Ghoreishi, F.S.; Hosseini-Nami, S.; Mehrzadi, S.; Shakeri-Zadeh, A.; Kamrava, S.K. Nanotechnology in hyperthermia cancer therapy: From fundamental principles to advanced applications. *J. Control. Release* **2016**, *235*, 205–221. [CrossRef]

45. Amini, S.M.; Kharrazi, S.; Rezayat, S.M.; Gilani, K. Radiofrequency electric field hyperthermia with gold nanostructures: Role of particle shape and surface chemistry. *Artif. Cells Nanomed. Biotechnol.* **2018**, *46*, 1452–1462. [CrossRef] [PubMed]

46. Postnikov, A.; Moldosanov, K. Phonon-Assisted Radiofrequency Absorption by Gold Nanoparticles Resulting in Hyperthermia. In *Fundamental and Applied Nano-Electromagnetics*; Springer: Dordrecht, The Netherlands, 2016; pp. 171–201. [CrossRef]

47. Connord, V.; Clerc, P.; Hallali, N.; Diab, D.E.H.; Fourmy, D.; Gigoux, V.; Carrey, J. Real-Time Analysis of Magnetic Hyperthermia Experiments on Living Cells under a Confocal Microscope. *Small* **2015**, *11*, 2437–2445. [CrossRef] [PubMed]

48. Blanco-Andujar, C.; Walter, A.; Cotin, G.; Bordeianu, C.; Mertz, D.; Felder-Flesch, D.; Begin-Colin, S. Design of iron oxide-based nanoparticles for MRI and magnetic hyperthermia. *Nanomedicine* **2016**, *11*, 1889–1910. [CrossRef]

49. Unkelbach, J.; Alber, M.; Bangert, M.; Bokrantz, R.; Chan, T.C.Y.; Deasy, J.O.; Fredriksson, A.; Gorissen, B.L.; van Herk, M.; Liu, W.; et al. Robust radiotherapy planning. *Phys. Med. Biol.* **2018**, *63*, 22TR02. [CrossRef]

50. Hergt, R.; Dutz, S.; Zeisberger, M. Validity limits of the Néel relaxation model of magnetic nanoparticles for hyperthermia. *Nanotechnology* **2010**, *21*, 015706. [CrossRef]

51. Patsula, V.; Horák, D.; Kučka, J.; Macková, H.; Lobaz, V.; Francová, P.; Herynek, V.; Heizer, T.; Páral, P.; Šefc, L. Synthesis and modification of uniform PEG-neridronate-modified magnetic nanoparticles determines prolonged blood circulation and biodistribution in a mouse preclinical model. *Sci. Rep.* **2019**, *9*, 10765. [CrossRef] [PubMed]

52. Ruiz, A.; Salas, G.; Calero, M.; Hernández, Y.; Villanueva, A.; Herranz, F.; Veintemillas-Verdaguer, S.; Martínez, E.; Barber, D.; Morales, M. Short-chain PEG molecules strongly bound to magnetic nanoparticle for MRI long circulating agents. *Acta Biomater.* **2013**, *9*, 6421–6430. [CrossRef]

53. Formica, D.; Silvestri, S. Biological effects of exposure to magnetic resonance imaging: An overview. *Biomed. Eng. Online* **2004**, *3*, 11. [CrossRef]

54. Dutz, S.; Hergt, R. Magnetic particle hyperthermia—A promising tumour therapy? *Nanotechnology* **2014**, *25*, 452001. [CrossRef]

55. Coffey, W. *Relaxation Phenomena in Condensed Matter*; John Wiley & Sons: Hoboken, NJ, USA, 1994; Volume 87, ISBN 0471303127.

56. Mamiya, H.; Jeyadevan, B. Hyperthermic effects of dissipative structures of magnetic nanoparticles in large alternating magnetic fields. *Sci. Rep.* **2011**, *1*, 157. [CrossRef]

57. Wang, C.; Hsu, C.-H.; Li, Z.; Hwang, L.-P.; Lin, Y.-C.; Chou, P.-T.; Lin, Y.-Y. Effective heating of magnetic nanoparticle aggregates for in vivo nano-theranostic hyperthermia. *Int. J. Nanomed.* **2017**, *12*, 6273–6287. [CrossRef] [PubMed]

58. Soukup, D.; Moise, S.; Céspedes, E.; Dobson, J.; Telling, N. In Situ Measurement of Magnetization Relaxation of Internalized Nanoparticles in Live Cells. *ACS Nano* **2015**, *9*, 231–240. [CrossRef]

59. Maier-Hauff, K.; Ulrich, F.; Nestler, D.; Niehoff, H.; Wust, P.; Thiesen, B.; Orawa, H.; Budach, V.; Jordan, A. Efficacy and safety of intratumoral thermotherapy using magnetic iron-oxide nanoparticles combined with external beam radiotherapy on patients with recurrent glioblastoma multiforme. *J. Neuro-Oncol.* **2011**, *103*, 317–324. [CrossRef]

60. Wang, X.; Law, J.; Luo, M.; Gong, Z.; Yu, J.; Tang, W.; Zhang, Z.; Mei, X.; Huang, Z.; You, L.; et al. Magnetic Measurement and Stimulation of Cellular and Intracellular Structures. *ACS Nano* **2020**, *14*, 3805–3821. [CrossRef]

61. Gahl, T.J.; Kunze, A. Force-Mediating Magnetic Nanoparticles to Engineer Neuronal Cell Function. *Front. Neurosci.* **2018**, *12*, 299. [CrossRef] [PubMed]

62. Singh, H.; Laibinis, P.E.; Hatton, T.A. Rigid, Superparamagnetic Chains of Permanently Linked Beads Coated with Magnetic Nanoparticles. Synthesis and Rotational Dynamics under Applied Magnetic Fields. *Langmuir* **2005**, *21*, 11500–11509. [CrossRef]

63. Serantes, D.; Chantrell, R.; Gavilán, H.; Morales, M.D.P.; Chubykalo-Fesenko, O.; Baldomir, D.; Satoh, A. Anisotropic magnetic nanoparticles for biomedicine: Bridging frequency separated AC-field controlled domains of actuation. *Phys. Chem. Chem. Phys.* **2018**, *20*, 30445–30454. [CrossRef]
64. Usov, N.A.; Serebryakova, O.; Gubanova, E.M. The heating of magnetic nanoparticles in a rotating magnetic field. *Nanoscale Microscale Thermophys. Eng.* **2020**, *24*, 20–28. [CrossRef]
65. Chen, M.; Wu, J.; Ning, P.; Wang, J.; Ma, Z.; Huang, L.; Plaza, G.R.; Shen, Y.; Xu, C.; Han, Y.; et al. Remote Control of Mechanical Forces via Mitochondrial-Targeted Magnetic Nanospinners for Efficient Cancer Treatment. *Small* **2020**, *16*, e1905424. [CrossRef]
66. Zhang, E.; Kircher, M.F.; Koch, M.; Eliasson, L.; Goldberg, S.N.; Renström, E. Dynamic Magnetic Fields Remote-Control Apoptosis via Nanoparticle Rotation. *ACS Nano* **2014**, *8*, 3192–3201. [CrossRef]
67. Hallali, N.; Clerc, P.; Fourmy, D.; Gigoux, V.; Carrey, J. Influence on cell death of high frequency motion of magnetic nanoparticles during magnetic hyperthermia experiments. *Appl. Phys. Lett.* **2016**, *109*, 032402. [CrossRef]
68. Cheng, D.; Li, X.; Zhang, G.; Shi, H. Morphological effect of oscillating magnetic nanoparticles in killing tumor cells. *Nanoscale Res. Lett.* **2014**, *9*, 195. [CrossRef]
69. Kim, D.-H.; Rozhkova, E.A.; Ulasov, I.; Bader, S.D.; Rajh, T.; Lesniak, M.S.; Novosad, V. Biofunctionalized magnetic-vortex microdiscs for targeted cancer-cell destruction. *Nat. Mater.* **2010**, *9*, 165–171. [CrossRef] [PubMed]
70. Golovin, Y.I.; Klyachko, N.L.; Gribanovskii, S.L.; Golovin, D.Y.; Samodurov, A.A.; Majouga, A.G.; Sokolsky-Papkov, M.; Kabanov, A. Nanomechanical control of properties of biological membranes achieved by rodlike magnetic nanoparticles in a superlow-frequency magnetic field. *Tech. Phys. Lett.* **2015**, *41*, 455–457. [CrossRef]
71. Wells, J.; Ortega, D.; Steinhoff, U.; Dutz, S.; Garaio, E.; Sandre, O.; Natividad, E.; Cruz, M.M.; Brero, F.; Southern, P.; et al. Challenges and recommendations for magnetic hyperthermia characterization measurements. *Int. J. Hyperth.* **2021**, *38*, 447–460. [CrossRef] [PubMed]
72. Hand, J.; Li, Y.; Thomas, E.; Rutherford, M.; Hajnal, J. Prediction of specific absorption rate in mother and fetus associated with MRI examinations during pregnancy. *Magn. Reson. Med.* **2006**, *55*, 883–893. [CrossRef]
73. Kallumadil, M.; Tada, M.; Nakagawa, T.; Abe, M.; Southern, P.; Pankhurst, Q.A. Suitability of commercial colloids for magnetic hyperthermia. *J. Magn. Magn. Mater.* **2009**, *321*, 1509–1513. [CrossRef]
74. Huang, S.; Wang, S.-Y.; Gupta, A.; Borca-Tasciuc, D.-A.; Salon, S.J. On the measurement technique for specific absorption rate of nanoparticles in an alternating electromagnetic field. *Meas. Sci. Technol.* **2012**, *23*, 035701. [CrossRef]
75. Rosensweig, R. Heating magnetic fluid with alternating magnetic field. *J. Magn. Magn. Mater.* **2002**, *252*, 370–374. [CrossRef]
76. Prigo, E.A.; Hemery, G.; Sandre, O.; Ortega, D.; Garaio, E.; Plazaola, F.; Teran, F.J. Fundamentals and advances in magnetic hyperthermia. *Appl. Phys. Rev.* **2015**, *2*, 041302. [CrossRef]
77. Garaio, E.; Collantes, J.M.; Plazaola, F.; Garcia, J.A.; Castellanos-Rubio, I. A multifrequency eletromagnetic applicator with an integrated AC magnetometer for magnetic hyperthermia experiments. *Meas. Sci. Technol.* **2014**, *25*, 115702. [CrossRef]
78. Atkinson, W.J.; Brezovich, I.A.; Chakraborty, D.P. Usable Frequencies in Hyperthermia with Thermal Seeds. *IEEE Trans. Biomed. Eng.* **1984**, *31*, 70–75. [CrossRef] [PubMed]
79. Thiesen, B.; Jordan, A. Clinical applications of magnetic nanoparticles for hyperthermia. *Int. J. Hyperth.* **2008**, *24*, 467–474. [CrossRef] [PubMed]
80. Kossatz, S.; Ludwig, R.L.; Dähring, H.; Ettelt, V.; Rimkus, G.; Marciello, M.; Salas, G.; Patel, V.; Teran, F.J.; Hilger, I. High Therapeutic Efficiency of Magnetic Hyperthermia in Xenograft Models Achieved with Moderate Temperature Dosages in the Tumor Area. *Pharm. Res.* **2014**, *31*, 3274–3288. [CrossRef]
81. Mamiya, H. Recent Advances in Understanding Magnetic Nanoparticles in AC Magnetic Fields and Optimal Design for Targeted Hyperthermia. *J. Nanomater.* **2013**, *2013*, 752973. [CrossRef]
82. Ovejero, J.G.; Yoon, S.J.; Li, J.; Mayoral, A.; Gao, X.; O'Donnell, M.; García, M.A.; Herrasti, P.; Hernando, A. Synthesis of hybrid magneto-plasmonic nanoparticles with potential use in photoacoustic detection of circulating tumor cells. *Microchim. Acta* **2018**, *185*, 130. [CrossRef]
83. Galanzha, E.I.; Zharov, V.P. Circulating Tumor Cell Detection and Capture by Photoacoustic Flow Cytometry in Vivo and ex Vivo. *Cancers* **2013**, *5*, 1691–1738. [CrossRef]
84. Sanz-Ortega, L.; Rojas, J.M.; Marcos, A.; Portilla, Y.; Stein, J.V.; Barber, D.F. T cells loaded with magnetic nanoparticles are retained in peripheral lymph nodes by the application of a magnetic field. *J. Nanobiotechnol.* **2019**, *17*, 14. [CrossRef]
85. Kilinc, D.; Lesniak, A.; Rashdan, S.A.; Gandhi, D.; Blasiak, A.; Fannin, P.C.; von Kriegsheim, A.; Kolch, W.; Lee, G.U. Mechanochemical Stimulation of MCF7 Cells with Rod-Shaped Fe-Au Janus Particles Induces Cell Death Through Paradoxical Hyperactivation of ERK. *Adv. Healthc. Mater.* **2014**, *4*, 395–404. [CrossRef]
86. Smith, B.R.; Gambhir, S.S. Nanomaterials for In Vivo Imaging. *Chem. Rev.* **2017**, *117*, 901–986. [CrossRef] [PubMed]
87. Tay, Z.W.; Chandrasekharan, P.; Chiu-Lam, A.; Hensley, D.W.; Dhavalikar, R.; Zhou, X.Y.; Yu, E.Y.; Goodwill, P.W.; Zheng, B.; Rinaldi, C.; et al. Magnetic Particle Imaging-Guided Heating in Vivo Using Gradient Fields for Arbitrary Localization of Magnetic Hyperthermia Therapy. *ACS Nano* **2018**, *12*, 3699–3713. [CrossRef] [PubMed]
88. Cornell, R.M.; Schwertmann, U. *The Iron Oxides: Structure, Properties, Reactions, Occurrences and Uses*, 2nd ed.; John Wiley & Sons: Hoboken, NJ, USA, 2003; pp. 32–34. ISBN 9783527302741.

89. Lasheras, X.; Insausti, M.; de la Fuente, J.M.; Gil de Muro, I.; Castellanos-Rubio, I.; Marcano, L.; Fernández-Gubieda, M.L.; Serrano, A.; Martín-Rodríguez, R.; Garaio, E.; et al. Mn-Doping level dependence on the magnetic response of MnxFe3−xO4 ferrite nanoparticles. *Dalton Trans.* **2019**, *48*, 11480–11491. [CrossRef] [PubMed]

90. Yang, L.; Ma, L.; Xin, J.; Li, A.; Sun, C.; Wei, R.; Ren, B.W.; Chen, Z.; Lin, H.; Gao, J. Composition Tunable Manganese Ferrite Nanoparticles for Optimized T2 Contrast Ability. *Chem. Mater.* **2017**, *29*, 3038–3047. [CrossRef]

91. Jang, J.-T.; Nah, H.; Lee, J.-H.; Moon, S.H.; Kim, M.-G.; Cheon, J. Critical Enhancements of MRI Contrast and Hyperthermic Effects by Dopant-Controlled Magnetic Nanoparticles. *Angew. Chem. Int. Ed.* **2009**, *48*, 1234–1238. [CrossRef]

92. Munoz-Menendez, C.; Serantes, D.; Ruso, J.; Baldomir, D. Towards improved magnetic fluid hyperthermia: Major-loops to diminish variations in local heating. *Phys. Chem. Chem. Phys.* **2017**, *19*, 14527–14532. [CrossRef]

93. Serantes, D.; Baldomir, D.; Martinez-Boubeta, C.; Simeonidis, K.; Angelakeris, M.; Natividad, E.; Castro, M.; Mediano, A.; Chen, D.-X.; Sanchez, A.; et al. Influence of dipolar interactions on hyperthermia properties of ferromagnetic particles. *J. Appl. Phys.* **2010**, *108*, 073918. [CrossRef]

94. Mazarío, E.; Herrasti, P.; Morales, M.D.P.; Menendez, N. Synthesis and characterization of CoFe2O4ferrite nanoparticles obtained by an electrochemical method. *Nanotechnology* **2012**, *23*, 355708. [CrossRef]

95. Coey, J.M.D. *Magnetism and Magnetic Materials*; Cambridge University Press: Cambridge, UK, 2004; ISBN 9780521016766.

96. Roca, A.G.; Gutiérrez, L.; Gavilán, H.; Fortes Brollo, M.E.; Veintemillas-Verdaguer, S.; del Morales, M.P. Design strategies for shape-controlled magnetic iron oxide nanoparticles. *Adv. Drug Deliv. Rev.* **2019**, *138*, 68–104. [CrossRef] [PubMed]

97. Guardia, P.; Di Corato, R.; Lartigue, L.; Wilhelm, C.; Espinosa, A.; Garcia-Hernandez, M.; Gazeau, F.; Manna, L.; Pellegrino, T. Water-Soluble Iron Oxide Nanocubes with High Values of Specific Absorption Rate for Cancer Cell Hyperthermia Treatment. *ACS Nano* **2012**, *6*, 3080–3091. [CrossRef]

98. Geng, S.; Yang, H.; Ren, X.; Liu, Y.; He, S.; Zhou, J.; Su, N.; Li, Y.; Xu, C.; Zhang, X.; et al. Anisotropic Magnetite Nanorods for Enhanced Magnetic Hyperthermia. *Chem.-Asian J.* **2016**, *11*, 2996–3000. [CrossRef]

99. Golovin, Y.I.; Klyachko, N.L.; Majouga, A.G.; Gribanovskii, S.L.; Golovin, D.Y.; Zhigachev, A.O.; Shuklinov, A.V.; Efremova, M.V.; Veselov, M.M.; Vlasova, K.Y.; et al. New Approaches to Nanotheranostics: Polyfunctional Magnetic Nanoparticles Activated by Non-Heating Low-Frequency Magnetic Field Control Biochemical System with Molecular Locality and Selectivity. *Nanotechnol. Russ.* **2018**, *13*, 215–239. [CrossRef]

100. Lisjak, D.; Mertelj, A. Anisotropic magnetic nanoparticles: A review of their properties, syntheses and potential applications. *Prog. Mater. Sci.* **2018**, *95*, 286–328. [CrossRef]

101. He, S.; Zhang, H.; Liu, Y.; Sun, F.; Yu, X.; Li, X.; Zhang, L.; Wang, L.; Mao, K.; Wang, G.; et al. Maximizing Specific Loss Power for Magnetic Hyperthermia by Hard-Soft Mixed Ferrites. *Small* **2018**, *14*, e1800135. [CrossRef]

102. Phan, M.-H.; Alonso, J.; Khurshid, H.; Lampen-Kelley, P.; Chandra, S.; Repa, K.S.; Nemati, Z.; Das, R.; Iglesias, Ó.; Srikanth, H. Exchange Bias Effects in Iron Oxide-Based Nanoparticle Systems. *Nanomaterials* **2016**, *6*, 221. [CrossRef]

103. Lewis, L.H.; Jiménez-Villacorta, F. Perspectives on Permanent Magnetic Materials for Energy Conversion and Power Generation. *Met. Mater. Trans. A* **2013**, *44*, 2–20. [CrossRef]

104. Lee, J.-H.; Jang, J.-T.; Choi, J.-S.; Moon, S.H.; Noh, S.-H.; Kim, J.-W.; Kim, J.-G.; Kim, I.-S.; Park, K.I.; Cheon, J. Exchange-coupled magnetic nanoparticles for efficient heat induction. *Nat. Nanotechnol.* **2011**, *6*, 418–422. [CrossRef]

105. Noh, S.-H.; Na, W.; Jang, J.-T.; Lee, J.-H.; Lee, E.J.; Moon, S.H.; Lim, Y.; Shin, J.-S.; Cheon, J. Nanoscale Magnetism Control via Surface and Exchange Anisotropy for Optimized Ferrimagnetic Hysteresis. *Nano Lett.* **2012**, *12*, 3716–3721. [CrossRef]

106. Nica, V.; Caro, C.; Páez-Muñoz, J.M.; Leal, M.P.; Garcia-Martin, M.L. Bi-Magnetic Core-Shell CoFe2O4@MnFe2O4 Nanoparticles for In Vivo Theranostics. *Nanomaterials* **2020**, *10*, 907. [CrossRef]

107. Hugounenq, P.; Levy, M.; Alloyeau, D.; Lartigue, L.; Dubois, E.; Cabuil, V.; Ricolleau, C.; Roux, S.; Wilhelm, C.; Gazeau, F.; et al. Iron Oxide Monocrystalline Nanoflowers for Highly Efficient Magnetic Hyperthermia. *J. Phys. Chem. C* **2012**, *116*, 15702–15712. [CrossRef]

108. Lartigue, L.; Hugounenq, P.; Alloyeau, D.; Clarke, S.P.; Lévy, M.; Bacri, J.-C.; Bazzi, R.; Brougham, D.F.; Wilhelm, C.; Gazeau, F. Cooperative Organization in Iron Oxide Multi-Core Nanoparticles Potentiates Their Efficiency as Heating Mediators and MRI Contrast Agents. *ACS Nano* **2012**, *6*, 10935–10949. [CrossRef]

109. Hemery, G.; Keyes, J.A.C.; Garaio, E.; Rodrigo, I.; Garcia, J.A.; Plazaola, F.; Garanger, E.; Sandre, O. Tuning Sizes, Morphologies, and Magnetic Properties of Monocore Versus Multicore Iron Oxide Nanoparticles through the Controlled Addition of Water in the Polyol Synthesis. *Inorg. Chem.* **2017**, *56*, 8232–8243. [CrossRef]

110. Jonasson, C.; Schaller, V.; Zeng, L.; Olsson, E.; Frandsen, C.; Castro, A.; Nilsson, L.; Bogart, L.K.; Southern, P.; Pankhurst, Q.A.; et al. Modelling the effect of different core sizes and magnetic interactions inside magnetic nanoparticles on hyperthermia performance. *J. Magn. Magn. Mater.* **2018**, *477*, 198–202. [CrossRef]

111. Fiorani, D.; Dormann, J.; Cherkaoui, R.; Tronc, E.; Lucari, F.; D'Orazio, F.; Spinu, L.; Nogues, M.; García-Santiago, A.; Testa, A.M. Collective magnetic state in nanoparticles systems. *J. Magn. Magn. Mater.* **1999**, *196*, 143–147. [CrossRef]

112. Ovejero, J.G.; Cabrera, D.; Carrey, J.; Valdivielso, T.; Salas, G.; Teran, F.J. Effects of inter- and intra-aggregate magnetic dipolar interactions on the magnetic heating efficiency of iron oxide nanoparticles. *Phys. Chem. Chem. Phys.* **2016**, *18*, 10954–10963. [CrossRef]

113. Sadat, M.; Patel, R.; Sookoor, J.; Bud'Ko, S.L.; Ewing, R.C.; Zhang, J.; Xu, H.; Wang, Y.; Pauletti, G.M.; Mast, D.B.; et al. Effect of spatial confinement on magnetic hyperthermia via dipolar interactions in Fe3O4 nanoparticles for biomedical applications. *Mater. Sci. Eng. C* **2014**, *42*, 52–63. [CrossRef]

114. Cabrera, D.; Coene, A.; Leliaert, J.; Artés-Ibáñez, E.J.; Dupré, L.; Telling, N.D.; Teran, F.J. Dynamical Magnetic Response of Iron Oxide Nanoparticles Inside Live Cells. *ACS Nano* **2018**, *12*, 2741–2752. [CrossRef]

115. Serantes, D.; Simeonidis, K.; Angelakeris, M.; Chubykalo-Fesenko, O.; Marciello, M.; Morales, M.D.P.; Baldomir, D.; Boubeta, C.M. Multiplying Magnetic Hyperthermia Response by Nanoparticle Assembling. *J. Phys. Chem. C* **2014**, *118*, 5927–5934. [CrossRef]

116. Patil, R.M.; Thorat, N.D.; Shete, P.B.; Bedge, P.A.; Gavde, S.; Joshi, M.G.; Tofail, S.A.; Bohara, R.A. Comprehensive cytotoxicity studies of superparamagnetic iron oxide nanoparticles. *Biochem. Biophys. Rep.* **2018**, *13*, 63–72. [CrossRef]

117. Levy, M.; Luciani, N.; Alloyeau, D.; Elgrabli, D.; Deveaux, V.; Pechoux, C.; Chat, S.; Wang, G.; Vats, N.; Gendron, F.; et al. Long term in vivo biotransformation of iron oxide nanoparticles. *Biomaterials* **2011**, *32*, 3988–3999. [CrossRef] [PubMed]

118. Mejías, R.; Gutirrez, L.; Salas, G.; Pérez-Yagüe, S.; Zotes, T.M.; Lázaro, F.J.; Morales, M.P.; Barber, D.F. Long term biotransformation and toxicity of dimercaptosuccinic acid-coated magnetic nanoparticles support their use in biomedical applications. *J. Control. Release* **2013**, *171*, 225–233. [CrossRef] [PubMed]

119. Rojas, J.M.; Gavilán, H.; del Dedo, V.; Lorente-Sorolla, E.; Sanz-Ortega, L.; da Silva, G.B.; Costo, R.; Perez-Yagüe, S.; Talelli, M.; Marciello, M.; et al. Time-course assessment of the aggregation and metabolization of magnetic nanoparticles. *Acta Biomater.* **2017**, *58*, 181–195. [CrossRef] [PubMed]

120. Mazuel, F.; Espinosa, A.; Luciani, N.; Reffay, M.; Le Borgne, R.; Motte, L.; Desboeufs, K.; Michel, A.; Pellegrino, T.; Lalatonne, Y.; et al. Massive Intracellular Biodegradation of Iron Oxide Nanoparticles Evidenced Magnetically at Single-Endosome and Tissue Levels. *ACS Nano* **2016**, *10*, 7627–7638. [CrossRef] [PubMed]

121. Anselmo, A.C.; Mitragotri, S. Nanoparticles in the clinic. *Bioeng. Transl. Med.* **2016**, *1*, 10–29. [CrossRef]

122. Soetaert, F.; Korangath, P.; Serantes, D.; Fiering, S.; Ivkov, R. Cancer therapy with iron oxide nanoparticles: Agents of thermal and immune therapies. *Adv. Drug Deliv. Rev.* **2020**, *163*, 65–83. [CrossRef]

123. Grüttner, C.; Müller, K.; Teller, J.; Westphal, F. Synthesis and functionalisation of magnetic nanoparticles for hyperthermia applications. *Int. J. Hyperth.* **2013**, *29*, 777–789. [CrossRef]

124. Gomes, R.C.; Silva, F.G.; Aquino, R.; Bonville, P.; Tourinho, F.; Perzynski, R.; Depeyrot, J. Exchange bias of MnFe2O4@γFe2O3 and CoFe2O4@γFe2O3 core/shell nanoparticles. *J. Magn. Magn. Mater.* **2014**, *368*, 409–414. [CrossRef]

125. Gavilán, H.; Sánchez, E.H.; Brollo, M.E.F.; Asín, L.; Moerner, K.K.; Frandsen, C.; Lázaro, F.J.; Serna, C.J.; Veintemillas-Verdaguer, S.; Morales, M.P.; et al. Formation Mechanism of Maghemite Nanoflowers Synthesized by a Polyol-Mediated Process. *ACS Omega* **2017**, *2*, 7172–7184. [CrossRef]

126. Ammar, S.; Fiévet, F. Polyol Synthesis: A Versatile Wet-Chemistry Route for the Design and Production of Functional Inorganic Nanoparticles. *Nanomaterials* **2020**, *10*, 1217. [CrossRef]

127. Gallo-Cordova, A.; Veintemillas-Verdaguer, S.; Tartaj, P.; Mazarío, E.; Morales, M.; Ovejero, J. Engineering Iron Oxide Nanocatalysts by a Microwave-Assisted Polyol Method for the Magnetically Induced Degradation of Organic Pollutants. *Nanomaterials* **2021**, *11*, 1052. [CrossRef]

128. Fdez-Gubieda, M.L.; Alonso, J.; García-Prieto, A.; García-Arribas, A.; Fernández Barquín, L.; Muela, A. Magnetotactic bacteria for cancer therapy. *J. Appl. Phys.* **2020**, *128*, 070902. [CrossRef]

129. Ovejero, J.G.; Morales, I.; De La Presa, P.; Mille, N.; Carrey, J.; Garcia, M.A.; Hernando, A.; Herrasti, P. Hybrid nanoparticles for magnetic and plasmonic hyperthermia. *Phys. Chem. Chem. Phys.* **2018**, *20*, 24065–24073. [CrossRef]

130. Sarkar, S.; Levi-Polyachenko, N. Conjugated polymer nano-systems for hyperthermia, imaging and drug delivery. *Adv. Drug Deliv. Rev.* **2020**, *163*, 40–64. [CrossRef]

131. Rodrigues, H.F.; Capistrano, G.; Bakuzis, A.F. In vivo magnetic nanoparticle hyperthermia: A review on preclinical studies, low-field nano-heaters, noninvasive thermometry and computer simulations for treatment planning. *Int. J. Hyperth.* **2020**, *37*, 76–99. [CrossRef] [PubMed]

132. Carrey, J.; Mehdaoui, B.; Respaud, M. Simple models for dynamic hysteresis loop calculations of magnetic single-domain nanoparticles: Application to magnetic hyperthermia optimization. *J. Appl. Phys.* **2011**, *109*, 083921. [CrossRef]

133. Roti, J.L.R. Cellular responses to hyperthermia (40–46 °C): Cell killing and molecular events. *Int. J. Hyperth.* **2008**, *24*, 3–15. [CrossRef] [PubMed]

134. Oei, A.L.; Vriend, L.E.M.; Crezee, J.; Franken, N.A.P.; Krawczyk, P.M. Effects of hyperthermia on DNA repair pathways: One treatment to inhibit them all. *Radiat. Oncol.* **2015**, *10*, 165. [CrossRef] [PubMed]

135. Song, C.W. Effect of local hyperthermia on blood flow and micro environment: A review. *Cancer Res.* **1984**, *44*, 4721s–7370s.

136. Wydra, R.J.; Oliver, C.E.; Anderson, K.W.; Dziubla, T.D.; Hilt, J.Z. Accelerated generation of free radicals by iron oxide nanoparticles in the presence of an alternating magnetic field. *RSC Adv.* **2015**, *5*, 18888–18893. [CrossRef] [PubMed]

137. Wydra, R.J.; Rychahou, P.; Evers, B.M.; Anderson, K.W.; Dziubla, T.D.; Hilt, J.Z. The role of ROS generation from magnetic nanoparticles in an alternating magnetic field on cytotoxicity. *Acta Biomater.* **2015**, *25*, 284–290. [CrossRef]

138. Lepock, J.R. Cellular effects of hyperthermia: Relevance to the minimum dose for thermal damage. *Int. J. Hyperth.* **2003**, *19*, 252–266. [CrossRef]

139. Asín, L.; Goya, G.F.; Tres, A.; Ibarra, M.R. Induced cell toxicity originates dendritic cell death following magnetic hyperthermia treatment. *Cell Death Dis.* **2013**, *4*, e596. [CrossRef]

140. Villanueva, A.; de la Presa, P.; Alonso, J.M.; Rueda, T.; Martínez, A.; Crespo, P.; Morales, M.D.P.; Gonzalez-Fernandez, M.A.; Valdés, J.; Rivero, G. Hyperthermia HeLa Cell Treatment with Silica-Coated Manganese Oxide Nanoparticles. *J. Phys. Chem. C* **2010**, *114*, 1976–1981. [CrossRef]

141. Creixell, M.; Bohórquez, A.; Torres-Lugo, M.; Rinaldi, C. EGFR-Targeted Magnetic Nanoparticle Heaters Kill Cancer Cells without a Perceptible Temperature Rise. *ACS Nano* **2011**, *5*, 7124–7129. [CrossRef] [PubMed]

142. Marcos-Campos, I.; Asin, L.; Torres, T.E.; Marquina, C.; Tres, A.; Ibarra, M.R.; Goya, G. Cell death induced by the application of alternating magnetic fields to nanoparticle-loaded dendritic cells. *Nanotechnology* **2011**, *22*, 205101. [CrossRef]

143. Domenech, M.; Marrero-Berrios, I.; Torres-Lugo, M.; Rinaldi, C. Lysosomal Membrane Permeabilization by Targeted Magnetic Nanoparticles in Alternating Magnetic Fields. *ACS Nano* **2013**, *7*, 5091–5101. [CrossRef]

144. Clerc, P.; Jeanjean, P.; Hallali, N.; Gougeon, M.; Pipy, B.; Carrey, J.; Fourmy, D.; Gigoux, V. Targeted Magnetic Intra-Lysosomal Hyperthermia produces lysosomal reactive oxygen species and causes Caspase-1 dependent cell death. *J. Control. Release* **2018**, *270*, 120–134. [CrossRef] [PubMed]

145. Master, A.M.; Williams, P.N.; Pothayee, N.; Pothayee, N.; Zhang, R.; Vishwasrao, H.; Golovin, Y.I.; Riffle, J.S.; Sokolsky, M.; Kabanov, A.V. Remote Actuation of Magnetic Nanoparticles For Cancer Cell Selective Treatment Through Cytoskeletal Disruption. *Sci. Rep.* **2016**, *6*, 33560. [CrossRef]

146. Pereira, M.; Oliveira, L.C.A.; Murad, E. Iron oxide catalysts: Fenton and Fentonlike reactions—A review. *Clay Miner.* **2012**, *47*, 285–302. [CrossRef]

147. Kobayashi, T.; Kakimi, K.; Nakayama, E.; Jimbow, K. Antitumor immunity by magnetic nanoparticle-mediated hyperthermia. *Nanomedicine* **2014**, *9*, 1715–1726. [CrossRef]

148. Ludwig, R.; Teran, F.J.; Teichgraeber, U.; Hilger, I. Nanoparticle-based hyperthermia distinctly impacts production of ROS, expression of Ki-67, TOP2A, and TPX2, and induction of apoptosis in pancreatic cancer. *Int. J. Nanomed.* **2017**, *12*, 1009–1018. [CrossRef] [PubMed]

149. Evans, S.S.; Repasky, E.A.; Fisher, D.T. Fever and the thermal regulation of immunity: The immune system feels the heat. *Nat. Rev. Immunol.* **2015**, *15*, 335–349. [CrossRef]

150. Mejías, R.; Hernández Flores, P.; Talelli, M.; Tajada-Herráiz, J.L.; Brollo, M.E.; Portilla, Y.; Morales, M.P.; Barber, D.F. Cell-Promoted Nanoparticle Aggregation Decreases Nanoparticle-Induced Hyperthermia under an Alternating Magnetic Field Independently of Nanoparticle Coating, Core Size, and Subcellular Localization. *ACS Appl. Mater. Interfaces* **2019**, *11*, 340–355. [CrossRef] [PubMed]

151. Moise, S.; Céspedes, E.; Soukup, D.; Byrne, J.M.; El Haj, A.J.; Telling, N.D. The cellular magnetic response and biocompatibility of biogenic zinc- and cobalt-doped magnetite nanoparticles. *Sci. Rep.* **2017**, *7*, 39922. [CrossRef] [PubMed]

152. Etheridge, M.L.; Hurley, K.R.; Zhang, J.; Jeon, S.; Ring, H.L.; Hogan, C.; Haynes, C.L.; Garwood, M.; Bischof, J.C. Accounting for biological aggregation in heating and imaging of magnetic nanoparticles. *Technology* **2014**, *2*, 214–228. [CrossRef] [PubMed]

153. Di Corato, R.; Espinosa, A.; Lartigue, L.; Tharaud, M.; Chat, S.; Pellegrino, T.; Ménager, C.; Gazeau, F.; Wilhelm, C. Magnetic hyperthermia efficiency in the cellular environment for different nanoparticle designs. *Biomaterials* **2014**, *35*, 6400–6411. [CrossRef] [PubMed]

154. Fortes Brollo, M.E.F.; Flores, P.H.; Gutierrez, L.; Johansson, C.; Barber, D.; Morales, M.D.P. Magnetic properties of nanoparticles as a function of their spatial distribution on liposomes and cells. *Phys. Chem. Chem. Phys.* **2018**, *20*, 17829–17838. [CrossRef]

155. Mehdaoui, B.; Meffre, A.; Carrey, J.; Lachaize, S.; Lacroix, L.-M.; Gougeon, M.; Chaudret, B.; Respaud, M. Optimal Size of Nanoparticles for Magnetic Hyperthermia: A Combined Theoretical and Experimental Study. *Adv. Funct. Mater.* **2011**, *21*, 4573–4581. [CrossRef]

156. Martinez-Boubeta, C.; Simeonidis, K.; Makridis, A.; Angelakeris, M.; Iglesias, O.; Guardia, P.; Cabot, A.; Yedra, L.; Estradé, S.; Peiró, F.; et al. Learning from Nature to Improve the Heat Generation of Iron-Oxide Nanoparticles for Magnetic Hyperthermia Applications. *Sci. Rep.* **2013**, *3*, 1652. [CrossRef]

157. Khandhar, A.; Ferguson, R.M.; Simon, J.A.; Krishnan, K.M. Enhancing cancer therapeutics using size-optimized magnetic fluid hyperthermia. *J. Appl. Phys.* **2012**, *111*. [CrossRef]

158. Santhosh, P.B.; Ulrih, N.P. Multifunctional superparamagnetic iron oxide nanoparticles: Promising tools in cancer theranostics. *Cancer Lett.* **2013**, *336*, 8–17. [CrossRef] [PubMed]

159. Sivakumar, B.; Aswathy, R.G.; Nagaoka, Y.; Suzuki, M.; Fukuda, T.; Yoshida, Y.; Maekawa, T.; Sakthikumar, D.N. Multifunctional Carboxymethyl Cellulose-Based Magnetic Nanovector as a Theragnostic System for Folate Receptor Targeted Chemotherapy, Imaging, and Hyperthermia against Cancer. *Langmuir* **2013**, *29*, 3453–3466. [CrossRef]

160. Castellanos-Rubio, I.; Rodrigo, I.; Olazagoitia-Garmendia, A.; Arriortua, O.; Gil De Muro, I.; Garitaonandia, J.S.; Bilbao, J.R.; Fdez-Gubieda, M.L.; Plazaola, F.; Orue, I.; et al. Highly Reproducible Hyperthermia Response in Water, Agar, and Cellular Environment by Discretely PEGylated Magnetite Nanoparticles. *ACS Appl. Mater. Interfaces* **2020**, *12*, 27917–27929. [CrossRef] [PubMed]
161. Del Sol-Fernández, S.; Portilla-Tundidor, Y.; Gutiérrez, L.; Odio, O.F.; Reguera, E.; Barber, D.F.; Morales, M.D.P. Flower-like Mn-Doped Magnetic Nanoparticles Functionalized with αvβ3-Integrin-Ligand to Efficiently Induce Intracellular Heat after Alternating Magnetic Field Exposition, Triggering Glioma Cell Death. *ACS Appl. Mater. Interfaces* **2019**, *11*, 26648–26663. [CrossRef] [PubMed]
162. Skandalakis, G.P.; Rivera, D.; Rizea, C.D.; Bouras, A.; Raj, J.G.J.; Bozec, D.; Hadjipanayis, C.G. Hyperthermia treatment advances for brain tumors. *Int. J. Hyperth.* **2020**, *37*, 3–19. [CrossRef]
163. Boateng, F.; Ngwa, W. Delivery of Nanoparticle-Based Radiosensitizers for Radiotherapy Applications. *Int. J. Mol. Sci.* **2020**, *21*, 273. [CrossRef]
164. Leal, M.P.; Rivera-Fernández, S.; Franco, J.M.; Pozo, D.; de la Fuente, J.M.; García-Martín, M.L. Long-circulating PEGylated manganese ferrite nanoparticles for MRI-based molecular imaging. *Nanoscale* **2015**, *7*, 2050–2059. [CrossRef]
165. Zhou, P.; Zhao, H.; Wang, Q.; Zhou, Z.; Wang, J.; Deng, G.; Wang, X.; Liu, Q.; Yang, H.; Yang, S. Photoacoustic-Enabled Self-Guidance in Magnetic-Hyperthermia Fe@Fe3 O4 Nanoparticles for Theranostics In Vivo. *Adv. Healthc. Mater.* **2018**, *7*, e1701201. [CrossRef]
166. Piehler, S.; Dähring, H.; Grandke, J.; Göring, J.; Couleaud, P.; Aires, A.; Cortajarena, A.L.; Courty, J.; Latorre, A.; Somoza, Á.; et al. Iron Oxide Nanoparticles as Carriers for DOX and Magnetic Hyperthermia after Intratumoral Application into Breast Cancer in Mice: Impact and Future Perspectives. *Nanomaterials* **2020**, *10*, 1016. [CrossRef]
167. García-Hevia, L.; Casafont, Í.; Oliveira, J.; Terán, N.; Fanarraga, M.L.; Gallo, J.; Bañobre-López, M. Magnetic lipid nanovehicles synergize the controlled thermal release of chemotherapeutics with magnetic ablation while enabling non-invasive monitoring by MRI for melanoma theranostics. *Bioact. Mater.* **2021**, *8*, 153–164. [CrossRef]
168. Xie, L.; Jin, W.; Zuo, X.; Ji, S.; Nan, W.; Chen, H.; Gao, S.; Zhang, Q. Construction of small-sized superparamagnetic Janus nanoparticles and their application in cancer combined chemotherapy and magnetic hyperthermia. *Biomater. Sci.* **2020**, *8*, 1431–1441. [CrossRef]
169. Mai, B.T.; Balakrishnan, P.B.; Barthel, M.J.; Piccardi, F.; Niculaes, D.; Marinaro, F.; Fernandes, S.; Curcio, A.; Kakwere, H.; Autret, G.; et al. Thermoresponsive Iron Oxide Nanocubes for an Effective Clinical Translation of Magnetic Hyperthermia and Heat-Mediated Chemotherapy. *ACS Appl. Mater. Interfaces* **2019**, *11*, 5727–5739. [CrossRef] [PubMed]
170. Vilas-Boas, V.; Carvalho, F.; Espiña, B. Magnetic Hyperthermia for Cancer Treatment: Main Parameters Affecting the Outcome of In Vitro and In Vivo Studies. *Molecules* **2020**, *25*, 2874. [CrossRef]
171. Etemadi, H.; Plieger, P.G. Magnetic Fluid Hyperthermia Based on Magnetic Nanoparticles: Physical Characteristics, Historical Perspective, Clinical Trials, Technological Challenges, and Recent Advances. *Adv. Ther.* **2020**, *3*, 2000061. [CrossRef]
172. van Landeghem, F.; Maier-Hauff, K.; Jordan, A.; Hoffmann, K.-T.; Gneveckow, U.; Scholz, R.; Thiesen, B.; Brück, W.; von Deimling, A. Post-mortem studies in glioblastoma patients treated with thermotherapy using magnetic nanoparticles. *Biomaterials* **2009**, *30*, 52–57. [CrossRef] [PubMed]
173. Jordan, A.; Wust, P.; Fählin, H.; John, W.; Hinz, A.; Felix, R. Inductive heating of ferrimagnetic particles and magnetic fluids: Physical evaluation of their potential for hyperthermia. *Int. J. Hyperth.* **1993**, *9*, 51–68. [CrossRef]
174. Jordan, A.; Scholz, R.; Maier-Hauff, K.; Johannsen, M.; Wust, P.; Nadobny, J.; Schirra, H.; Schmidt, H.; Deger, S.; Loening, S.; et al. Presentation of a new magnetic field therapy system for the treatment of human solid tumors with magnetic fluid hyperthermia. *J. Magn. Magn. Mater.* **2001**, *225*, 118–126. [CrossRef]
175. Gneveckow, U.; Jordan, A.; Scholz, R.; Brüß, V.; Waldöfner, N.; Ricke, J.; Feussner, A.; Hildebrandt, B.; Rau, B.; Wust, P. Description and characterization of the novel hyperthermia- and thermoablation-system MFH®300F for clinical magnetic fluid hyperthermia. *Med. Phys.* **2004**, *31*, 1444–1451. [CrossRef] [PubMed]
176. Sreenivasa, G.; Gellermann, J.; Rau, B.; Nadobny, J.; Schlag, P.; Deuflhard, P.; Felix, R.; Wust, P. Clinical use of the hyperthermia treatment planning system HyperPlan to predict effectiveness and toxicity. *Int. J. Radiat. Oncol.* **2003**, *55*, 407–419. [CrossRef]
177. Johannsen, M.; Gneveckow, U.; Eckelt, L.; Feussner, A.; Waldöfner, N.; Scholz, R.; Deger, S.; Wust, P.; Loening, S.A.; Jordan, A. Clinical hyperthermia of prostate cancer using magnetic nanoparticles: Presentation of a new interstitial technique. *Int. J. Hyperth.* **2005**, *21*, 637–647. [CrossRef] [PubMed]
178. Wust, P.; Gneveckow, U.; Johannsen, M.; Böhmer, D.; Henkel, T.; Kahmann, F.; Sehouli, J.; Felix, R.; Ricke, J.; Jordan, A. Magnetic nanoparticles for interstitial thermotherapy—Feasibility, tolerance and achieved temperatures. *Int. J. Hyperth.* **2006**, *22*, 673–685. [CrossRef]
179. Johannsen, M.; Gneveckow, U.; Thiesen, B.; Taymoorian, K.; Cho, C.H.; Waldöfner, N.; Scholz, R.; Jordan, A.; Loening, S.A.; Wust, P. Thermotherapy of Prostate Cancer Using Magnetic Nanoparticles: Feasibility, Imaging, and Three-Dimensional Temperature Distribution. *Eur. Urol.* **2007**, *52*, 1653–1662. [CrossRef] [PubMed]
180. Johannsen, M.; Gneveckow, U.; Taymoorian, K.; Thiesen, B.; Waldöfner, N.; Scholz, R.; Jung, K.; Jordan, A.; Wust, P.; Loening, S.A. Morbidity and quality of life during thermotherapy using magnetic nanoparticles in locally recurrent prostate cancer: Results of a prospective phase I trial. *Int. J. Hyperth.* **2007**, *23*, 315–323. [CrossRef] [PubMed]

181. Maier-Hauff, K.; Rothe, R.; Scholz, R.; Gneveckow, U.; Wust, P.; Thiesen, B.; Feussner, A.; von Deimling, A.; Waldoefner, N.; Felix, R.; et al. Intracranial Thermotherapy using Magnetic Nanoparticles Combined with External Beam Radiotherapy: Results of a Feasibility Study on Patients with Glioblastoma Multiforme. *J. Neuro-Oncol.* **2007**, *81*, 53–60. [CrossRef] [PubMed]
182. Magnetic Nanoparticle Thermoablation-Retention and Maintenance in the Prostate: A Phase 0 Study in Men (MAGNABLATE I). Available online: https://clinicaltrials.gov/ct2/show/nct02033447?Term=Hyperthermia&cond=Cancer&intr=magnetic&draw=2&rank=6 (accessed on 27 July 2021).
183. Rubia-Rodríguez, I.; Santana-Otero, A.; Spassov, S.; Tombácz, E.; Johansson, C.; De La Presa, P.; Teran, F.; Morales, M.D.P.; Veintemillas-Verdaguer, S.; Thanh, N.; et al. Whither Magnetic Hyperthermia? A Tentative Roadmap. *Materials* **2021**, *14*, 706. [CrossRef] [PubMed]

Review

Magnetic Particle Imaging: An Emerging Modality with Prospects in Diagnosis, Targeting and Therapy of Cancer

Zhi Wei Tay [1,*], Prashant Chandrasekharan [2], Benjamin D. Fellows [2], Irati Rodrigo Arrizabalaga [2], Elaine Yu [2], Malini Olivo [1] and Steven M. Conolly [2]

[1] Institute of Bioengineering and Bioimaging, Agency for Science, Technology and Research (A*STAR), 11 Biopolis Way, #02-02 Helios Building, Singapore 138667, Singapore; malini_olivo@ibb.a-star.edu.sg

[2] Department of Bioengineering, 340 Hearst Memorial Mining Building, University of California Berkeley, Berkeley, CA 94720-1762, USA; prashantc@berkeley.edu (P.C.); bdfello@berkeley.edu (B.D.F.); irati.rodrigo@berkeley.edu (I.R.A.); elaineyu@berkeley.edu (E.Y.); sconolly@berkeley.edu (S.M.C.)

* Correspondence: tay_zhi_wei@ibb.a-star.edu.sg

Simple Summary: Magnetic Particle Imaging (MPI) is an emerging imaging technique that provides quantitative direct imaging of superparamagnetic iron oxide nanoparticles. In the last decade, MPI has shown great prospects as one of the magnetic methods other than Magnetic Resonance Imaging with applications covering cancer diagnosis, targeting enhancement, actuating cancer therapy, and post-therapy monitoring. Working on different physical principles from Magnetic Resonance Imaging, MPI benefits from ideal image contrast with zero background tissue signal, enabling hotspot-type images similar to Nuclear Medicine scans but using magnetic agents rather than radiotracers. In this review, we discussed the relevance of MPI to cancer diagnostics and image-guided therapy as well as recent progress to clinical translation.

Abstract: Background: Magnetic Particle Imaging (MPI) is an emerging imaging modality for quantitative direct imaging of superparamagnetic iron oxide nanoparticles (SPION or SPIO). With different physics from MRI, MPI benefits from ideal image contrast with zero background tissue signal. This enables clear visualization of cancer with image characteristics similar to PET or SPECT, but using radiation-free magnetic nanoparticles instead, with infinite-duration reporter persistence in vivo. MPI for cancer imaging: demonstrated months of quantitative imaging of the cancer-related immune response with in situ SPION-labelling of immune cells (e.g., neutrophils, CAR T-cells). Because MPI suffers absolutely no susceptibility artifacts in the lung, immuno-MPI could soon provide completely noninvasive early-stage diagnosis and treatment monitoring of lung cancers. MPI for magnetic steering: MPI gradients are ~150 × stronger than MRI, enabling remote magnetic steering of magneto-aerosol, nanoparticles, and catheter tips, enhancing therapeutic delivery by magnetic means. MPI for precision therapy: gradients enable focusing of magnetic hyperthermia and magnetic-actuated drug release with up to 2 mm precision. The extent of drug release from the magnetic nanocarrier can be quantitatively monitored by MPI of SPION's MPS spectral changes within the nanocarrier. Conclusion: MPI is a promising new magnetic modality spanning cancer imaging to guided-therapy.

Keywords: magnetic particle imaging; magnetic nanoparticles; magnetic hyperthermia; magnetic drug delivery

Citation: Tay, Z.W.; Chandrasekharan, P.; Fellows, B.D.; Arrizabalaga, I.R.; Yu, E.; Olivo, M.; Conolly, S.M. Magnetic Particle Imaging: An Emerging Modality with Prospects in Diagnosis, Targeting and Therapy of Cancer. *Cancers* **2021**, *13*, 5285. https://doi.org/10.3390/cancers13215285

Academic Editors: Moriaki Kusakabe and Akihiro Kuwahata

Received: 31 July 2021
Accepted: 19 October 2021
Published: 21 October 2021

Publisher's Note: MDPI stays neutral with regard to jurisdictional claims in published maps and institutional affiliations.

1. Introduction

Magnetic Particle Imaging (MPI) is an emerging magnetics-based imaging technique first introduced by Philips, Hamburg in 2005 [1]. While the name is very similar to Magnetic Resonance Imaging (MRI), it operates on very different physical principles. Unlike MRI, where the signal comes from the precession of nuclear spin magnetic moments of the target nuclei (e.g., 1H, 2H, 13C, 17O, 19F, 23Na, 31P), the MPI signal is obtained from the ensemble magnetization of superparamagnetic iron oxide nanoparticles (SPION) as described by the Langevin model [2]. Because there are no SPIONs found in native biological tissue unlike the ^{1}H in water and biological tissue sensed by MRI, MPI benefits from zero tissue background signal and achieves excellent image contrast comparable to tracer images typical of nuclear medicine scans such as positron emission tomography (PET) or single-photon emission computerized tomography (SPECT), which are the gold standard for diagnostic cancer imaging [3–5]. Since only SPIONs produce signal in an MPI scan, the MPI images obtained are fully quantitative in a linear fashion and are robust to minute changes in susceptibility. In comparison, the same SPIONs in an MRI scan are typically semi-quantitative as they produce contrast changes via susceptibility differences (Figure 1a), yielding a non-linear indirect effect on the ^{1}H signal [2]. MPI operates in the kilohertz frequency range where magnetic fields fully penetrate tissue, bone, and air with negligible attenuation and reflection differences. Thus, MPI does not have any view limitations and works robustly even in lungs [6–9] and bones, which are challenging for MRI [2,10] and ultrasound.

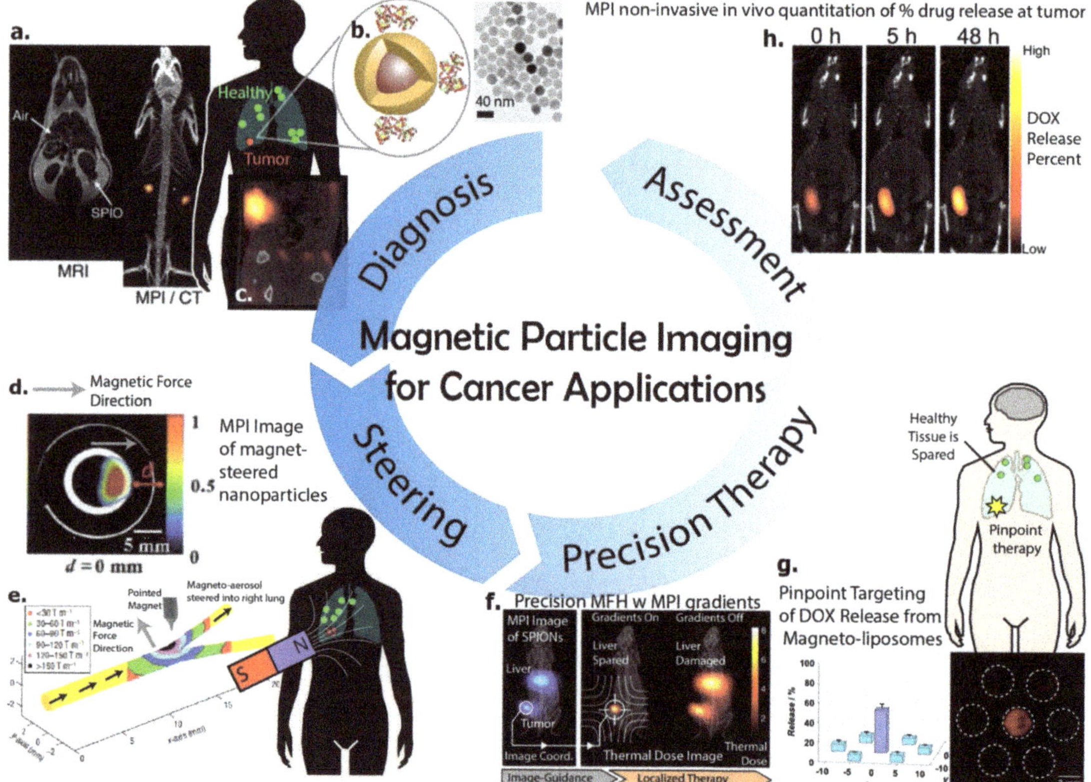

Figure 1. Overview figure for Magnetic Particle Imaging's prospects in diagnosis, targeting, and therapy of cancer. Magnetic Particle Imaging (MPI) is an emerging magnetic imaging technology that works completely differently from MRI, providing radiation-free tracer-like contrast and linear quantitation with nanogram sensitivity to superparamagnetic iron-oxide nanoparticles (SPIONs). While most MPI research is still preclinical, MPI hardware has recently reached clinical scale scanners and is en route to clinical translation. (**a**) Zheng et al. 2016 [8]—Comparison of MRI SPION contrast to MPI contrast of the same SPIONs. (**b**) Arami et al. 2017 [11] — schematic of cancer-targeted magnetic nanoparticles. (**c**) Yu et al. 2017 [12]—MPI image of SPIONs accumulated within a tumor. (**d**) Banura et al. 2017 [13]—magnetic steering of magneto-aerosol with MPI as post-event verification. (**e**) Dames et al. 2007 [14]—magnetic tip for targeted delivery of magnetic

aerosol to lung. (**f**) Tay et al. 2018 [15]—MPI scanner's gradients enable pinpoint heating at user-selected locations, heating SPIONs at the tumor without heating off-target SPIONs. (**g**) Liu et al. 2018 [16] — magnetic gradients enable pinpoint drug release at target location without triggering release from adjacent nanocarriers 2 mm away. (**h**) Zhu et al. 2019 [17] — MPI in vivo non-invasive quantification of the percentage of release of drug from nanocarriers in mice tumors, enabling real-time assessment of the success of drug delivery for cancer. Figures within insets reproduced with permission from respective authors cited in reference numbers and publishers.

Besides the excellent image contrast, one of the other key benefits of Magnetic Particle Imaging for cancer imaging is the relatively high sensitivity of the modality. The electronic magnetization of SPIONs sensed by MPI is 22 million times stronger than that of the nuclear magnetization of water (^{1}H) at 7 Tesla [2]. Furthermore, the dose limit of iron oxide is 510 mg according to Lu et al. 2010, which is 25 million times higher than the 2 ng dose limit of PET [18,19]. The 2 ng value was calculated from the 370 MBq FDA-mandated dose limit for 18-FDG divided by the specific (radio)activity of 1ng of 18-fluorine (in MBq, averaged specific activity value). This implies that MPI can increase the systemic administered dosage to compensate and ensure tumor detection at 260 nM Fe sensitivity, even though this is poorer sensitivity than the 2 pM of PET [20]. Therefore, if SPIONs can achieve similar targeting efficiencies to tumors as 18-FDG, MPI can be expected to be competitive with PET on a dose-limited comparison, and thus help avoid radiation dose (especially important for pediatrics). Other advantages include the near-infinite signal half-life of SPIONs enabling longer time for circulation and binding to tumors while the 110 min half-life of 18-FDG necessitates a PET scan merely 40 min after injection [19]. Convenience is also improved as SPIONs can be used off-the-shelf, thus avoiding cyclotron facility overheads and the radiation safety measures for hot chemistry preparations.

Regarding the imaging agent, the SPION magnetic "tracers" used in MPI will differ depending on the application. For stem cell or immune cell labeling, carboxydextran-based SPIONs have shown good labeling efficiency, likely due to the affinity of the dextran coating to cell uptake and internalization. For vascular imaging or tumor targeting, long-circulating stealth SPIONs with PEG-based coatings are ideal due to the enhancement of blood circulation half-life, allowing more time for the tracer to remain in circulation for vascular imaging or for the tracer to aggregate into the tumors. For magnetic hyperthermia applications where the magnetic nanoparticles are heated via external alternating magnetic fields (AMF), the magnetic core of the SPIONs must demonstrate high heating perfor-mance, i.e., good specific absorption rate (SAR) values at typical hyperthermia frequencies (100–1000 kHz) [21]. It is important to note that these qualities are not mutually exclusive and an SPION can be designed with multiple of these qualities such as a high heating performance magnetic core with stealth PEG-coating.

2. Physical Mechanisms Underlying Magnetic Particle Imaging

In brief, MPI performs spatial encoding, signal detection, and image reconstruction based on very different magnetic principles from MRI. From the high-contrast and spatial resolution characteristics, MPI is more similar to PET and SPECT, although it uses non-radioactive SPIONs at 20–100 nm sizes rather than small-molecule radiotracers. This section will explain the mechanism of MPI and its spatial and temporal resolution.

2.1. Localization and Collection of Signal from a Specific Slice or Volume

Magnetic Particle Imaging has two methods of localization of signal and thereby achieves spatial encoding in order to reconstruct an image. For the system matrix method, a point source of SPIONs is physically moved to every voxel in the field-of-view (FOV) and the MPI harmonic signal recorded as a calibration to determine the system matrix transfer function. In order to encode a different MPI harmonic signal at every voxel, a static background selection field is applied. The selection field is defined as a gradient field (magnetic field strength varies spatially) with a zero-field region at a central point (defined as field-free-point system—FFP) or a zero-field in a line geometry (defined as field-free-line system—FFL) as shown in Figure 2a,b. The differing background field strength as a function of position changes the SPION magnetization and results in a different MPI harmonic signal depending on position in the FOV. The static selection field alone cannot excite an MPI signal, and thus a time-varying drive field of 25 kHz and 16 mT is applied. The definition of a drive field is a monotonal excitation magnetic field operating in the kilohertz frequency range that aims to generate rapid magnetization changes in SPIONs as the MPI signal. When superimposed on top of the selection field, the result is the motion of the FFP in a Lissajous trajectory so as to pass near every voxel in the FOV at least once during the scan. In conclusion, spatial encoding for the system matrix meth-od uses the combination of the selection field gradient and a Lissajous (rather than raster) trajectory to determine a unique MPI harmonic signature for each and every voxel in the FOV [5].

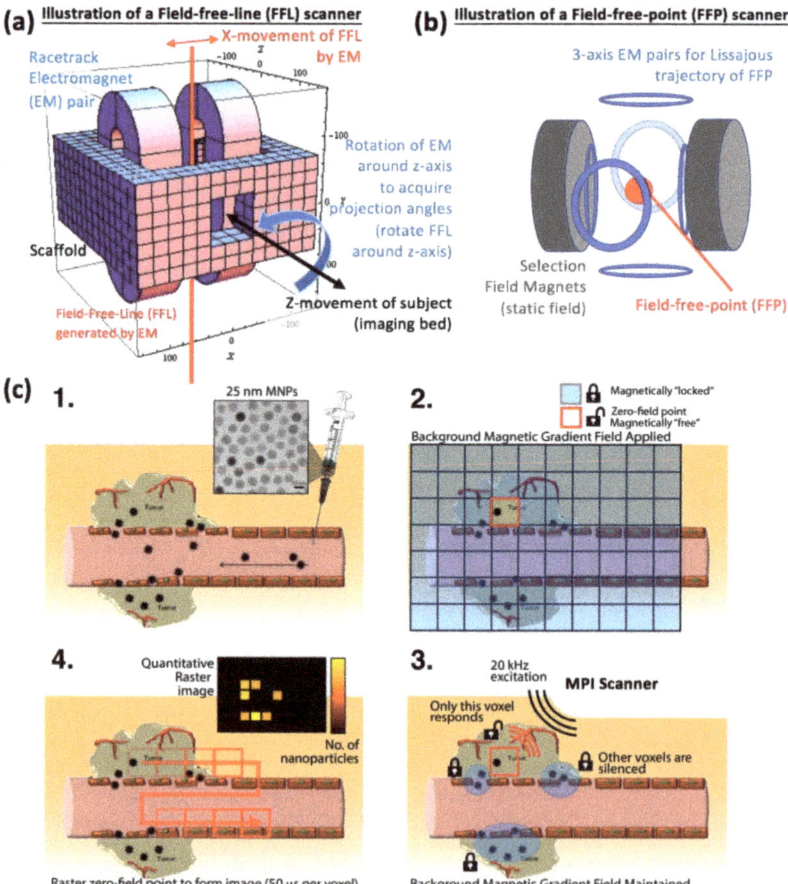

Figure 2. Physical mechanisms underlying how Magnetic Particle Imaging (MPI) scans and produces an image. (**a**) Scanner configuration for a field-free-line (FFL) MPI scanner. (**b**) Scanner configuration for a field-free-point (FFP) scanner. (**c**) Illustrative workflow diagram on how the previously defined MPI-related magnetic fields can be applied to SPIONs that have accumulated within a tumor in terms of how to spatially encode, signal detect, and image reconstruct the MPI image. The background tumor image is adapted from Jhaveri AM, Torchilin VP. Multifunctional polymeric micelles for delivery of drugs and siRNA. Front Pharmacol 2014; 5:77 under a CC by 3.0 creative commons license [22].

For the x-space method, there is no pre-calibration step. Spatial encoding relies on the fact that SPIONs physically at the location of the FFP or FFL give the largest amount of signal and SPIONs away from the FFP give less and less signal in accordance to the point-spread-function (PSF) of that specific MPI scanner and SPION combination. This is basically predicated on the PSF as imaging spot size, with a smaller spot size enabling greater precision in discerning at-FFP signal and suppressing off-FFP signal. To acquire the entire FOV, the FFP or FFL is rastered across the FOV, usually in a cartesian trajectory, although the Lissajous trajectory used in the system matrix method could work too. The x-space drive field is usually in the same-axis as the detector coil (single-axis) and uses typical values of 20 kHz and 20 mT [3].

2.2. Signal Detection and Image Reconstruction Approach for MPI

MPI uses receive coils, defined as inductive solenoid or saddle-shaped wire coil sensors, which detect the SPION signal based off the time-varying magnetization changes of the SPION in response to the drive field. As a result, the signal strength is proportional to dM/dt and frequencies around 20 kHz are preferred as a trade-off between high signal strength from dM/dt and minimizing relaxation-induced blurring when the SPION cannot keep up with the drive field switching [3]. Unlike MRI where the readout timing is usually delayed after the excitation, the MPI signal is read out at the same time as the drive field (excitation) application. There is thus a large amount of direct feedthrough of the drive field into the inductive received signal. This is mitigated by high-pass or band-pass filters as well as gradiometric sensing coil design.

Image reconstruction for the system matrix method solves an inverse problem using the calibrated system function (3D matrix) achieved by recording the MPI harmonics at each and every voxel in the FOV. Image reconstruction for the x-space method relies on knowledge of the instantaneous position of the FFP and FFL in 3D space. The current instantaneous MPI signal is directly gridded to the known FFP/FFL location [3].

2.3. Spatial Resolution and Time Requirements for MPI

The current spatial resolution for MPI is around 0.5–2 mm depending on the magnetization characteristics of the SPION used as well as the gradient strength of the scanner [2]. Figure 2 compares the resolution and sensitivity of MPI to other imaging modalities. The temporal resolution of MPI can be relatively good at 45 frames per second as achieved by system matrix MPI due to the speed of the Lissajous trajectory [5].

3. Imaging Cancer Using Magnetic Particle Imaging

In brief, MPI is similar in image-quality to PET because of its zero-background, high-contrast, and ~1 mm spatial resolution. However, MPI's imaging agent of SPIONs cannot utilize the Warburg effect to target tumors and must rely on other mechanisms such as targeting of cancer cell receptors or cancer-specific proteases. This section discusses the various imaging studies on cancer that have been performed with MPI.

The gold standard for clinical cancer imaging is Positron Emission Tomography (PET). The ability of 18-FDG to selectively accumulate in even small metastatic tumors due to the Warburg effect and the tracer-nature and positive contrast of PET scans allow for unambiguous diagnosis of tumor presence and location [20]. Coverage of the whole-body is possible except for the brain or bladder due to the low contrast caused by high background 18-FDG uptake by healthy tissue. However, PET scans still have a non-negligible radiation dose and are not recommended for pediatric imaging. MRI and CT are also widely used for cancer imaging but often require tumors to be relatively large for reliable detection on scans. In this context, Magnetic Particle Imaging is promising as it provides tracer-like contrast (see Figure 1a–c) without any radiation dose due to the use of magnetic "tracers" as opposed to radiotracers. Figure 3 summarizes the differences of MPI from other imaging modalities for cancer imaging. In practice however, the SPIONs used in MPI still need to make progress towards matching 18-FDG's high affinity to cancerous tissue in order to be competitive with PET.

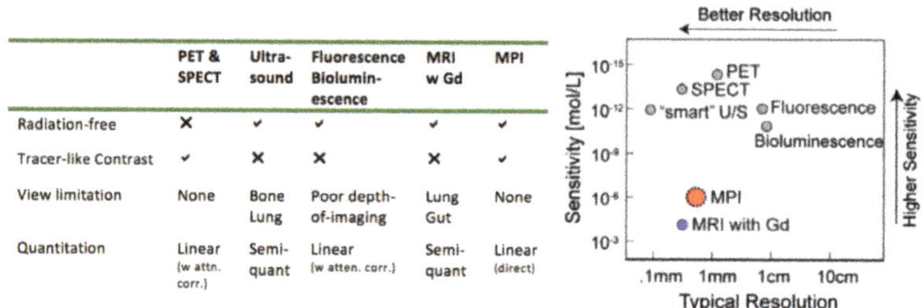

	PET & SPECT	Ultra-sound	Fluorescence Biolumin-escence	MRI w Gd	MPI
Radiation-free	✗	✔	✔	✔	✔
Tracer-like Contrast	✔	✗	✗	✗	✔
View limitation	None	Bone Lung	Poor depth-of-imaging	Lung Gut	None
Quantitation	Linear (w attn. corr.)	Semi-quant	Linear (w atten. corr.)	Semi-quant	Linear (direct)

Figure 3. Comparison of MPI to other molecular imaging modalities. Figure on right panel adapted with permission from Saritas et al. J. Magn. Reson. 229 [2]. Copyright 2013 Elsevier.

The earliest demonstrated application of Magnetic Particle Imaging towards cancer was a 2014 in vitro study by the University Hospital of Schleswig Holstein and Institute of Medical Engineering at the University of Luebeck, Germany [23]. Custom dextran-coated SPIONs (UL-D) were synthesized and demonstrated good internalization by head and neck squamous cancer cells as well as significant MPI signal via in vitro measurements using a Magnetic Particle Spectrometer. Although in vivo images were not demonstrated, the authors comprehensively characterized the labeled cells showing that their MPI-suitable SPIONs did not impact cell mitochondrial activity (MTT assay), cell viability (annexin V-APC-Propidium Iodide flow cytometry), cell proliferation (xCELLigence DP), cytokine secretion (Bead-based immunoassays for IL-6, IL-8, IL-1β and TNF-α), and reactive oxygen species generation (ROS assay by Dichlorofluorescein diacetate). These assays suggest that labeling of the cancer cells should not negatively impact tumor behavior such as increased tumor invasion or metastases.

In 2016, another in vitro study demonstrated the detection of cancer-specific proteases using changes in Magnetic Particle Spectrum (MPS) of MPI-compatible monodisperse iron oxide nanoparticles [24]. The linker-peptide-aggregated nanoparticles demonstrated a significant change in their spectrum when exposed to cancer-specific proteases. Although this assay was not verified for in vivo MPI, since MPI can be calibrated to tune specifically to a designated MPS (color MPI), this strategy could be promising to increase MPI image-specificity to cancer cells. Sensitivity can be improved by optimization of the magnetic core size [25] as well as designing contrast-enhancing MPI pulsed excitation rather than continuous-wave excitation [26].

The earliest full study of in vivo Magnetic Particle Imaging of cancer (Figure 4a) was demonstrated in 2017 by the University of California Berkeley on their academic MPI scanner using long-circulating SPIONs (LS-008) from Lodespin Laboratories [12].

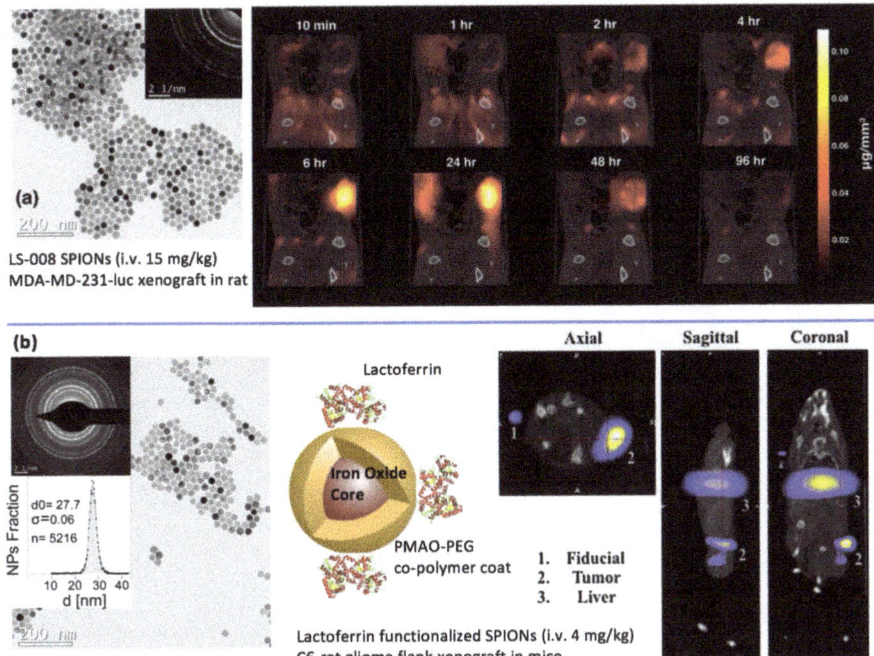

Figure 4. Imaging Cancer using Magnetic Particle Imaging. Figures adapted with permission from Yu et al. Nano Lett. 17. Copyright 2017 American Chemical Society. (**a**) Yu et al. 2017 [12] used LS-008 SPIONs injected i.v. at 15 mg/kg into an MDA-MD-231-luc flank xenograft in rat. The long-circulating SPIONs were non-targeted and after a few hours of systemic circulation, accumulated by the enhanced permeability and retention effect in the tumor. The image time course showcases the benefits of MPI with tracer-like contrast and direct linear quantitation, enabling clear visualization of the particle EPR dynamics with initial rim enhancement, accumulation, and then wash-out. (**b**) Arami et al. 2017 [11] used Lactoferrin functionalized SPIONs injected i.v. at 4 mg/kg for targeting of a C6-rat glioma flank xenograft in mice. After 2 h post-injection, the MPI image showed accumulation in the tumor together with substantial clearance to the liver. Figures reproduced from Arami et al. Nanoscale. 9 with permission from the Royal Society of Chemistry.

The cancer model used seven athymic nude rats bearing flank xenografts of MDA-MB-231-luc breast tumor cells. This work emphasized some of the inherent advantages of MPI for cancer imaging such as excellent image contrast and full quantitation of the tracer dynamics from administration to initial rim enhancement of the tumor, accumulation within the tumor between 1–24 h (peaking at 6 h), and then slow clearance to the liver over a period of 96 h. The signal-to-background ratio of the tumor was very high (>100) as there was no background uptake of SPIONs by biological tissue unlike 18-FDG. Because attenuation correction and signal half-life compensation is not required in MPI, the image quantification was demonstrated to be facile and straightforward. The tracer accumulation in the tumor occurred via enhanced permeability and retention effect (EPR) as this was an untargeted SPION study without cancer-targeting functionalization.

Later in 2017, the first in vivo MPI image of cancer using targeted SPIONs (Figure 4b) was demonstrated by the Stanford School of Medicine Department of Radiology in mice [11]. This study improved targeting to the flank xenograft of C6 brain cancer cells via surface functionalization with lactoferrin and also by placing a permanent magnet on the rodent flank. The SPIONs were multi-modal with Cy5.5 NIRF and 67-Ga radiolabel for near-infrared and SPECT imaging respectively. This study further demonstrated the excellent image contrast of MPI as compared with near-infrared imaging and showed that it approaches the image contrast achievable by 67-Ga SPECT images.

Imaging Cell Therapy for Cancer Immunotherapy Using Magnetic Particle Imaging

In brief, MPI has many advantages for monitoring of labeled adoptive cell transfer immune cells such as long-lasting magnetic label that does not lose signal over time by radioactive decay and MPI's high-contrast yet quantitative nature. This section introduces cancer immunotherapy and recent efforts to image immunotherapy with MPI.

In the last twenty years, immunotherapy for cancer has steadily gained traction in clinical practice. There are five major types of cancer immunotherapy: (1) cancer vaccines, (2) cytokine therapies, (3) adoptive cell transfer (ACT), (4) immune checkpoint inhibitors, and (5) oncolytic virus therapies [27–31]. Of all these categories, MPI is well-poised to contribute in the adoptive cell transfer category and oncolytic virus category. This is because the magnetic nanoparticles used in MPI have optimal core sizes of 20–30 nm and can thus label immune cells (micron-sized) or oncolytic viruses (150–240 nm). For the adoptive cell transfer category, there is a need to verify that the transferred cells have arrived at and remain at the target cancer site throughout the course of therapy. Furthermore, it is essential to monitor the viability and functionality of the cells to ensure the success of the therapy [32]. These requirements are similar to the imaging requirements for stem cell therapy. Since MPI has been validated in many stem cell therapy studies [33–37], we anticipate MPI's advantages to be applicable to the adoptive cell therapy application as well. The main benefits of MPI in stem cells are innately transferable to the adoptive cell therapy application, such as (1) no loss in signal over time from magnetic cell labels enabling >90% of signal left over 89 days in vivo [33], (2) no radiation dose that will limit the length of a longitudinal study, (3) direct and quantitative measurement of magnetic label that is unaffected by changes in subject anatomy background over time [34], and (4) potential for assessment of viability of labeled cells via color MPI spectroscopic techniques demonstrated in various MPI studies that leverages microenvironment sensitivity for color/contrast change or for multi-contrast multiplexing [38–41]. These initial stem cell studies have demonstrated that the magnetic label remains internalized within the cell population of interest, and that any released label is rapidly cleared to the liver and does not confound the quantitation [33].

ACT has shown the greatest success in "liquid" malignancies such as B lymphocyte leukemia and lymphoma. However, ACT as an immunotherapy for solid tumors has been hampered by an inability to adequately manipulate infused T cells to efficiently traffic into and specifically target deep-seated tumors for destruction, while minimizing immune-related adverse events (irAEs) caused by low-level recognition of antigen on surrounding healthy tissues [42]. Clinicians thus require real-time information on the biodistribution of ACT products in patients for accurate prognosis and treatment success [43]. MPI of SPION-labeled ACT immune cells can provide high-contrast, sensitive visualization of biodistribution and are thus ideal for this unmet need. The same SPIONs also appear on MRI scans (albeit lower contrast), thus allowing MPI's quantitative nature to complement the high-resolution anatomic MRI scans [44,45]. Rivera-Rodriguez et al. recently demonstrated MPI of ACT immune cells in a mouse model and showed that labeled immune cells showed up in the brain of C57BL/6 mice bearing intracranial KLuc-gp100 tumors 24 h after ACT infusion [46].

Furthermore, ideally immune cells should demonstrate native magnetic signal in order to prevent under-counting that occurs when in vitro magnetic labels are diluted by cell division. Recent efforts tried to overcome this limitation by genetically modifying cells with genes from magnetotactic bacteria [47–49], in order to produce magnetic crystals to enable label-free native magnetic contrast, but this has not been widely implemented on different mammalian cell types yet.

Other than ACT, Magnetic Particle Imaging has also been demonstrated to be helpful in other immunology studies that help advance the field of cancer immunotherapy. For example, the tumor microenvironment is known to greatly impact the success rate and thus a better understanding will help decipher the mechanisms of immunotherapies, define predictive biomarkers, and identify novel therapeutic targets. Figure 5 showcases recent work on MPI to track tumor-associated macrophages (TAMs). [50,51] Aptly named "Magnetic Particle Imaging of Macrophages Associated with Cancer: Filling the Voids Left by Iron-Based Magnetic Resonance Imaging.", the study showcased how MPI's positive contrast and quantitative nature complements the traditional MRI images of TAMs. In addition to this, MPI can also image inflammation by in situ labeling of inflammatory immune cells [52]. Although this study did not target cancer cells per se, the same in situ labeling concept could be used to image the inflammatory tumor microenvironment.

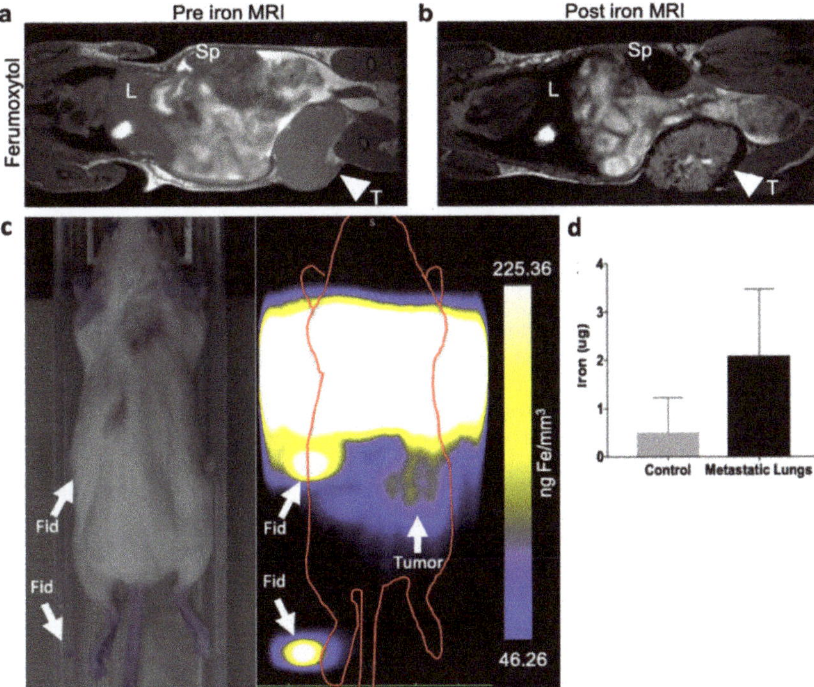

Figure 5. Imaging of Tumor-Associated Macrophages (TAMs) using MRI and MPI. Adapted with permission from Makela et al. Mol Imaging Biol 22. Copyright 2020 Springer Nature. [51]. (**a**) MRI image before non-targeted systemic i.v. injection of 0.5 mmol/kg Ferumoxytol (n = 8). L denotes liver, Sp denotes spleen, and T denotes tumor. (**b**) MRI image at 24 h post-injection of Ferumoxytol, where TAMs were seen to have iron uptake due to signal voids visible within the tumor. (**c**) MPI image at 24 h post-injection of Ferumoxytol. The image resolution was lower because Ferumoxytol is not optimal for MPI due to 7-fold worse spatial resolution of the nanoparticle than MPI standard Ferucarbotran. The MPI signal, while visible in the tumor, was not visible in the lung, as the lungs could not be spatially resolved from the liver due to the poor Ferumoxytol spatial resolution. (**d**) Ex vivo MPI of the lungs showed a significantly higher amount of iron in lung metastases compared with healthy control, indicating the presence of iron-labeled TAMs. This study demonstrated that not all MRI iron contrast work well for MPI due to differences in the physical principles of the imaging signal between MRI and MPI.

4. Magnetic-Based Steering and Targeting Strategies Using MPI Hardware

In brief, MPI use of the strongest magnetic gradients in the imaging field (up to 7 Tesla per meter) equips the MPI scanner to perform magnetic steering of magnetic agents to target tumor sites. This section elaborates on recent efforts to demonstrate this.

To introduce this topic, we must first note that one of the key benefits of using a magnetic imaging agent is the fact that magnetism remains the strongest force-from-a-distance method for remote steering or targeting [53–55]. There have been many studies of targeting of magnetic entities to a desired in vivo location using strong magnets [56–58]. Both MRI and MPI can benefit from these targeting strategies to enhance the concentration of imaging agent in a region-of-interest for increased binding probability to targets resulting in better imaging or localization of dose for better therapy. For example, Dames et al. 2007 demonstrated the use of a shaped magnetic tip for targeted delivery of magnetic aerosol droplets to the lung [14] (Figure 1e) and Banura et al. 2017 conducted a similar study with the addition of MPI to image the final biodistribution in the lungs after targeting [13] (Figure 6a). Other than the lungs, permanent magnets have been used to enhance delivery to tumors in other parts of the body. Arami et al. 2017 was able to enhance delivery to a flank tumor using an external permanent magnet affixed to the rodent flank [11].

One limitation of these single-magnet strategies is that targeting is only efficient at regions close to the body surface (Figure 1d). Other than embedding a sufficiently strong magnetic dipole source deep within the body, there is no method to magnetically attract towards an arbitrary point in 3D space. However, the hardware of MPI is able to generate a "repulsive" point at an arbitrary point in 3D space. This is because MPI uses a field-free-point or field-free-line gradient architecture with rapidly increasing magnetic field strength away from the zero-field-region which implies that magnetic material moves towards the edges of the gradient away from the zero-field-point. Magnetic steering is not unique to MPI, and while MRI has been used to steer large magnetic millimeter-sized ferromagnetic beads in vivo before [59], the weaker gradients used in MRI limit the particle size to about 0.2 mm as smaller objects do not have sufficient magnetic mass for MRI gradients to control [60]. The magnetic force $F = \tilde{N}(m \cdot B)$, where $\tilde{N}$ denotes the change of $(m \cdot B)$ per unit distance with units of m^{-1}. Assuming a magnetically saturated magnetic moment m (constant) as the object with units of Am^2, and B as our applied field with base SI units of $N\ A^{-1}\ m^{-1}$ (*note Tesla = $N\ A^{-1}\ m^{-1}$), then in general the magnetic force F scales linearly with the applied field (gradient) strength. MPI's 7 T/m gradients [12] are much stronger than the 0.045 T/m gradients used in MRI [2] and can provide much larger magnetic forces for targeting. This has resulted in the capability of MPI scanners to remotely steer catheter tips [61], remotely manipulate an iron screw [62], and in theory also steer particles of sub-micrometer scale. Specifically, magnetic catheter steering has seen clinical usage such as the NIOBE® ES Remote Magnetic Navigation (RMN) System (Stereotaxis, St. Louis, MO, USA) albeit with fluoroscopy imaging. In that clinical application, remote magnetic catheter navigation was performed to guide the catheter through the four heart chambers in order to locally perform atrial fibrillation ablation. Over 200 patients were tested, and it was shown that magnetic steering significantly reduced total fluoroscopy time (10.4 ± 6.4 vs. 16.3 ± 10.9 min) and thus lowered radiation dose to the patient when compared with manual pull-wire catheter navigation [63]. Recent preliminary work in the MPI field has shown some promise to completely replace the fluoroscopy aspect of catheter navigation with non-radioactive magnetic imaging by using one MPI "tracer" to mark the catheter tip and a second MPI "tracer" to replace the iodine contrast that shows the blood vessel size, shape, and branching. By distinguishing the magnetic signatures of the two different "tracers", it enabled interactive magnetic catheter steering with 3D real-time image feedback via "multi-color" MPI [61].

This can be combined with MPI's relatively high temporal resolution of up to 45 fps [64] to enable scan+steer sequences where an image is taken of a volume within 1/45th of a second every second for real-time image feedback of magnetic targeting while dedicating the 44 other frames to holding the magnetically repulsive point in 3D space. With real-time feedback, this can dynamically target the magnetic material towards an arbitrary region in 3D space despite only using a magnetically repulsive point. Proof-of-concept of this simultaneous imaging and MPI-steering of nanoparticles in Figure 6b–d was demonstrated by Griese et al. 2019 in vitro in a bifurcation flow phantom [65].

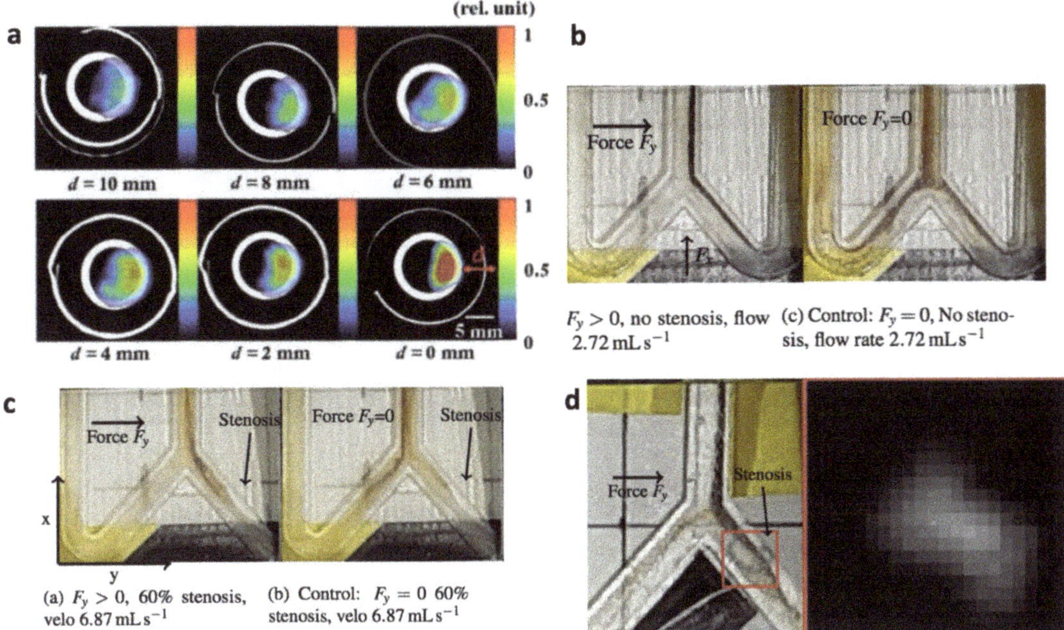

Figure 6. MPI image-guided magnetic steering of magnetic nanoparticles. Figures adapted with permission from Banura et al. Jpn. J. Appl. Phys 2017 [13] under Creative Commons 4.0. (**a**) Magnetic steering of aerosolized magnetic nanoparticles to deposit in a user-selected side of a lung imaging phantom. With closer distance d of the magnetic steering point to the lung, the stronger the accumulation of the magnetic nanoparticles. (**b**) Griese et al. 2020 [65] demonstrated magnetic steering of magnetic nanoparticles in a bifurcation flow phantom. Figures (**b**)–(**d**) adapted with permission Griese et al. J. Magn. Magn. Mater. 498. Copyright 2020 Elsevier. Without the magnetic steering in the control experiment, the dark brown nanoparticle stream bifurcated evenly. Once magnetic steering was turned on, 100% of the particles flowed into the selected right stream. (**c**) One important proof this study demonstrated was that magnetic steering can be performed against a strong flow rate of 6.87 mL/s, as shown by steering of particles into the right arm with a 60% stenosis, although the control clearly shows the flow rate favoring the left arm by order-of-magnitude. (**d**) Seamless switching between steer and image mode was shown in steering particles into an 100% stenosis arm, demonstrating that it is possible to maintain sufficient magnetic force to steer while performing a quick MPI image scan with a time ratio of 20:1 for steer:image.

As testament to the much stronger gradient strengths used in MPI versus MRI, Griese demonstrated that steering against the flow direction is possible by showing steering of nanoparticles into the arm with a 100% stenosis, although the control experiment showed the flow directs the nanoparticles to the non-stenosed arm when the MPI magnetic force is absent. The concept of seamless switching between "steer" and "image" modes was shown too. With a time ratio of 20:1 for force and imaging mode, the induced magnetic force acts for sufficient durations to maneuver the particles towards the stenosis, although no force is acting on the particles during the short time of the imaging mode [65]. In addition, multiple other studies have shown the feasibility of remote magnetic steering of micron-sized objects or synthetic bacteria in vivo [66,67].

Other than using magnetic forces for targeting, some groups have utilized anaerobic magnetotactic bacteria's natural tendency to migrate towards hypoxic regions for targeting hypoxic tumor regions [68]. In this case, the targeting depends on the bacteria, but because the bacteria natively produce magnetic crystals, this can be easily imaged with MPI or MRI. However, other groups have used magnetotactic bacteria for magnetic-field controlled manipulation and actuation of micro-objects [69]. Other strategies do not use the magnetic field for attraction forces but mainly for alignment of travel axis. While the bulk of the propulsion comes from micro-turbines or flagella [70–72], these micro-swimmers possess a magnetic axis that can be aligned to an external magnetic field for directionality. Unlike MRI having a fixed direction B0 field that limits alignment to the B0 field axis only, MPI's hardware is well-suited here because the "felt" magnetic field lines can be directed in any arbitrary direction by simply shifting the field-free-point gradient field around since the flux lines just surrounding the field-free-point are directed from every direction towards or away from the point.

5. Magnetic Methods for Cancer Therapy in Context of Magnetic Particle Imaging

Magnetic methods for cancer therapy generally fall into a few categories: (1) Hyperthermia methods that raise the temperature of the cancer cells ranging from mild heating to ablative levels via magnetic nanoparticles, (2) Magnetically actuated drug release from cancer-targeted nanocarriers, or (3) Magnetically actuated mechanical disruption of cancer cells by magnetic particles or magnetic micro-/nano-robots. In this section, each category is discussed and the benefits and relevance of Magnetic Particle Imaging towards these methods is explained.

5.1. Magnetic Hyperthermia Therapy (MHT)

In brief, MPI unique scanner architecture gives it the potential to be integrated with the alternating magnetic field (AMF) coils used for magnetic hyperthermia, enabling seamless image-guided therapy workflows. Another unique point is that the pre-existing gradients on the MPI scanner can be used to focus magnetic hyperthermia solely at the field-free-point (FFP) or field-free-line (FFL), enabling unprecedented targetable precision at-depth and in a 3D manner. This section explains the background of hyperthermia and recent MPI efforts to synergize with MHT.

The general principle of hyperthermia is based on increasing the temperature of a tissue of interest above 40 degrees Celsius [73,74]. While there are several methods to increase the temperature in hyperthermia, including microwaves, ultrasound, and laser, we focused on radiofrequency magnetic hyperthermia in this article. Magnetic hyperthermia (MH) is a promising cancer therapy that is induced by applying an alternating magnetic field (AMF) of frequencies ranging between 100 kHz and 1 MHz into magnetic nanoparticles targeted in the tumor area [75]. Under such conditions, magnetic nanoparticles act as very local heat sources, which are capable of raising the temperature of cancer tissues and consequently destroying the tumor in a localized and effective way. The heat generated by both superparamagnetic or ferromagnetic nanoparticles is originated from hysteresis losses and is proportional to the area of the hysteresis loop described by the magnetic nanoparticles during the application of the AMF [76,77].

The key advantages of MH are (1) the ability to treat at deeper regions of the body where other surface methods like microwaves, ultrasound, and radiation cannot, (2) negligible energy dose is deposited in healthy tissue en route to the target site as almost all the heat dose comes from the magnetic material on-site, (3) the magnetic material is not consumed by the therapy and allows for multiple treatment sessions per injection, and (4) the thermal dose is externally controlled by the AMF applicator which can compensate for variability in magnetic material accumulation at cancer site to ensure correct thermal dosing [78].

The first application of Magnetic Hyperthermia was in 1957 in dogs, where it aimed to treat cancers that had metastasized to the lymph nodes [79,80]. Most of the subsequent studies relied on direct injection of magnetic material into the tumor [81,82] rather than systemic delivery. To address this issue, Ivkov et al., in 2005, utilized monoclonal antibody targeting to cancer tissue [83]. Various other groups used magnetic nanoparticles within cationic liposomes for efficient accumulation into tumors and demonstrated therapeutic effect in rat glioma [84–86], melanoma [87,88], and prostate [89] animal tumor models. In recent years, an increasing number of in vivo and in vitro works have been reported in the literature [21,90]. In 2001, Jordan et al. showed the treatment of human solid tumors with MFH [91]. Due to the obtained promising results, several clinical trials have been carried out for the treatment of glioblastoma multiforme and prostate cancer. In 2003, the first phase I clinical trial was performed on 14 patients with glioblastoma multiforme (GBM) at the Charité Hospital in Berlin (MagForce Nanotechnologies) [77,91–93]. In 2005, Johannsen et al. reposted the first phase I clinical trial carried out in 10 patients with locally recurrent prostate cancer [94–96]. In 2010, MagForce AG obtained European Union Regulatory Approval (10/2011) for its the Nanotherm® therapy and later in 2013 started a clinical study in current gliobastoma with Nanotherm® therapy after receiving approval from the German Federal Institute for Drugs and Medical Devices. Recently, the FDA approved a single-arm study of NanoTherm (R) therapy system for intermediate-risk prostate cancer [97].

Despite all these clinical trials, there are several challenges that need to be addressed. One issue of MH is related to low accumulation of magnetic nanoparticles at the tumor site [98]. In order to achieve an efficient magnetic hyperthermia treatment, the heating efficiency (also known as Specific Absorption Rate (SAR)) of magnetic nanoparticles needs to be as high as possible in order to destroy the cancer with the low amount of magnetic nanoparticle available in the target site. The SAR greatly depends on the physicochemical properties of the nanoparticles such as composition, size, shape, crystallinity, and saturation magnetization [99,100]. Additionally, interparticle magnetic interactions, the interplay between particles and biological systems, and AMF parameters also affect the heating performance of magnetic nanoparticles [101,102].

Currently, different approaches have been proposed in the literature to design magnetic nanoparticles that exhibit high SAR values. Tailoring the shape of the magnetic nanoparticles can provide an effective strategy to increase their heating efficiency. For instance, Guardia et al. showed that the 19 nm iron oxide nanocubes possess very high SAR values (up to 2452 W/g at 29,000 A/m and 520 kHz) compared with spherical particles of similar size [100]. Other promising designs include magnetic vortex nanorings reaching 3000 W/g (at 64,000 A/m and 400 kHz) with demonstrated efficacy in vivo [103]. Some studies also use exchange-coupling between a magnetically hard core and magnetically soft shell to enhance SAR values (3886 W/g at 37,000 A/m and 500 kHz) to an order-of-magnitude greater than conventional iron-oxide nanoparticles, with superior therapeutic effectiveness in mice tumor models over chemotherapeutic drugs [104]. In addition, tuning the arrangements formed by dipolar interactions can also help enhance the heating efficiency of magnetic nanoparticles. Some works in literature have reported that specific arrangements formed by dipolar interaction, like chain-like structures, increase the SAR due to their ability to mechanically orient along the field lines [105,106]. Gandia et al. [107] proved that magnetotactic bacteria of the species M. gryphiswaldense, which internally biomineralized magnetosome chains, give rise to very high SAR values, up to 2400 W/g at 28,000 A/m and 300 kHz. Table 1 provides a summary and key characteristics of these MHT agents.

Table 1. Summary of studies demonstrating MHT agents with high SAR values.

MHT Agent	Characteristics	SAR (W/g)	Ref.
Iron Oxide Nanocubes	Size: 19 nm $\pm$ 3 nm, Msat: 80 emu/g	2452	[100]
Magnetic Vortex Nanorings	Size: 42/70 nm (ID/OD), 50 nm thick K_1: 135,000 erg/cc, Msat: 77 emu/g	~3000	[103]
Core-shell $ZnCoFe_2O_4$ @$ZnMnFe_2O_4$	Size: 15 nm K: 15,000 J/m^3 Msat: 125 emu/g	3886	[104]
Magnetite nanoparticle assembled chains	Size: 44 nm, sigma = 0.17 Msat: 87 emu/g	4.3-fold SAR w chaining	[105]
Magnetotactic bacteria M. gryphiswaldense	Size: 45 nm in 1 micron chain Msat: ~90 emu/g	2400	[107]

Magnetic Particle Imaging provides key benefits for MHT such as image-guidance (Figure 1f), quantitation of magnetic material on-site, which is essential for MHT thermal dose planning, and also the ability to select which magnetic nanoparticles to heat with pinpoint precision as low as a few millimeters [15,108–110]. This precision capability is a novel benefit in the field of MHT for cancer therapy. To explain further, consider chemotherapy and radiotherapy, which benefit from a significant differential cytotoxicity between cancerous and healthy cells [111]. Despite this, significant side-effects still exist due to collateral damage to healthy tissue. A similar issue exists for MHT where nanoparticle targeting/trafficking to tumors is not perfect and healthy tissue also accumulate nanoparticles. The additional precision in magnetic excitation enabled by MPI thus greatly mitigates collateral thermal damage to off-target healthy tissues frequently caused by magnetic particles biodistributed to other sites in the body, especially clearance organs such as the liver or spleen. This indirectly increases the therapeutic ratio to allow higher nanoparticle dosage as the side-effects to healthy cells are minimized. This concept is also seen for targeted nanocarriers for drug delivery, where precision of drug release enables higher doses while having lesser side-effects. Details are shown in the next section of this article. Conventional external AMF applicators used in MHT are unable to target magnetic excitation and heating only to the tumor because the long wavelength of the AMF at about ~50 m precludes the possibility of lens-based focusing of the magnetic field at a distance [15]. Other attempts using an array-based synthesization technique were able to project a focal point AMF at a distance of 10 cm but precision remained low with spot sizes of 5 cm [112,113]. Improving the precision to 2.5 cm required exponentially high currents in the kilo-ampere range [113]. In contrast, the mechanism for MPI's precision heating relies not on "focusing" the AMF into a narrow spot, but rather it suppresses the heating capability of off-target magnetic material by magnetically saturating off-target material so that it cannot respond and get heated by the AMF [15,114–116]. This can be achieved by MPI's field-free-line or field-free-point gradient hardware where the precision linearly scales with the gradient strength [115,116]. For example, in Figure 7b, at a gradient strength of 2.35 T/m, precision of 7 mm was achieved [15]. The field-free-region (zero field point) was simply placed over the target spot, enabling only that point-in-space to respond to AMF while magnetically saturating all other regions in space. Because the hardware for precision targeting exists within the MPI scanner and because the MPI scan at 20 kHz is demonstrated to have zero heating of particles [15], MPI is innately suited for image-guided precision MHT by simply imaging at 20 kHz then switching to a ~300 kHz for gradient-targeted precision MHT. Considering MPI's fully quantitative imaging of magnetic nanoparticle mass, it is possible to develop the ideal MHT workflow of (1) image, (2) quantitate, (3) dose planning, (4) target positioning, and (5) precision MHT all within a single MPI scanner. This ideal workflow was demonstrated in a rodent cancer model by Tay et al. 2018 (Figure 7a), where the efficacy of precision MHT and mitigation of collateral thermal damage to the liver was validated in vivo [15] (Figure 7c–e).

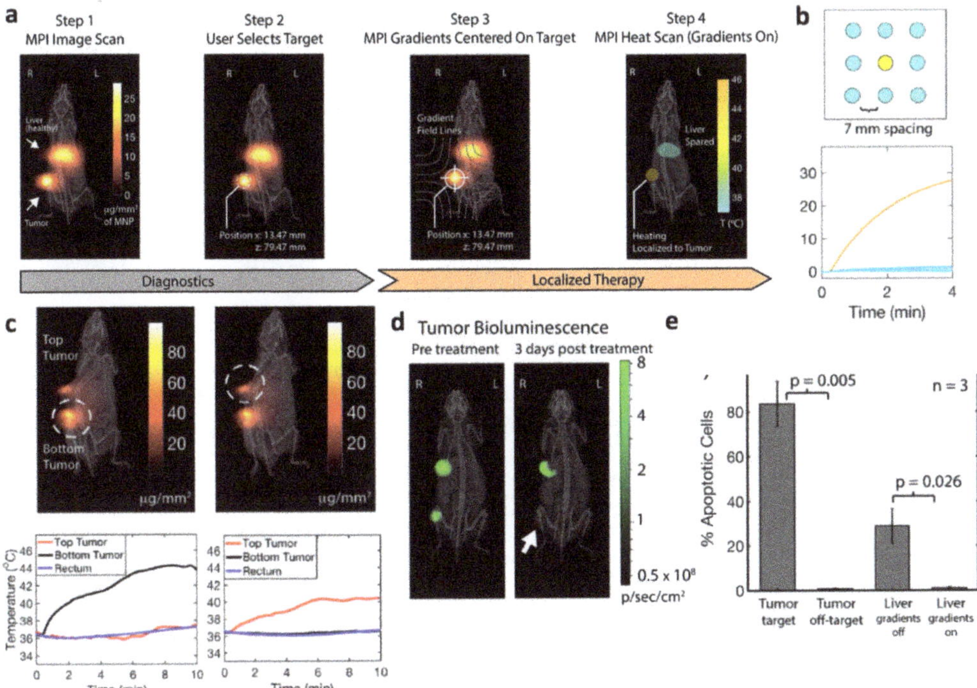

Figure 7. MPI image-guided precision Magnetic Hyperthermia with non-invasive pinpoint heating to 2–7 mm precision at a depth of 4 cm. Adapted with permission from Tay et al. 2018 [15]. Copyright 2018 American Chemical Society. (**a**) Theranostic workflow with MPI image guidance, selection of tumor target, and pinpoint localized heating. (**b**) Precision of at least 7 mm was demonstrated where any user-selected well in a custom 9-well plate can be heated to a 30 degrees increase in temperature with negligible heating in all adjacent targets 7 mm away. (**c**) In vivo results show user-selected precision heating of one of two adjacent tumors to the exclusion of the other. (**d**) Bioluminescence as a viability marker of luc-competent tumor confirms pinpoint therapy of one tumor with negligible impact on off-target tumor. (**e**) Although the mouse liver received a significant nanoparticle dose as shown in a, apoptosis assay showed that MPI hyperthermia (third column) improved precision-to-tumor over conventional hyperthermia (fourth column) which, due to its wide area magnetic excitation, collaterally damaged the liver while treating the tumor. This precision capability can reduce the side-effects of damage to healthy tissue in clinical settings.

5.2. Magnetically Actuated Drug Release

In brief, MPI can provide image-tracking of the magnetic-labeled drug delivery platform in vivo, ensuring that arrival at the target tumor site has occurred before triggering drug release by magneto-mechanical or magnetic heating in the case of thermosensitive liposomes. Similar to Section 5.1, MPI's selection field (FFP or FFL) can localize the triggering to only the field-free-region, increasing precision of therapy and further reducing drug side-effects. Finally, MPI images during therapy provide real-time feedback on the extent of drug released from the carrier. This section reviews recent efforts of MPI in the drug delivery field.

Chemotherapy has been one of the mainstays of cancer therapy and there has been much work in developing targeted nanocarriers with controlled release of chemotherapeutic drugs at the tumor to reduce systemic toxicity while maximizing the drug dosage at close proximity to the tumor to improve the therapeutic index [117,118]. Several methods to actuate the release of the chemotherapeutic have been developed and can be widely classified into external stimuli (magnetic, ultrasound, electric field, thermosensitive, UV–vis light, etc.) or endogeneous stimuli (pH-sensitive release, cancer-linked enzyme cleavage reactions, redox reactions, etc.) [119]. Magnetic methods to actuate drug release have several benefits over other methods such as (1) the ability to access deeper regions of the body with no view limitations and (2) the relative safety of magnetic fields compared with other methods for external stimuli that may affect healthy tissue en route to the target [120]. There have been many studies detailed below showing the efficacy of magnetic actuation for controlled chemotherapeutic release. The mechanism relies on a magnetic force to mechanical energy conversion and in many cases there is no detectable temperature rise, although it is also possible to combine both mechanical and MHT heating to doubly trigger release. In 2012, Peiris et al. developed a multi-component iron oxide nanochain with radiofrequency-tunable drug release [121]. The magnetic nanochain efficiently converts magnetic energy from a 10 kHz, 1–50 W external magnetic field into mechanical vibrations that trigger drug release from the attached DOX-loaded liposome. The release rate could be modulated by the operating parameters of the magnetic field. A temperature-sensitive fluorophore attached to the chain acted as a thermometer to verify the absence of local heating. In 2013, Oliveira et al. showed magnetic field triggered drug release (14 mT 750 kHz) from polymersomes, which are notable for their ability to load both hydrophilic and hydrophobic drugs [122]. In 2018, Nardoni et al. used pulsed magnetic fields (20 kHz, 60 A/m) to actuate drug release from high transition-temperature (Tm = 52 °C) magnetoliposomes [123]. The transient increase in membrane permeability upon actuation allowed on-demand drug release while ensuring negligible leakage and safety at all other times. For magneto-thermal mechanisms of drug release, Fuller et al. 2019 demonstrated nanocarriers with a hydrophobic core of superparamagnetic iron oxide nanoparticles that released heat upon AMF to actuate release of drug cargo from a thermoresponsive polymer based on thermally labile Diels-Alder bonds [124].

Magnetic Particle Imaging provides several key benefits for magnetic drug release—(1) Image-guidance and quantitative assessment of nanocarrier accumulation at target tumor site, (2) pinpoint precision of a few millimeters in actuating drug release while suppressing drug release from off-target nanocarriers (Figure 1g), and (3) real-time feedback on the amount of drug released from the magnetic nanocarrier via changes in the magnetic component's MPI spectrum. Similar to MHT, (1) is crucial for dose planning, especially when the amount of drug release is tunable such as in the study led by Peiris [121] (Figure 8a,b). Benefit (2) works on similar principles to that earlier described for MHT, where the suppression of off-target magnetic entities via magnetic saturation also works to suppress the induced mechanical forces. In other words, the magnetic components are overwhelmed by the background gradient magnetic field and are thus aligned and locked to the directionality of the background magnetic field. MPI's most unique benefit can be considered to be the real-time feedback on the amount of drug release. The mechanism of this depends on the different microenvironment around the nanoparticles within the nanocarrier as opposed to free nanoparticles after rupture of the nanocarrier. The particles report the change in microenvironment by a quantitative shift in the MPI spectrum. There has been much work in the MPI field to make these "color MPI" algorithms robust and quantitative to microenvironment factors, i.e., viscosity, pH, and inter-molecular binding [38–41].

Combining benefit (3) with the ability to switch between imaging 20 kHz and actuation ~300 kHz on the same MPI scanner, it is possible to develop the ideal therapy workflow within a single scanner.

The ideal workflow of (1) image, (2) quantitate, (3) dose planning, (4) target positioning, (5) precision drug release, and (6) real-time feedback (Figure 1h) on the amount of drug released would be desirable for controlled drug release applications. This workflow is theoretically feasible, although no one group has demonstrated the entirety of this workflow. Separate groups have proof-of-concept studies on each step of the workflow. Maruyama et al. 2016 demonstrated MPI quantitation of magnetic nanocarriers based on a thermoresponsive liposome design [125]. Liu et al. 2018 demonstrated target positioning and precision drug release (Figure 8c,d) at millimeter-scale precision from magnetic nanoliposomes with MPI-like gradient fields (referred to as static gating fields in this paper) [16]. Zhu et al. 2019 used MPI for in vivo quantitative drug release monitoring in tumors of a murine breast cancer model to measure in real-time the amount of drug release [17] (Figure 9a–c). Finally, MPI can be used to monitor apoptosis in tumors post-treatment. Using an apoptosis-specific tracer, MPI can accurately quantify apoptosis as the imaging signal was almost proportional to the number of apoptotic cells [126] (Figure 9d).

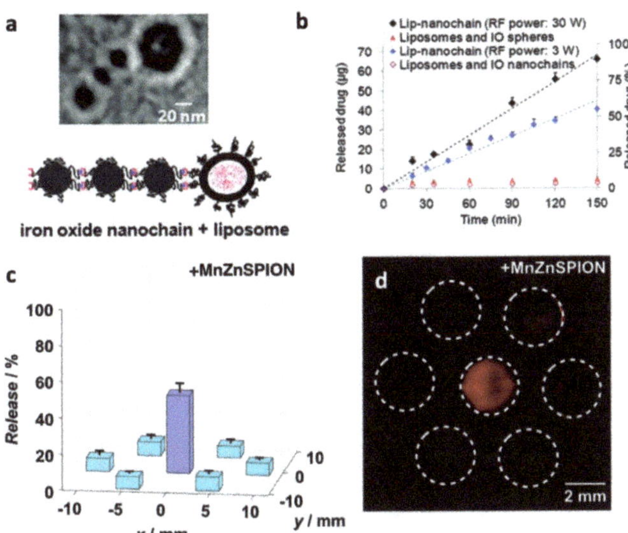

Figure 8. MPI for magnetically-actuated drug release [121]. Top row figures adapted with permission from Peiris et al. ACS Nano (2012) [121]. Copyright 2012 American Chemical Society (**a**) Liposome attached to a nanochain of three SPIONs that mechanically oscillates when exposed to an alternating magnetic field. Rather than using thermo-sensitive release, this work shows the feasibility of mechanical energy for rupturing the attached liposome. (**b**) The extent of drug release can be finely-tuned and controlled by the AMF frequency and power. Bottom row figures adapted with permission from Liu et al. Small (2018) [16]. Copyright 2018 WILEY-VCH Verlag GmbH. (**c**) Magnetic gradients (also termed static gating field in this article) can be used to target drug release to selected locations with 2 mm precision while suppressing release from other neighboring nanocarriers. Since MPI have the strongest magnetic gradients in imaging, the same concept demonstrated in Figure 7 can be replicated here for image-guided targeting of drug release. (**d**) Fluorescence imaging of released DOX from thermosensitive liposomes verifies that only the targeted well triggered drug release.

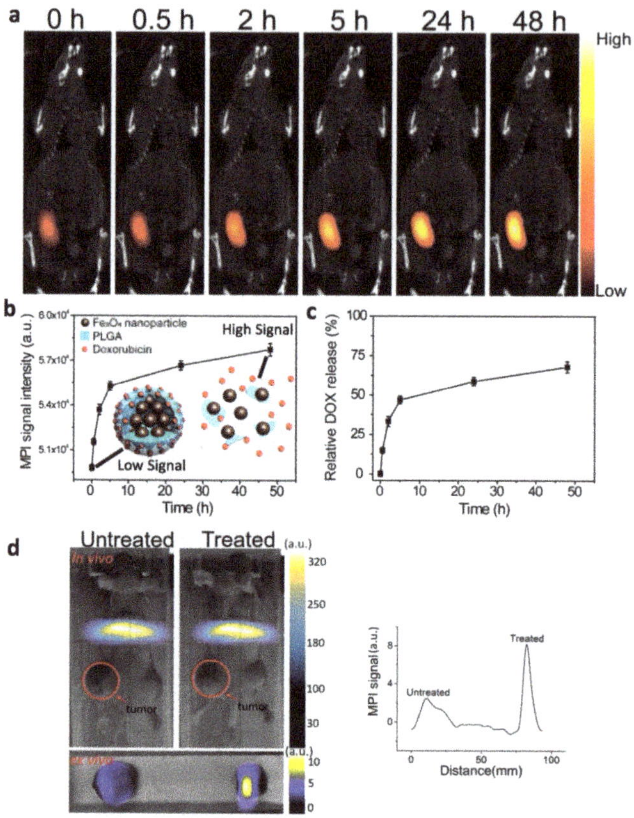

Figure 9. MPI for monitoring of the percentage of drug release from nanocarrier in vivo. Adapted with permission from Zhu et al. Nano Lett (2019) [17]. Copyright 2019 American Chemical Society. MPI enables non-invasive imaging assessment of the extent of drug release via MPI signal differences when SPIONs are encapsulated within a pH-sensitive nanocarrier and after their release together with drug upon nanocarrier rupture. (**a**) In vivo MPI images of nanoparticle-PLGA-Doxorubicin nanocarriers in vivo showing increasing MPI signal over a span of 48 h. (**b,c**) Quantification of MPI image intensity and percentage of DOX release from the nanocarrier shows a good correlation between MPI signal intensity and the percentage of release, verifying that the designed nanocarrier works as intended to have low MPI signal pre-release and high MPI signal after the nanoparticles are freed from the nanocarrier together with the Doxorubicin. (**d**) MPI of AnnexinV-SPION that binds to apoptotic cells in mouse xenograft model post-therapy, showing that MPI can evaluate the anti-tumor efficacy of cancer therapy. Figure (**d**) adapted with permission from Liang et al. Phys Med Biol. 2020. [126]. Copyright 2020 by Institute of Physics and Engineering in Medicine. Reproduced by permission of IOP Publishing. All rights reserved.

5.3. Magnetically Actuated Mechanical Disruption of Cancer Cells

In brief, other than magnetic hyperthermia and actuation of drug release, MPI can provide the magnetic energy and control necessary to actuate mechanical disruption of cancer cells. MPI's unique scanner architecture is well-suited for this because it already has 3-axis drive coils capable of up to 25 kHz and 25 mT field strength to power the magnetic actuation in various directions or to produce rotating magnetic fields. Most importantly, unlike MRI, which has an "always-on" main field forcing a fixed alignment of all magnetic material axes within the MRI scanner, MPI can turn off the electromagnet selection field and allow the drive fields to establish magnetic control of mechanical agents in vivo. This section reviews magneto-mechanical methods for cancer and explains how MPI's electromagnets can specifically achieve magnetic actuation for these methods.

To introduce the magneto-mechanical approach, we must first note that the magnetic forces incident on magnetic particles can be translated into mechanical energy that directly destroys cancer cells. Creixell et al. 2011 demonstrated that EGFR-targeted magnetic nanoparticles under AMF excitation were able to kill cancer cells (at a 99.9% loss in viability) without a perceptible temperature rise [127]. This runs contrary to the expectation that a temperature rise of up to 43–46 °C is needed to kill the cells under AMF. Because binding and subsequent activation of EGFR is implicated in cancer cell apoptosis, the EGF-nanoparticles without application of AMF already demonstrate some toxicity to the cancer cells. However, the application of AMF significantly increased the toxicity and suggests that magneto-mechanical stimulation of the EGFR via the attached EGF-nanoparticle greatly upregulates the relevant apoptotic pathways. This showcases the feasibility of magneto-mechanical actuation of apoptotic pathways in cancer. Other than EGFR, overactivation of ERK proteins via magnetic particles was also investigated to stop the cancer cell cycle of replication.

Besides mechanical activation of receptor-linked pathways, a more direct method is the mechanical disruption of cellular structures. Externally-bound magnetic particles can compromise cell membrane integrity promoting cell lysis, while internalized magnetic particles cause perturbations in lysozymes leading to enzyme-based cell suicide or damage the cytoskeletal integrity of the cell [128]. Liu et al. 2012 used magnetic carbon nanotubes at 75 mT 16 Hz for magnetoporation of the cancer cell membrane, as measured by increased membrane roughness by Atomic Force Microscopy and Scanning Electron Microscopy [129]. Wong et al. reported similar membrane integrity alteration with magnetic NiFe nanowires at 14 mT 5 Hz via ethidium bromide staining [130]. Domenech et al. showed lysosomal membrane permeabilization in cells that internalized iron oxide magnetic particles with increased the release of proteolytic cathepsin B activity leading to the cancer cell to self-digest [131]. Zhang et al. and Shen et al. reported similar results [132,133], but Master et al. reported negative results where lysosomal disturbance was not observed [134]. Master et al. also targeted the cell cytoskeletal component actin, harnessing the observation that cancer cells are less rigid than healthy cells. The results showed cancer cells were more susceptible to cytoskeletal disruption by actin-targeted magnetic particles under AMF [134]. Additionally, disk-shaped magnetic particles have also been used to magnetomechanical damage cancer cell integrity [135]. The disks can be actuated by an external magnetic field to exercise mechanical force on the cancer cell. Kim et al. used Ni80Fe20 microdisks with magnetic vortex configuration for in vitro experiments with glioma cancer cells. In this work, a 90% of cell death was reported after applying 9 mT and 10–20 Hz during 10 min [136]. Goriena et al. used Ni80Fe20 vortex configuration nanodisks [137] almost ten times smaller than those used by Kim et al. to destroy lung cancer cells [135]. The application of a 10 Hz oscillating magnetic field of 10 mT during 30 min reduced the cell by 30%. Beside disk-shaped magnetic particles with vortex state, perpendicularly magnetized synthetic antiferromagnetic (P-SAF) disks have also been used for cancer treatment through mechanical cell disruption [138].

Magnetic actuation was also demonstrated to be useful in a more macroscopic scale. For example, magnetic microbots at micron-level sizes [139–141]. Lee et al. 2020 demonstrated a micron-sized nickel-based magnetic corkscrew that is actuated by an external rotating magnetic field to "corkscrew" itself onto the cancer cell [142]. The microrobot then releases chemotherapeutic drugs after affixing itself to the target cell. Vyskocil et al. 2020 developed Au/Ag/Ni microrobotic scalpels that enter and exit an individual cancer cell and cut the cancer cell under actuation by an external rotating magnetic field [143]. This is relevant to MPI scanners because these rotating magnetic fields can be achieved with the 3-axis electromagnets used to produce the Lissajous trajectory for the FFP. Betal et al. 2018 developed a core-shell magnetoelectric nanorobot that uses DC magnetic gradients for navigation and steering to the target cell. This is relevant to MPI scanners because the

DC magnetic gradients needed can be fulfilled by the MPI selection field gradients. The same nanorobot under AMF actuation transforms into a localized electric-pulse generator for targeted cell electroporation. This can directly kill the cancer cells or increase their susceptibility to chemotherapeutics. [144].

As mentioned in previous chapters, MPI systems can provide image-guidance for this type of cancer strategy while offering much stronger AMF (25 mT at 20 kHz) and background gradient capabilities (up to 7 T/m) than MRI systems that typically have a 0.045 T/m gradient and microtesla RF excitation as AMF. While MRI systems can be enhanced with in-bore additions such as the shaped soft-iron core used in dipole-field navigation [145], the tradeoff in image quality due to susceptibility artifacts is usually substantial [60]. As such, MPI's unique hardware makeup as an imaging modality utilizing strong AMF and strong gradients enables its theranostic capabilities too, leading to better magnetic targeting, control, and actuation for magneto-mechanical strategies to kill cancer cells.

6. Safety of MPI and Current Status of Clinical Translation

MPI has comparable safety to MRI, which has been widely recognized as a safe medical imaging modality because it utilizes safe magnetic fields for excitation and signaling. Magnetic fields are non-ionizing and pass through the human body safely without attenuation or any mechanical tissue destruction. There are only two main safety considerations: (1) magnetic stimulation of peripheral nerves causing tingling sensations at the body peripheries when the alternating magnetic field strength is too high and (2) eddy-current induced warming of tissue when the magnetic field is at radio-frequency operating range, which is also known as SAR-related safety limits. For MPI, the drive field operating frequency is relatively low between 1–50 kHz and therefore close to the 42 kHz junction of magnetic stimulation (magstim) safety limits and SAR limit dominance as outlined by the safety study on human volunteers performed by Saritas et al. [146]. This MPI tailored safety study by Saritas et al. is most suitable for MPI's 20 kHz drive fields that finds no precedence in MRI safety standards. The results showed a limit of 15 mT peak-to-peak for the drive field, which is amenable to MPI imaging-parameters and therefore there are no fundamental safety concerns for MPI.

We can also evaluate MPI safety from the viewpoint of international commission on non-ionizing radiation protection (ICNIRP) standards. For instance, a maximal value of 2.7 Tesla per second for dB/dt (1 Hz–3 kHz, applicable to gradient fields MRI/MPI) was recommended to prevent any magnetostimulation or magnetophosphor effects on patients [147]. MPI's selection gradient fields, which are shifted by mechanical motion of the patient or by electromagnets at the range of 1 Hz to 3 kHz, stay within the limits proposed. Even for a very strong 7 Tesla per meter selection gradient field, the shift rate can be as high as 0.38 m per second, sufficient to raster the FFP or FFL across a typical clinical FOV dimension of 38 cm in 1 s. For a weaker selection field, this can be proportionally faster. Thus, MPI's gradient fields have no safety issues under the ICNIRP standards applied to MRI gradient fields.

For Magnetic Hyperthermia, the limits are dominated by the SAR-related safety regime and the safety limits are well-defined by general Atkinson-Brezovich limit ($H \times f <= 5 \times 10^8$ Am^{-1} s^{-1}) [148] and the Hergt criterion [149], which is less rigid and tailored to the area of application of the body ($H \times f <= 5 \times 10^9$ Am^{-1} s^{-1}). To give examples of existing devices in Johns Hopkins University, Attaluri et al. 2020 constructed a Maxwell-type induction coil prototype for magnetic nanoparticle hyperthermia in phantoms and large animals. The prototype was designed to be scalable to a human-sized system (60 cm diameter) [150]. For hyperthermia devices in Berlin, MagForce AG obtained European Union Regulatory Approval (10/2011) for its Nanotherm® therapy, and clinical studies for glioblastoma therapy were performed in 2013 with this approved instrument. More recently, MagForce received 2020 FDA approval for use in intermediate-risk prostate cancer. These examples and associated references were previously discussed in detail in Section 5.1.

The iron oxide nanoparticles used in MPI have a long history of safe usage in medicine, both as MRI tracer agents and for treating anemia. Iron oxide nanoparticle is considered safe and specific anaphylactic reaction observed when used is often associated with the parenteral formulation (can be made safer) and not the magnetic core that produces the MPI signal [151]. The nanoparticles used in MPI do not contain toxic magnetic elements like cobalt and are entirely iron-oxide based for biocompatibility. Some examples of clinically approved iron oxide are Ferumoxytol (USA) and Ferucarbotran/Resovist$^{®}$ (Japan) and the latter has been shown to work well for MPI [37,152,153]. Regarding clearance, iron oxide nanoparticles are easily assimilated by the liver and spleen and cleared by the hepatobiliary system [154]. Any digested iron from the particles becomes incorporated in the porphyrin rings of hemoglobin [151], replenishing the blood with iron-rich hemoglobin and forming the basis for treating anemia [155].

Although the MPI field has mostly shown preclinical studies, recent work has shown that the imaging technology can be scaled-up to clinical scale. Graeser et al. 2019 showcased a human-sized MPI scanner for brain imaging applications [156]. Mason et al. 2020 showcased an MPI design for clinical intraoperative applications [157].

7. Conclusions

Magnetic Particle Imaging is an emerging imaging modality with numerous complementary aspects to the more established MRI in the field of magnetic methods for cancer. Other than direct imaging of tumors, MPI has shown promise to value-add to passive nanocarriers [22] in other aspects such as targeting enhancement, actuating therapy, and post-therapy monitoring. Existing magnetic nanoparticles have mostly been optimized for MRI imaging, but with recent work on optimizing nanoparticles for MPI and heating theranostics, we hope that the capabilities of MPI can be significantly enhanced in the near future by these new classes of nanoparticles. The MPI engineering field has also shown great progress towards clinical translation with recent work showcasing a human head-sized MPI scanner. Overall, MPI has demonstrated its potential in a wide range of applications from tumor imaging to magnetically-actuated in situ drug release. With good compatibility for immunotherapy cell labeling, intrinsic high gradient strengths for magnetic steering and targeting, and finally the capability for spatially precise AMF magnetic heating/actuation, MPI shows great promise as a magnetic platform technology for cancer theranostics.

Author Contributions: Conceptualization, Z.W.T. and S.M.C.; writing—original draft preparation, Z.W.T.; writing—review and editing, Z.W.T., P.C., B.D.F., I.R.A., E.Y., M.O., S.M.C.; supervision, M.O., S.M.C.; funding acquisition, M.O, S.M.C. All authors have read and agreed to the published version of the manuscript.

Funding: For affiliation #2, we gratefully acknowledge funding support from National Institutes of Health, NIH1R01EB024578-03, 3R01EB024578-03S1, R44EB029877-01A1 and 1R24MH106053-01. For affiliation #1, we acknowledge funding support from Agency for Science, Technology and Research's (A*STAR) BMRC Central Research Fund (CRF, UIBR) Award and CDA grant 202D800036.

Conflicts of Interest: The authors declare no conflict of interest.

References

1. Gleich, B.; Weizenecker, J. Tomographic Imaging Using the Nonlinear Response of Magnetic Particles. *Nature* **2005**, *435*, 1214–1217. [CrossRef] [PubMed]
2. Saritas, E.U.; Goodwill, P.W.; Croft, L.R.; Konkle, J.J.; Lu, K.; Zheng, B.; Conolly, S.M. Magnetic Particle Imaging (MPI) for NMR and MRI Researchers. *J. Magn. Reson.* **2013**, *229*, 116–126. [CrossRef] [PubMed]
3. Zheng, B.; Yu, E.; Orendorff, R.; Lu, K.; Konkle, J.J.; Tay, Z.W.; Hensley, D.; Zhou, X.Y.; Chandrasekharan, P.; Saritas, E.U.; et al. Seeing SPIOs Directly In Vivo with Magnetic Particle Imaging. *Mol. Imaging Biol.* **2017**, *19*, 385–390. [CrossRef] [PubMed]

4. Zhou, X.Y.; Tay, Z.W.; Chandrasekharan, P.; Yu, E.Y.; Hensley, D.W.; Orendorff, R.; Jeffris, K.E.; Mai, D.; Zheng, B.; Goodwill, P.W.; et al. Magnetic Particle Imaging for Radiation-Free, Sensitive and High-Contrast Vascular Imaging and Cell Tracking. *Curr. Opin. Chem. Biol.* **2018**, *45*, 131–138. [CrossRef]

5. Chandrasekharan, P.; Tay, Z.W.; Zhou, X.Y.; Yu, E.; Orendorff, R.; Hensley, D.; Huynh, Q.; Fung, K.L.B.; VanHook, C.C.; Goodwill, P.; et al. A Perspective on a Rapid and Radiation-Free Tracer Imaging Modality, Magnetic Particle Imaging, with Promise for Clinical Translation. *Br. J. Radiol.* **2018**, *91*, 20180326. [CrossRef]

6. Tay, Z.W.; Chandrasekharan, P.; Zhou, X.Y.; Yu, E.; Zheng, B.; Conolly, S. In Vivo Tracking and Quantification of Inhaled Aerosol Using Magnetic Particle Imaging towards Inhaled Therapeutic Monitoring. *Theranostics* **2018**, *8*, 3676–3687. [CrossRef]

7. Zhou, X.Y.; Jeffris, K.E.; Yu, E.Y.; Zheng, B.; Goodwill, P.W.; Nahid, P.; Conolly, S.M. First in Vivo Magnetic Particle Imaging of Lung Perfusion in Rats. *Phys. Med. Biol.* **2017**, *62*, 3510–3522. [CrossRef]

8. Zheng, B.; von See, M.P.; Yu, E.; Gunel, B.; Lu, K.; Vazin, T.; Schaffer, D.V.; Goodwill, P.W.; Conolly, S.M. Quantitative Magnetic Particle Imaging Monitors the Transplantation, Biodistribution, and Clearance of Stem Cells In Vivo. *Theranostics* **2016**, *6*, 291–301. [CrossRef]

9. Nishimoto, K.; Mimura, A.; Aoki, M.; Banura, N.; Murase, K. Application of Magnetic Particle Imaging to Pulmonary Imaging Using Nebulized Magnetic Nanoparticles. *Open J. Med. Imaging* **2015**, *5*, 49. [CrossRef]

10. Oakes, J.M.; Breen, E.C.; Scadeng, M.; Tchantchou, G.S.; Darquenne, C. MRI-Based Measurements of Aerosol Deposition in the Lung of Healthy and Elastase-Treated Rats. *J. Appl. Physiol.* **2014**, *116*, 1561–1568. [CrossRef]

11. Arami, H.; Teeman, E.; Troksa, A.; Bradshaw, H.; Saatchi, K.; Tomitaka, A.; Gambhir, S.S.; Häfeli, U.O.; Liggitt, D.; Krishnan, K.M. Tomographic Magnetic Particle Imaging of Cancer Targeted Nanoparticles. *Nanoscale* **2017**, *9*, 18723–18730. [CrossRef]

12. Yu, E.Y.; Bishop, M.; Zheng, B.; Ferguson, R.M.; Khandhar, A.P.; Kemp, S.J.; Krishnan, K.M.; Goodwill, P.W.; Conolly, S.M. Magnetic Particle Imaging: A Novel in Vivo Imaging Platform for Cancer Detection. *Nano Lett.* **2017**, *17*, 1648–1654. [CrossRef]

13. Banura, N.; Murase, K. Magnetic Particle Imaging for Aerosol-Based Magnetic Targeting. *Jpn. J. Appl. Phys.* **2017**, *56*, 088001. [CrossRef]

14. Dames, P.; Gleich, B.; Flemmer, A.; Hajek, K.; Seidl, N.; Wiekhorst, F.; Eberbeck, D.; Bittmann, I.; Bergemann, C.; Weyh, T.; et al. Targeted Delivery of Magnetic Aerosol Droplets to the Lung. *Nat. Nanotechnol.* **2007**, *2*, 495–499. [CrossRef]

15. Tay, Z.W.; Chandrasekharan, P.; Chiu-Lam, A.; Hensley, D.W.; Dhavalikar, R.; Zhou, X.Y.; Yu, E.Y.; Goodwill, P.W.; Zheng, B.; Rinaldi, C.; et al. Magnetic Particle Imaging-Guided Heating in Vivo Using Gradient Fields for Arbitrary Localization of Magnetic Hyperthermia Therapy. *ACS Nano* **2018**, *12*, 3699–3713. [CrossRef]

16. Liu, J.F.; Neel, N.; Dang, P.; Lamb, M.; McKenna, J.; Rodgers, L.; Litt, B.; Cheng, Z.; Tsourkas, A.; Issadore, D. Radiofrequency-Triggered Drug Release from Nanoliposomes with Millimeter-Scale Resolution Using a Superimposed Static Gating Field. *Small* **2018**, *14*, e1802563. [CrossRef]

17. Zhu, X.; Li, J.; Peng, P.; Hosseini Nassab, N.; Smith, B.R. Quantitative Drug Release Monitoring in Tumors of Living Subjects by Magnetic Particle Imaging Nanocomposite. *Nano Lett.* **2019**, *19*, 6725–6733. [CrossRef]

18. Lu, M.; Cohen, M.H.; Rieves, D.; Pazdur, R. FDA Report: Ferumoxytol for Intravenous Iron Therapy in Adult Patients with Chronic Kidney Disease. *Am. J. Hematol.* **2010**, *85*, 315–319. [CrossRef]

19. Cherry, S.R.; Sorenson, J.A.; Phelps, M.E. *Physics in Nuclear Medicine*; Elsevier Health Sciences: Amsterdam, The Netherlands, 2012.

20. James, M.L.; Gambhir, S.S. A Molecular Imaging Primer: Modalities, Imaging Agents, and Applications. *Physiol. Rev.* **2012**, *92*, 897–965. [CrossRef]

21. Périgo, E.A.; Hemery, G.; Sandre, O.; Ortega, D.; Garaio, E.; Plazaola, F.; Teran, F.J. Fundamentals and Advances in Magnetic Hyperthermia. *Appl. Phys. Rev.* **2015**, *2*, 041302. [CrossRef]

22. Jhaveri, A.M.; Torchilin, V.P. Multifunctional polymeric micelles for delivery of drugs and siRNA. *Front. Pharmacol.* **2014**, *5*, 77. [CrossRef]

23. Lindemann, A.; Lüdtke-Buzug, K.; Fräderich, B.M.; Gräfe, K.; Pries, R.; Wollenberg, B. Biological Impact of Superparamagnetic Iron Oxide Nanoparticles for Magnetic Particle Imaging of Head and Neck Cancer Cells. *Int. J. Nanomed.* **2014**, *9*, 5025–5040. [CrossRef]

24. Gandhi, S.; Arami, H.; Krishnan, K.M. Detection of Cancer-Specific Proteases Using Magnetic Relaxation of Peptide-Conjugated Nanoparticles in Biological Environment. *Nano Lett.* **2016**, *16*, 3668–3674. [CrossRef]

25. Tay, Z.W.; Hensley, D.W.; Vreeland, E.C.; Zheng, B.; Conolly, S.M. The Relaxation Wall: Experimental Limits to Improving MPI Spatial Resolution by Increasing Nanoparticle Core Size. *Biomed. Phys. Eng Express* **2017**, *3*. [CrossRef]

26. Tay, Z.W.; Hensley, D.; Ma, J.; Chandrasekharan, P.; Zheng, B.; Goodwill, P.; Conolly, S. Pulsed Excitation in Magnetic Particle Imaging. *IEEE Trans. Med. Imaging* **2019**, *38*, 2389–2399. [CrossRef]

27. Alard, E.; Butnariu, A.-B.; Grillo, M.; Kirkham, C.; Zinovkin, D.A.; Newnham, L.; Macciochi, J.; Pranjol, M.Z.I. Advances in Anti-Cancer Immunotherapy: Car-T Cell, Checkpoint Inhibitors, Dendritic Cell Vaccines, and Oncolytic Viruses, and Emerging Cellular and Molecular Targets. *Cancers* **2020**, *12*, 1826. [CrossRef]

28. Sivanandam, V.; LaRocca, C.J.; Chen, N.G.; Fong, Y.; Warner, S.G. Oncolytic Viruses and Immune Checkpoint Inhibition: The Best of Both Worlds. *Mol. Ther. Oncolytics* **2019**, *13*, 93–106. [CrossRef]

29. Russell, S.J.; Barber, G.N. Oncolytic Viruses as Antigen-Agnostic Cancer Vaccines. *Cancer Cell* **2018**, *33*, 599–605. [CrossRef]

30. Rosenberg, S.A.; Restifo, N.P. Adoptive Cell Transfer as Personalized Immunotherapy for Human Cancer. *Science* **2015**, *348*, 62–68. [CrossRef]

31. Branca, M.A. Rekindling cancer vaccines. *Nat. Biotechnol.* **2016**, *34*, 1292. [CrossRef]

32. Lewin, M.; Carlesso, N.; Tung, C.H.; Tang, X.W.; Cory, D.; Scadden, D.T.; Weissleder, R. Tat Peptide-Derivatized Magnetic Nanoparticles Allow In Vivo Tracking and Recovery of Progenitor Cells. *Nat. Biotechnol.* **2000**, *18*, 410–414. [CrossRef] [PubMed]

33. Zheng, B.; Vazin, T.; Goodwill, P.W.; Conway, A.; Verma, A.; Saritas, E.U.; Schaffer, D.; Conolly, S.M. Magnetic Particle Imaging Tracks the Long-Term Fate of in Vivo Neural Cell Implants with High Image Contrast. *Sci. Rep.* **2015**, *5*, 14055. [CrossRef] [PubMed]

34. Sehl, O.C.; Gevaert, J.J.; Melo, K.P.; Knier, N.N.; Foster, P.J. A Perspective on Cell Tracking with Magnetic Particle Imaging. *Tomography* **2020**, *6*, 315–324. [CrossRef]

35. Lemaster, J.E.; Chen, F.; Kim, T.; Hariri, A.; Jokerst, J.V. Development of a Trimodal Contrast Agent for Acoustic and Magnetic Particle Imaging of Stem Cells. *ACS Appl. Nano Mater.* **2018**, *1*, 1321–1331. [CrossRef] [PubMed]

36. Kratz, H.; Taupitz, M.; Ariza de Schellenberger, A.; Kosch, O.; Eberbeck, D.; Wagner, S.; Trahms, L.; Hamm, B.; Schnorr, J. Novel Magnetic Multicore Nanoparticles Designed for MPI and Other Biomedical Applications: From Synthesis to First in Vivo Studies. *PLoS ONE* **2018**, *13*, e0190214. [CrossRef] [PubMed]

37. Nejadnik, H.; Pandit, P.; Lenkov, O.; Lahiji, A.P.; Yerneni, K.; Daldrup-Link, H.E. Ferumoxytol Can Be Used for Quantitative Magnetic Particle Imaging of Transplanted Stem Cells. *Mol. Imaging Biol.* **2018**, *21*, 465–472. [CrossRef] [PubMed]

38. Fidler, F.; Steinke, M.; Kraupner, A.; Gruttner, C.; Hiller, K.-H.; Briel, A.; Westphal, F.; Walles, H.; Jakob, P.M. Stem Cell Vitality Assessment Using Magnetic Particle Spectroscopy. *IEEE Trans. Magn.* **2015**, *51*, 1–4. [CrossRef]

39. Utkur, M.; Muslu, Y.; Saritas, E.U. Relaxation-Based Color Magnetic Particle Imaging for Viscosity Mapping. *Appl. Phys. Lett.* **2019**, *115*, 152403. [CrossRef]

40. Khurshid, H.; Friedman, B.; Berwin, B.; Shi, Y.; Ness, D.B.; Weaver, J.B. Blood Clot Detection Using Magnetic Nanoparticles. *AIP Adv.* **2017**, *7*, 056723. [CrossRef]

41. Szwargulski, P.; Wilmes, M.; Javidi, E.; Thieben, F.; Graeser, M.; Koch, M.; Gruettner, C.; Adam, G.; Gerloff, C.; Magnus, T.; et al. Monitoring Intracranial Cerebral Hemorrhage Using Multicontrast Real-Time Magnetic Particle Imaging. *ACS Nano* **2020**, *14*, 13913–13923. [CrossRef]

42. Tey, S.-K. Adoptive T-Cell Therapy: Adverse Events and Safety Switches. *Clin. Transl. Immunol.* **2014**, *3*, e17. [CrossRef]

43. Kalos, M.; June, C.H. Adoptive T Cell Transfer for Cancer Immunotherapy in the Era of Synthetic Biology. *Immunity* **2013**, *39*, 49–60. [CrossRef]

44. Ahrens, E.T.; Bulte, J.W.M. Tracking Immune Cells In Vivo Using Magnetic Resonance Imaging. *Nat. Rev. Immunol.* **2013**, *13*, 755–763. [CrossRef]

45. Cromer Berman, S.M.; Walczak, P.; Bulte, J.W.M. Tracking Stem Cells Using Magnetic Nanoparticles. *Wiley Interdiscip. Rev. Nanomed. Nanobiotechnol.* **2011**, *3*, 343–355. [CrossRef]

46. Rivera-Rodriguez, A.; Hoang-Minh, L.B.; Chiu-Lam, A.; Sarna, N.; Marrero-Morales, L.; Mitchell, D.A.; Rinaldi-Ramos, C.M. Tracking adoptive T cell immunotherapy using magnetic particle imaging. *Nanotheranostics* **2021**, *5*, 431–444. [CrossRef]

47. Kolinko, I.; Lohße, A.; Borg, S.; Raschdorf, O.; Jogler, C.; Tu, Q.; Pósfai, M.; Tompa, E.; Plitzko, J.M.; Brachmann, A.; et al. Biosynthesis of Magnetic Nanostructures in a Foreign Organism by Transfer of Bacterial Magnetosome Gene Clusters. *Nat. Nanotechnol.* **2014**, *9*, 193–197. [CrossRef]

48. Zhang, X.-Y.; Robledo, B.N.; Harris, S.S.; Hu, X.P. A Bacterial Gene, mms6, as a New Reporter Gene for Magnetic Resonance Imaging of Mammalian Cells. *Mol. Imaging* **2014**, *13*, 1–12. [CrossRef]

49. Kraupner, A.; Eberbeck, D.; Heinke, D.; Uebe, R.; Schüler, D.; Briel, A. Bacterial Magnetosomes—Nature's Powerful Contribution to MPI Tracer Research. *Nanoscale* **2017**, *9*, 5788–5793. [CrossRef]

50. Makela, A.V.; Gaudet, J.M.; Foster, P.J. Quantifying Tumor Associated Macrophages in Breast Cancer: A Comparison of Iron and Fluorine-Based MRI Cell Tracking. *Sci. Rep.* **2017**, *7*, 42109. [CrossRef]

51. Makela, A.V.; Gaudet, J.M.; Schott, M.A.; Sehl, O.C.; Contag, C.H.; Foster, P.J. Magnetic Particle Imaging of Macrophages Associated with Cancer: Filling the Voids Left by Iron-Based Magnetic Resonance Imaging. *Mol. Imaging Biol.* **2020**, *22*, 958–968. [CrossRef]

52. Chandrasekharan, P.; Fung, K.L.B.; Zhou, X.Y.; Cui, W.; Colson, C.; Mai, D.; Jeffris, K.; Huynh, Q.; Saayujya, C.; Kabuli, L.; et al. Non-Radioactive and Sensitive Tracking of Neutrophils towards Inflammation Using Antibody Functionalized Magnetic Particle Imaging Tracers. *Nanotheranostics* **2021**, *5*, 240–255. [CrossRef]

53. Ilami, M.; Ahmed, R.J.; Petras, A.; Beigzadeh, B.; Marvi, H. Magnetic Needle Steering in Soft Phantom Tissue. *Sci. Rep.* **2020**, *10*, 2500. [CrossRef]

54. Lalande, V.; Gosselin, F.P.; Vonthron, M.; Conan, B.; Tremblay, C.; Beaudoin, G.; Soulez, G.; Martel, S. In Vivo Demonstration of Magnetic Guidewire Steerability in a MRI System with Additional Gradient Coils. *Med. Phys.* **2015**, *42*, 969–976. [CrossRef]

55. Heunis, C.; Sikorski, J.; Misra, S. Flexible Instruments for Endovascular Interventions: Improved Magnetic Steering, Actuation, and Image-Guided Surgical Instruments. *IEEE Robot. Autom. Mag.* **2018**, *25*, 71–82. [CrossRef]

56. Muthana, M.; Kennerley, A.J.; Hughes, R.; Fagnano, E.; Richardson, J.; Paul, M.; Murdoch, C.; Wright, F.; Payne, C.; Lythgoe, M.F.; et al. Directing Cell Therapy to Anatomic Target Sites in Vivo with Magnetic Resonance Targeting. *Nat. Commun.* **2015**, *6*, 8009. [CrossRef]

57. Pouponneau, P.; Leroux, J.-C.; Soulez, G.; Gaboury, L.; Martel, S. Co-Encapsulation of Magnetic Nanoparticles and Doxorubicin into Biodegradable Microcarriers for Deep Tissue Targeting by Vascular MRI Navigation. *Biomaterials* **2011**, *32*, 3481–3486. [CrossRef]

58. Carpi, F.; Kastelein, N.; Talcott, M.; Pappone, C. Magnetically Controllable Gastrointestinal Steering of Video Capsules. *IEEE Trans. Biomed. Eng.* **2011**, *58*, 231–234. [CrossRef]

59. Felfoul, O.; Becker, A.T.; Fagogenis, G.; Dupont, P.E. Simultaneous Steering and Imaging of Magnetic Particles Using MRI toward Delivery of Therapeutics. *Sci. Rep.* **2016**, *6*, 33567. [CrossRef]

60. Martel, S. Magnetic Microbots to Fight Cancer. IEEE Spectrum Feature Article on Medical Robots. 2012. Available online: https://spectrum.ieee.org/magnetic-microbots-to-fight-cancer (accessed on 21 October 2021).

61. Rahmer, J.; Wirtz, D.; Bontus, C.; Borgert, J.; Gleich, B. Interactive Magnetic Catheter Steering With 3-D Real-Time Feedback Using Multi-Color Magnetic Particle Imaging. *IEEE Trans. Med. Imaging* **2017**, *36*, 1449–1456. [CrossRef] [PubMed]

62. Rahmer, J.; Stehning, C.; Gleich, B. Remote Magnetic Actuation Using a Clinical Scale System. *PLoS ONE* **2018**, *13*, e0193546. [CrossRef] [PubMed]

63. Yuan, S.; Holmqvist, F.; Kongstad, O.; Jensen, S.M.; Wang, L.; Ljungström, E.; Hertervig, E.; Borgquist, R. Long-Term Outcomes of the Current Remote Magnetic Catheter Navigation Technique for Ablation of Atrial Fibrillation. *Scand. Cardiovasc. J.* **2017**, *51*, 308–315. [CrossRef] [PubMed]

64. Bakenecker, A.C.; Ahlborg, M.; Debbeler, C.; Kaethner, C.; Buzug, T.M.; Lüdtke-Buzug, K. Magnetic Particle Imaging in Vascular Medicine. *Innov. Surg. Sci.* **2018**, *3*, 179–192. [CrossRef]

65. Griese, F.; Knopp, T.; Gruettner, C.; Thieben, F.; Müller, K.; Loges, S.; Ludewig, P.; Gdaniec, N. Simultaneous Magnetic Particle Imaging and Navigation of Large Superparamagnetic Nanoparticles in Bifurcation Flow Experiments. *J. Magn. Magn. Mater.* **2020**, *498*, 166206. [CrossRef]

66. Buss, M.T.; Ramesh, P.; English, M.A.; Lee-Gosselin, A.; Shapiro, M.G. Biomagnetic Materials: Spatial Control of Probiotic Bacteria in the Gastrointestinal Tract Assisted by Magnetic Particles (adv. Mater. 17/2021). *Adv. Mater.* **2021**, *33*, 2170134. [CrossRef]

67. Zhang, H.; Li, Z.; Gao, C.; Fan, X.; Pang, Y.; Li, T.; Wu, Z.; Xie, H.; He, Q. Dual-Responsive Biohybrid Neutrobots for Active Target Delivery. *Sci. Robot.* **2021**, *6*, eaaz9519. [CrossRef]

68. Felfoul, O.; Mohammadi, M.; Taherkhani, S.; de Lanauze, D.; Zhong Xu, Y.; Loghin, D.; Essa, S.; Jancik, S.; Houle, D.; Lafleur, M.; et al. Magneto-Aerotactic Bacteria Deliver Drug-Containing Nanoliposomes to Tumour Hypoxic Regions. *Nat. Nanotechnol.* **2016**, *11*, 941–947. [CrossRef]

69. Martel, S.; Tremblay, C.C.; Ngakeng, S. Controlled Manipulation and Actuation of Micro-Objects with Magnetotactic Bacteria. *J. Phys. D Appl. Phys.* **2006**, *89*, 233904. [CrossRef]

70. Carlsen, R.W.; Edwards, M.R.; Zhuang, J.; Pacoret, C.; Sitti, M. Magnetic Steering Control of Multi-Cellular Bio-Hybrid Microswimmers. *Lab. Chip* **2014**, *14*, 3850–3859. [CrossRef]

71. Hoshiar, A.K.; Le, T.-A.; Amin, F.U.; Kim, M.O.; Yoon, J. Studies of Aggregated Nanoparticles Steering during Magnetic-Guided Drug Delivery in the Blood Vessels. *J. Magn. Magn. Mater.* **2017**, *427*, 181–187. [CrossRef]

72. Gagné, K.; Tremblay, C.; Majedi, Y.; Mohammadi, M.; Martel, S. Indirect MPI-Based Detection of Superparamagnetic Nanoparticles Transported by Computer-Controlled Magneto-Aerotactic Bacteria. In Proceedings of the 2017 International Conference on Manipulation, Automation and Robotics at Small Scales (MARSS), Montreal, QC, Canada, 17–21 July 2017; pp. 1–5.

73. Chang, D.; Lim, M.; Goos, J.A.C.M.; Qiao, R.; Ng, Y.Y.; Mansfeld, F.M.; Jackson, M.; Davis, T.P.; Kavallaris, M. Biologically Targeted Magnetic Hyperthermia: Potential and Limitations. *Front. Pharmacol.* **2018**, *9*, 831. [CrossRef]

74. Hildebrandt, B.; Wust, P.; Ahlers, O.; Dieing, A.; Sreenivasa, G.; Kerner, T.; Felix, R.; Riess, H. The cellular and molecular basis of hyperthermia. *Crit. Rev. Oncol. Hematol.* **2002**, *43*, 33–56. [CrossRef]

75. Ortega, D.; Pankhurst, Q.A. Magnetic hyperthermia. In *Nanoscience*; RSC Publishing: London, UK, 2012; pp. 60–88.

76. Dennis, C.L.; Ivkov, R. Physics of heat generation using magnetic nanoparticles for hyperthermia. *Int. J. Hyperth.* **2013**, *29*, 715–729. [CrossRef]

77. Carrey, J.; Mehdaoui, B.; Respaud, M. Simple Models for Dynamic Hysteresis Loop Calculations of Magnetic Single-Domain Nanoparticles: Application to Magnetic Hyperthermia Optimization. *J. Appl. Phys.* **2011**, *109*, 083921. [CrossRef]

78. Jordan, A.; Scholz, R.; Wust, P.; Fahling, H.; Felix, R. Magnetic fluid hyperthermia (MFH): Cancer treatment with AC magnetic field induced excitation of biocompatible superparamagnetic nanoparticles. *J. Magn. Magn. Mater.* **1999**, *201*, 413–419. [CrossRef]

79. Gilchrist, R.K.; Medal, R.; Shorey, W.D.; Hanselman, R.C.; Parrott, J.C.; Taylor, C.B. Selective Inductive Heating of Lymph Nodes. *Ann. Surg.* **1957**, *146*, 596–606. [CrossRef]

80. Thiesen, B.; Jordan, A. Clinical Applications of Magnetic Nanoparticles for Hyperthermia. *Int. J. Hyperthermia* **2008**, *24*, 467–474. [CrossRef]

81. Jordan, A.; Scholz, R.; Wust, P.; Fähling, H.; Krause, J.; Wlodarczyk, W.; Sander, B.; Vogl, T.; Felix, R. Effects of Magnetic Fluid Hyperthermia (MFH) on C3H Mammary Carcinoma in Vivo. *Int. J. Hyperthermia* **1997**, *13*, 587–605. [CrossRef]

82. Hilger, I.; Hiergeist, R.; Hergt, R.; Winnefeld, K.; Schubert, H.; Kaiser, W.A. Thermal Ablation of Tumors Using Magnetic Nanoparticles: An in Vivo Feasibility Study. *Invest. Radiol.* **2002**, *37*, 580–586. [CrossRef] [PubMed]

83. Ivkov, R.; De Nardo, S.J.; Daum, W.; Foreman, A.R.; Goldstein, R.C.; Nemkov, V.S.; De Nardo, G.L. Application of High Amplitude Alternating Magnetic Fields for Heat Induction of Nanoparticles Localized in Cancer. *Clin. Cancer Res.* **2005**, *11*, 7093s–7103s. [CrossRef] [PubMed]

84. Yanase, M.; Shinkai, M.; Honda, H.; Wakabayashi, T.; Yoshida, J.; Kobayashi, T. Antitumor Immunity Induction by Intracellular Hyperthermia Using Magnetite Cationic Liposomes. *Jpn. J. Cancer Res.* **1998**, *89*, 775–782. [CrossRef]

85. Shinkai, M.; Yanase, M.; Suzuki, M.; Honda, H.; Wakabayashi, T.; Yoshida, J.; Kobayashi, T. Intracellular Hyperthermia for Cancer Using Magnetite Cationic Liposomes. *J. Magn. Magn. Mater.* **1999**, *194*, 176–184. [CrossRef]

86. Le, B.; Shinkai, M.; Kitade, T.; Honda, H.; Yoshida, J.; Wakabayashi, T.; Kobayashi, T. Preparation of Tumor-Specific Magnetoliposomes and Their Application for Hyperthermia. *J. Chem. Eng. Jpn.* **2001**, *34*, 66–72. [CrossRef]

87. Ito, A.; Tanaka, K.; Kondo, K.; Shinkai, M.; Honda, H.; Matsumoto, K.; Saida, T.; Kobayashi, T. Tumor Regression by Combined Immunotherapy and Hyperthermia Using Magnetic Nanoparticles in an Experimental Subcutaneous Murine Melanoma. *Cancer Sci.* **2003**, *94*, 308–313. [CrossRef] [PubMed]

88. Tanaka, K.; Ito, A.; Kobayashi, T.; Kawamura, T.; Shimada, S.; Matsumoto, K.; Saida, T.; Honda, H. Intratumoral Injection of Immature Dendritic Cells Enhances Antitumor Effect of Hyperthermia Using Magnetic Nanoparticles. *Int. J. Cancer* **2005**, *116*, 624–633. [CrossRef] [PubMed]

89. Kawai, N.; Ito, A.; Nakahara, Y.; Futakuchi, M.; Shirai, T.; Honda, H.; Kobayashi, T.; Kohri, K. Anticancer Effect of Hyperthermia on Prostate Cancer Mediated by Magnetite Cationic Liposomes and Immune-Response Induction in Transplanted Syngeneic Rats. *Prostate* **2005**, *64*, 373–381. [CrossRef] [PubMed]

90. Kozissnik, B.; Bohorquez, A.C.; Dobson, J.; Rinaldi, C. Magnetic Fluid Hyperthermia: Advances, Challenges, and Opportunity. *Int. J. Hyperthermia* **2013**, *29*, 706–714. [CrossRef]

91. Jordan, A.; Scholz, R.; Maier-Hauff, K.; Johannsen, M.; Wust, P.; Nadobny, J.; Schirra, H.; Schmidt, H.; Deger, S.; Loening, S.; et al. Presentation of a new magnetic field therapy system for the treatment of human solid tumors with magnetic fluid hyperthermia. *J. Magn. Magn. Mater.* **2001**, *225*, 118–126. [CrossRef]

92. Maier-Hauff, K.; Rothe, R.; Scholz, R.; Gneveckow, U.; Wust, P.; Thiesen, B.; Feussner, A.; von Deimling, A.; Waldoefner, N.; Felix, R.; et al. Intracranial Thermotherapy Using Magnetic Nanoparticles Combined with External Beam Radiotherapy: Results of a Feasibility Study on Patients with Glioblastoma Multiforme. *J. Neurooncol.* **2007**, *81*, 53–60. [CrossRef]

93. Gneveckow, U.; Jordan, A.; Scholz, R.; Brüß, V.; Waldöfner, N.; Ricke, J.; Feussner, A.; Hildebrandt, B.; Rau, B.; Wust, P. Description and characterization of the novel hyperthermia and thermoablation-system MFH®300F for clinical magnetic fluid hyperthermia. *Med. Phys.* **2004**, *31*, 1444–1451. [CrossRef]

94. Johannsen, M.; Gneveckow, U.; Eckelt, L.; Feussner, A.; Waldöfner, N.; Scholz, R.; Deger, S.; Wust, P.; Loening, S.A.; Jordan, A. Clinical Hyperthermia of Prostate Cancer Using Magnetic Nanoparticles: Presentation of a New Interstitial Technique. *Int. J. Hyperthermia* **2005**, *21*, 637–647. [CrossRef]

95. Johannsen, M.; Gneveckow, U.; Thiesen, B.; Taymoorian, K.; Cho, C.H.; Waldöfner, N.; Scholz, R.; Jordan, A.; Loening, S.A.; Wust, P. Thermotherapy of Prostate Cancer Using Magnetic Nanoparticles: Feasibility, Imaging, and Three-Dimensional Temperature Distribution. *Eur. Urol.* **2007**, *52*, 1653–1662. [CrossRef]

96. Johannsen, M.; Gneveckow, U.; Taymoorian, K.; Thiesen, B.; Waldöfner, N.; Scholz, R.; Jung, K.; Jordan, A.; Wust, P.; Loening, S.A. Morbidity and Quality of Life during Thermotherapy Using Magnetic Nanoparticles in Locally Recurrent Prostate Cancer: Results of a Prospective Phase I Trial. *Int. J. Hyperthermia* **2007**, *23*, 315–323. [CrossRef]

97. Liu, X.; Zhang, Y.; Wang, Y.; Zhu, W.; Li, G.; Ma, X.; Zhang, Y.; Chen, S.; Tiwari, S.; Shi, K.; et al. Comprehensive Understanding of Magnetic Hyperthermia for Improving Antitumor Therapeutic Efficacy. *Theranostics* **2020**, *10*, 3793–3815. [CrossRef]

98. Wilhelm, S.; Tavares, A.J.; Dai, Q.; Ohta, S.; Audet, J.; Dvorak, H.F.; Chan, W.C.W. Analysis of Nanoparticle Delivery to Tumours. *Nat. Rev. Mater.* **2016**, *1*, 16014. [CrossRef]

99. Nemati, Z.; Alonso, J.; Rodrigo, I.; Das, R.; Garaio, E.; García, J.Á.; Orue, I.; Phan, M.-H.; Srikanth, H. Improving the Heating Efficiency of Iron Oxide Nanoparticles by Tuning Their Shape and Size. *J. Phys. Chem. C* **2018**, *122*, 2367–2381. [CrossRef]

100. Guardia, P.; Di Corato, R.; Lartigue, L.; Wilhelm, C.; Espinosa, A.; Garcia-Hernandez, M.; Gazeau, F.; Manna, L.; Pellegrino, T. Water-Soluble Iron Oxide Nanocubes with High Values of Specific Absorption Rate for Cancer Cell Hyperthermia Treatment. *ACS Nano* **2012**, *6*, 3080–3091. [CrossRef]

101. Castellanos-Rubio, I.; Rodrigo, I.; Olazagoitia-Garmendia, A.; Arriortua, O.; Gil de Muro, I.; Garitaonandia, J.S.; Bilbao, J.R.; Fdez-Gubieda, M.L.; Plazaola, F.; Orue, I.; et al. Highly Reproducible Hyperthermia Response in Water, Agar, and Cellular Environment by Discretely PEGylated Magnetite Nanoparticles. *ACS Appl. Mater. Interfaces* **2020**, *12*, 27917–27929. [CrossRef]

102. Cabrera, D.; Coene, A.; Leliaert, J.; Artés-Ibáñez, E.J.; Dupré, L.; Telling, N.D.; Teran, F.J. Dynamical Magnetic Response of Iron Oxide Nanoparticles Inside Live Cells. *ACS Nano* **2018**, *12*, 2741–2752. [CrossRef]

103. Liu, X.L.; Yang, Y.; Ng, C.T.; Zhao, L.Y.; Zhang, Y.; Bay, B.H.; Fan, H.M.; Ding, J. Magnetic Vortex Nanorings: A New Class of Hyperthermia Agent for Highly Efficient in Vivo Regression of Tumors. *Adv. Mater.* **2015**, *27*, 1939–1944. [CrossRef]

104. Lee, J.-H.; Jang, J.-T.; Choi, J.-S.; Moon, S.H.; Noh, S.-H.; Kim, J.-W.; Kim, J.-G.; Kim, I.-S.; Park, K.I.; Cheon, J. Exchange-Coupled Magnetic Nanoparticles for Efficient Heat Induction. *Nat. Nanotechnol.* **2011**, *6*, 418–422. [CrossRef]

105. Serantes, D.; Simeonidis, K.; Angelakeris, M.; Chubykalo-Fesenko, O.; Marciello, M.; del Puerto Morales, M.; Baldomir, D.; Martinez-Boubeta, C. Multiplying Magnetic Hyperthermia Response by Nanoparticle Assembling. *J. Phys. Chem. C* **2014**, *118*, 5927–5934. [CrossRef]

106. Martinez-Boubeta, C.; Simeonidis, K.; Makridis, A.; Angelakeris, M.; Iglesias, O.; Guardia, P.; Cabot, A.; Yedra, L.; Estradé, S.; Peiró, F.; et al. Learning from Nature to Improve the Heat Generation of Iron-Oxide Nanoparticles for Magnetic Hyperthermia Applications. *Sci. Rep.* **2013**, *3*, 1–8. [CrossRef] [PubMed]

107. Gandia, D.; Gandarias, L.; Rodrigo, I.; Robles-García, J.; Das, R.; Garaio, E.; García, J.Á.; Phan, M.-H.; Srikanth, H.; Orue, I.; et al. Unlocking the Potential of Magnetotactic Bacteria as Magnetic Hyperthermia Agents. *Small* **2019**, *15*, e1902626. [CrossRef] [PubMed]

108. Chandrasekharan, P.; Tay, Z.W.; Hensley, D.; Zhou, X.Y.; Fung, B.K.; Colson, C.; Lu, Y.; Fellows, B.D.; Huynh, Q.; Saayujya, C.; et al. Using magnetic particle imaging systems to localize and guide magnetic hyperthermia treatment: Tracers, hardware, and future medical applications. *Theranostics* **2020**, *10*, 2965–2981. [CrossRef] [PubMed]

109. Chandrasekharan, P.; Tay, Z.W.; Zhou, X.Y.; Yu, E.Y.; Fung, B.K.L.; Colson, C.; Lu, Y.; Fellows, B.D.; Huynh, Q.; Saayujya, C.; et al. Chapter 15—Magnetic Particle Imaging for Vascular, Cellular and Molecular Imaging. In *Molecular Imaging*, 2nd ed.; Ross, B.D., Gambhir, S.S., Eds.; Academic Press: Cambridge, MA, USA, 2021; pp. 265–282.

110. Lu, Y.; Rivera-Rodriguez, A.; Tay, Z.W.; Hensley, D.; Fung, K.L.B.; Colson, C.; Saayujya, C.; Huynh, Q.; Kabuli, L.; Fellows, B.; et al. Combining Magnetic Particle Imaging and Magnetic Fluid Hyperthermia for Localized and Image-Guided Treatment. *Int. J. Hyperthermia* **2020**, *37*, 141–154. [CrossRef] [PubMed]

111. Giustini, A.J.; Petryk, A.A.; Cassim, S.M.; Tate, J.A.; Baker, I.; Hoopes, P.J. Magnetic nanoparticle hyperthermia in cancer treatment. *Nano Life* **2010**, *1*, 17–32. [CrossRef] [PubMed]

112. Choi, B.H.; Kim, J.H.; Cheon, J.P.; Rim, C.T. Synthesized Magnetic Field Focusing Using a Current-Controlled Coil Array. *IEEE Magn. Lett.* **2016**, *7*, 1–4. [CrossRef]

113. Cubero, D.; Marmugi, L.; Renzoni, F. Exploring the Limits of Magnetic Field Focusing: Simple Planar Geometries. *Results Phys.* **2020**, *19*, 103562. [CrossRef]

114. Bauer, L.M.; Situ, S.F.; Griswold, M.A.; Samia, A.C.S. High-Performance Iron Oxide Nanoparticles for Magnetic Particle Imaging—Guided Hyperthermia (hMPI). *Nanoscale* **2016**, *8*, 12162–12169. [CrossRef]

115. Hensley, D.; Tay, Z.W.; Dhavalikar, R.; Zheng, B.; Goodwill, P.; Rinaldi, C.; Conolly, S. Combining Magnetic Particle Imaging and Magnetic Fluid Hyperthermia in a Theranostic Platform. *Phys. Med. Biol.* **2017**, *62*, 3483–3500. [CrossRef]

116. Murase, K.; Takata, H.; Takeuchi, Y.; Saito, S. Control of the Temperature Rise in Magnetic Hyperthermia with Use of an External Static Magnetic Field. *Phys. Med.* **2013**, *29*, 624–630. [CrossRef]

117. Kumari, P.; Ghosh, B.; Biswas, S. Nanocarriers for Cancer-Targeted Drug Delivery. *J. Drug Target.* **2016**, *24*, 179–191. [CrossRef]

118. Peer, D.; Karp, J.M.; Hong, S.; Farokhzad, O.C.; Margalit, R.; Langer, R. Nanocarriers as an Emerging Platform for Cancer Therapy. *Nat. Nanotechnol.* **2007**, *2*, 751–760. [CrossRef]

119. Pérez-Herrero, E.; Fernández-Medarde, A. Advanced Targeted Therapies in Cancer: Drug Nanocarriers, the Future of Chemotherapy. *Eur. J. Pharm. Biopharm.* **2015**, *93*, 52–79. [CrossRef]

120. Yang, Y.; Aw, J.; Chen, K.; Liu, F.; Padmanabhan, P.; Hou, Y.; Cheng, Z.; Xing, B. Enzyme-Responsive Multifunctional Magnetic Nanoparticles for Tumor Intracellular Drug Delivery and Imaging. *Chem. Asian J.* **2011**, *6*, 1381–1389. [CrossRef]

121. Peiris, P.M.; Bauer, L.; Toy, R.; Tran, E.; Pansky, J.; Doolittle, E.; Schmidt, E.; Hayden, E.; Mayer, A.; Keri, R.A.; et al. Enhanced Delivery of Chemotherapy to Tumors Using a Multicomponent Nanochain with Radio-Frequency-Tunable Drug Release. *ACS Nano* **2012**, *6*, 4157–4168. [CrossRef]

122. Oliveira, H.; Pérez-Andrés, E.; Thevenot, J.; Sandre, O.; Berra, E.; Lecommandoux, S. Magnetic Field Triggered Drug Release from Polymersomes for Cancer Therapeutics. *J. Control. Release* **2013**, *169*, 165–170. [CrossRef]

123. Nardoni, M.; Della Valle, E.; Liberti, M.; Relucenti, M.; Casadei, M.A.; Paolicelli, P.; Apollonio, F.; Petralito, S. Can Pulsed Electromagnetic Fields Trigger On-Demand Drug Release from High-Tm Magnetoliposomes? *Nanomaterials* **2018**, *8*, 196. [CrossRef]

124. Fuller, E.G.; Sun, H.; Dhavalikar, R.D.; Unni, M.; Scheutz, G.M.; Sumerlin, B.S.; Rinaldi, C. Externally Triggered Heat and Drug Release from Magnetically Controlled Nanocarriers. *ACS Appl. Polym. Mater.* **2019**, *1*, 211–220. [CrossRef]

125. Maruyama, S.; Shimada, K.; Enmeiji, K.; Murase, K. Development of Magnetic Nanocarriers Based on Thermosensitive Liposomes and Their Visualization Using Magnetic Particle Imaging. *Int J. Nanomed. Nanosurg.* **2016**, *2*. [CrossRef]

126. Liang, X.; Wang, K.; Du, J.; Tian, J.; Zhang, H. The First Visualization of Chemotherapy-Induced Tumor Apoptosis via Magnetic Particle Imaging in a Mouse Model. *Phys. Med. Biol.* **2020**, *65*, 195004. [CrossRef] [PubMed]

127. Creixell, M.; Bohórquez, A.C.; Torres-Lugo, M.; Rinaldi, C. EGFR-Targeted Magnetic Nanoparticle Heaters Kill Cancer Cells without a Perceptible Temperature Rise. *ACS Nano* **2011**, *5*, 7124–7129. [CrossRef] [PubMed]

128. Naud, C.; Thébault, C.; Carrière, M.; Hou, Y.; Morel, R.; Berger, F.; Diény, B.; Joisten, H. Cancer Treatment by Magneto-Mechanical Effect of Particles, a Review. *Nanoscale Adv.* **2020**, *2*, 3632–3655. [CrossRef]

129. Liu, D.; Wang, L.; Wang, Z.; Cuschieri, A. Magnetoporation and Magnetolysis of Cancer Cells via Carbon Nanotubes Induced by Rotating Magnetic Fields. *Nano Lett.* **2012**, *12*, 5117–5121. [CrossRef]

130. Wong, D.W.; Gan, W.L.; Liu, N.; Lew, W.S. Magneto-Actuated Cell Apoptosis by Biaxial Pulsed Magnetic Field. *Sci. Rep.* **2017**, *7*, 10919. [CrossRef]

131. Domenech, M.; Marrero-Berrios, I.; Torres-Lugo, M.; Rinaldi, C. Lysosomal Membrane Permeabilization by Targeted Magnetic Nanoparticles in Alternating Magnetic Fields. *ACS Nano* **2013**, *7*, 5091–5101. [CrossRef]

132. Zhang, E.; Kircher, M.F.; Koch, M.; Eliasson, L.; Goldberg, S.N.; Renström, E. Dynamic Magnetic Fields Remote-Control Apoptosis via Nanoparticle Rotation. *ACS Nano* **2014**, *8*, 3192–3201. [CrossRef]

133. Shen, Y.; Wu, C.; Uyeda, T.Q.P.; Plaza, G.R.; Liu, B.; Han, Y.; Lesniak, M.S.; Cheng, Y. Elongated Nanoparticle Aggregates in Cancer Cells for Mechanical Destruction with Low Frequency Rotating Magnetic Field. *Theranostics* **2017**, *7*, 1735–1748. [CrossRef]

134. Master, A.M.; Williams, P.N.; Pothayee, N.; Pothayee, N.; Zhang, R.; Vishwasrao, H.M.; Golovin, Y.I.; Riffle, J.S.; Sokolsky, M.; Kabanov, A.V. Remote Actuation of Magnetic Nanoparticles for Cancer Cell Selective Treatment Through Cytoskeletal Disruption. *Sci. Rep.* **2016**, *6*, 33560. [CrossRef]

135. Goiriena-Goikoetxea, M.; Muñoz, D.; Orue, I.; Fernández-Gubieda, M.L.; Bokor, J.; Muela, A.; García-Arribas, A. Disk-Shaped Magnetic Particles for Cancer Therapy. *Appl. Phys. Rev.* **2020**, *7*, 011306. [CrossRef]

136. Kim, D.-H.; Rozhkova, E.A.; Ulasov, I.V.; Bader, S.D.; Rajh, T.; Lesniak, M.S.; Novosad, V. Biofunctionalized Magnetic-Vortex Microdiscs for Targeted Cancer-Cell Destruction. *Nat. Mater.* **2009**, *9*, 165–171. [CrossRef]

137. Goiriena-Goikoetxea, M.; García-Arribas, A.; Rouco, M.; Svalov, A.V.; Barandiaran, J.M. High-Yield Fabrication of 60 Nm Permalloy Nanodiscs in Well-Defined Magnetic Vortex State for Biomedical Applications. *Nanotechnology* **2016**, *27*, 175302. [CrossRef]

138. Mansell, R.; Vemulkar, T.; Petit, D.C.M.C.; Cheng, Y.; Murphy, J.; Lesniak, M.S.; Cowburn, R.P. Magnetic Particles with Perpendicular Anisotropy for Mechanical Cancer Cell Destruction. *Sci. Rep.* **2017**, *7*, 1–7. [CrossRef]

139. Medina-Sánchez, M.; Schmidt, O.G. Medical Microbots Need Better Imaging and Control. *Nature* **2017**, *545*, 406–408. [CrossRef]

140. Li, J.; Esteban-Fernández de Ávila, B.; Gao, W.; Zhang, L.; Wang, J. Micro/Nanorobots for Biomedicine: Delivery, Surgery, Sensing, and Detoxification. *Sci. Robot.* **2017**, *2*, eaam6431. [CrossRef]

141. Chen, X.-Z.; Hoop, M.; Mushtaq, F.; Siringil, E.; Hu, C.; Nelson, B.J.; Pané, S. Recent Developments in Magnetically Driven Micro- and Nanorobots. *Appl. Mater. Today* **2017**, *9*, 37–48. [CrossRef]

142. Lee, S.; Kim, J.-Y.; Kim, J.; Hoshiar, A.K.; Park, J.; Lee, S.; Kim, J.; Pané, S.; Nelson, B.J.; Choi, H. A Needle-Type Microrobot for Targeted Drug Delivery by Affixing to a Microtissue. *Adv. Healthc. Mater.* **2020**, *9*, e1901697. [CrossRef]

143. Vyskočil, J.; Mayorga-Martinez, C.C.; Jablonská, E.; Novotný, F.; Ruml, T.; Pumera, M. Cancer Cells Microsurgery via Asymmetric Bent Surface Au/Ag/Ni Microrobotic Scalpels through a Transversal Rotating Magnetic Field. *ACS Nano* **2020**, *14*, 8247–8256. [CrossRef]

144. Betal, S.; Saha, A.K.; Ortega, E.; Dutta, M.; Ramasubramanian, A.K.; Bhalla, A.S.; Guo, R. Core-Shell Magnetoelectric Nanorobot—A Remotely Controlled Probe for Targeted Cell Manipulation. *Sci. Rep.* **2018**, *8*, 1–9. [CrossRef]

145. Latulippe, M.; Martel, S. Evaluation of the Potential of Dipole Field Navigation for the Targeted Delivery of Therapeutic Agents in a Human Vascular Network. *IEEE Trans. Magn.* **2018**, *54*, 1–12. [CrossRef]

146. Saritas, E.U.; Goodwill, P.W.; Zhang, G.Z.; Conolly, S.M. Magnetostimulation Limits in Magnetic Particle Imaging. *IEEE Trans. Med. Imaging* **2013**, *32*, 1600–1610. [CrossRef] [PubMed]

147. Acri, G.; Inferrera, P.; Denaro, L.; Sansotta, C.; Ruello, E.; Anfuso, C.; Salmeri, F.M.; Garreffa, G.; Vermiglio, G.; Testagrossa, B. dB/dt Evaluation in MRI Sites: Is ICNIRP Threshold Limit Exceeded? *Int. J. Environ. Res. Public Health* **2018**, *15*, 1298. [CrossRef] [PubMed]

148. Atkinson, W.J.; Brezovich, I.A.; Chakraborty, D.P. Usable Frequencies in Hyperthermia with Thermal Seeds. *IEEE Trans. Biomed. Eng.* **1984**, *31*, 70–75. [CrossRef] [PubMed]

149. Hergt, R.; Dutz, S.; Müller, R.; Zeisberger, M. Magnetic Particle Hyperthermia: Nanoparticle Magnetism and Materials Development for Cancer Therapy. *J. Phys. Condens. Matter* **2006**, *18*, S2919. [CrossRef]

150. Attaluri, A.; Jackowski, J.; Sharma, A.; Kandala, S.K.; Nemkov, V.; Yakey, C.; De Weese, T.L.; Kumar, A.; Goldstein, R.C.; Ivkov, R. Design and construction of a Maxwell-type induction coil for magnetic nanoparticle hyperthermia. *Int. J. Hyperthermia* **2020**, *37*, 1–14. [CrossRef] [PubMed]

151. Weissleder, R.; Stark, D.D.; Engelstad, B.L.; Bacon, B.R.; Compton, C.C.; White, D.L.; Jacobs, P.; Lewis, J. Superparamagnetic iron oxide: Pharmacokinetics and toxicity. *AJR Am. J. Roentgenol.* **1989**, *152*, 167–173. [CrossRef]

152. Yoshida, T.; Enpuku, K.; Ludwig, F.; Dieckhoff, J.; Wawrzik, T.; Lak, A.; Schilling, M. Characterization of Resovist®Nanoparticles for Magnetic Particle Imaging. In *Magnetic Particle Imaging*; Buzug, T.M., Borgert, J., Eds.; Springer Proceedings in Physics; Springer: Berlin/Heidelberg, Germany, 2012; pp. 3–7. ISBN 9783642241321.

153. Haegele, J.; Cremers, S.; Rahmer, J.; Franke, J.; Vaalma, S.; Heidenreich, M.; Borgert, J.; Borm, P.; Barkhausen, J.; Vogt, F. Magnetic Particle Imaging: A Resovist Based Marking Technology for Guide Wires and Catheters for Vascular Interventions. *IEEE Trans. Med. Imaging* **2016**, *35*, 2312–2318. [CrossRef]

154. Keselman, P.; Yu, E.Y.; Zhou, X.Y.; Goodwill, P.W.; Chandrasekharan, P.; Ferguson, R.M.; Khandhar, A.P.; Kemp, S.J.; Krishnan, K.M.; Zheng, B.; et al. Tracking short-term biodistribution and long-term clearance of SPIO tracers in magnetic particle imaging. *Phys. Med. Biol.* **2017**, *62*, 3440–3453. [CrossRef]

155. Spinowitz, B.S.; Kausz, A.T.; Baptista, J.; Noble, S.D.; Sothinathan, R.; Bernardo, M.V.; Brenner, L.; Pereira, B.J.G. Ferumoxytol for treating iron deficiency anemia in CKD. *J. Am. Soc. Nephrol.* **2008**, *19*, 1599–1605. [CrossRef]

156. Graeser, M.; Thieben, F.; Szwargulski, P.; Werner, F.; Gdaniec, N.; Boberg, M.; Griese, F.; Möddel, M.; Ludewig, P.; van de Ven, D.; et al. Human-Sized Magnetic Particle Imaging for Brain Applications. *Nat. Commun.* **2019**, *10*, 1–9. [CrossRef]

157. Mason, E.; Mattingly, E.; Herb, K.; Franconi, S.; Sliwiak, M.; Cooley, C.; Wald, L. Magnetic Particle Imaging for Intraoperative Margin Analysis in Breast-Conserving Surgery. *Int. J. Magn. Part. Imaging* **2020**, *6*, 2.

Article

Citrate-Coated Superparamagnetic Iron Oxide Nanoparticles Enable a Stable Non-Spilling Loading of T Cells and Their Magnetic Accumulation

Philipp Boosz [1,2,†], Felix Pfister [1,†], Rene Stein [1], Bernhard Friedrich [1], Lars Fester [3], Julia Band [1], Marina Mühlberger [1], Eveline Schreiber [1], Stefan Lyer [1], Diana Dudziak [4,5,6,7], Christoph Alexiou [1,†] and Christina Janko [1,*,†]

[1] Department of Otorhinolaryngology, Head and Neck Surgery, Section of Experimental Oncology and Nanomedicine (SEON), Else Kröner-Fresenius-Stiftung Professorship, Universitätsklinikum Erlangen, 91054 Erlangen, Germany; philipp.boosz@fau.de (P.B.); felix.pfister@uk-erlangen.de (F.P.); rene.stein@uk-erlangen.de (R.S.); bernhard.friedrich@uk-erlangen.de (B.F.); julia.band@uk-erlangen.de (J.B.); marina.muehlberger@uk-erlangen.de (M.M.); eveline.schreiber@uk-erlangen.de (E.S.); stefan.lyer@uk-erlangen.de (S.L.); christoph.alexiou@uk-erlangen.de (C.A.)
[2] Friedrich-Alexander-Universität Erlangen-Nürnberg (FAU), 91054 Erlangen, Germany
[3] Institute of Anatomy and Cell Biology, Friedrich-Alexander-Universität Erlangen-Nürnberg (FAU), 91054 Erlangen, Germany; lars.fester@uk-erlangen.de
[4] Laboratory of Dendritic Cell Biology, Department of Dermatology, Universitätsklinikum Erlangen, 91054 Erlangen, Germany; diana.dudziak@uk-erlangen.de
[5] Deutsches Zentrum Immuntherapie (DZI), 91054 Erlangen, Germany
[6] Comprehensive Cancer Center Erlangen-EMN (CCC ER-EMN), 91054 Erlangen, Germany
[7] Medical Immunology Campus Erlangen, 91054 Erlangen, Germany
[*] Correspondence: christina.janko@uk-erlangen.de
[†] These authors contributed equally to this paper.

Citation: Boosz, P.; Pfister, F.; Stein, R.; Friedrich, B.; Fester, L.; Band, J.; Mühlberger, M.; Schreiber, E.; Lyer, S.; Dudziak, D.; et al. Citrate-Coated Superparamagnetic Iron Oxide Nanoparticles Enable a Stable Non-Spilling Loading of T Cells and Their Magnetic Accumulation. *Cancers* **2021**, *13*, 4143. https://doi.org/10.3390/cancers13164143

Academic Editors: Moriaki Kusakabe and Akihiro Kuwahata

Received: 14 June 2021
Accepted: 13 August 2021
Published: 17 August 2021

Publisher's Note: MDPI stays neutral with regard to jurisdictional claims in published maps and institutional affiliations.

Simple Summary: In cancer patients, adoptive T cell transfer shall increase the number of circulating cytotoxic T cells to foster anti-tumor immune responses. In solid tumors, however, lack of lymphocyte infiltration into the tumor impairs treatment efficacy due to the immune-suppressive tumor microenvironment. To make cells controllable by external forces, we loaded primary human T cells with citrate-coated superparamagnetic iron oxide nanoparticles (SPIONs). SPIONs were tightly attached to the plasma membrane and also taken up intracellularly into vesicles. With their nanoparticle cargo, we were able to magnetically accumulate them, which is a promising finding for future magnetic delivery of immune cells after adoptive transfer.

Abstract: T cell infiltration into a tumor is associated with a good clinical prognosis of the patient and adoptive T cell therapy can increase anti-tumor immune responses. However, immune cells are often excluded from tumor infiltration and can lack activation due to the immune-suppressive tumor microenvironment. To make T cells controllable by external forces, we loaded primary human CD3+ T cells with citrate-coated superparamagnetic iron oxide nanoparticles (SPIONs). Since the efficacy of magnetic targeting depends on the amount of SPION loading, we investigated how experimental conditions influence nanoparticle uptake and viability of cells. We found that loading in the presence of serum improved both the colloidal stability of SPIONs and viability of T cells, whereas stimulation with CD3/CD28/CD2 and IL-2 did not influence nanoparticle uptake. Furthermore, SPION loading did not impair cytokine secretion after polyclonal stimulation. We finally achieved 1.4 pg iron loading per cell, which was both located intracellularly in vesicles and bound to the plasma membrane. Importantly, nanoparticles did not spill over to non-loaded cells. Since SPION-loading enabled efficient magnetic accumulation of T cells in vitro under dynamic conditions, we conclude that this might be a good starting point for the investigation of in vivo delivery of immune cells.

Keywords: nanomedicine; superparamagnetic iron oxide nanoparticles (SPIONs); adoptive T cell transfer; immune therapy; targeted transport; solid tumor; magnetic targeting

1. Introduction

Cancer is the second leading cause of mortality worldwide with over 8.7 million deaths in 2016 and is expected to become the leading one by 2060 [1]. Despite death rates have been steadily declining in the last years, treatment is still complex. Surgery, chemotherapy, radiotherapy, alone or in combination, have represented the main treatment options for a long time. Since the immune system itself can recognize and eliminate tumor cells, immune therapy has evolved. Infiltration of solid tumors with immune cells has turned out to be an important criterion for therapy as well as for patient prognosis [2,3]. Especially CD8+ T cells among the tumor-infiltrating lymphocytes (TILs) have a favorable impact on disease burden and progression [4,5]. Adoptive transfer of immune effector cells shall increase the amount of tumor-infiltrating lymphocytes to boost the anti-tumor immune response. This procedure utilizes the isolation of peripheral T cells or TILs from a patient and the expansion and modification to increase the immunogenic anti-tumor activity [6]. However, it has been shown that after the transfer of the T cells only a small number infiltrated into the tumor, resulting in only minor anti-tumor effects [7,8]. Irregular tumor vasculature, barrier function of the tumor epithelium, low expression of adhesion molecules, as well as low chemokine expression in the tumor and/or aberrant chemokine receptor expression on the T cells are discussed to hinder the infiltration of effector cells [9]. To overcome these issues, several very specific strategies dependent on the tumor and T cell type have been developed such as the use of antibodies or homing peptides, which increased activation of the cells but also induced severe adverse events such as cytokine release syndrome or seizures and cerebellar effects [10–12]. In addition, their specificity, short half-time, and high costs may represent an obstacle for wider use [13]. Thus, a safe and targeted approach, applicable to various effector cell types and tumors is of urgent need.

Superparamagnetic iron oxide nanoparticles (SPIONs) have come into focus as magnetically controllable shuttles to deliver drugs specifically to the desired region and sparing healthy tissues, referred to as "Magnetic Drug Targeting (MDT)" [14]. For this, SPIONs loaded with a (chemo-)therapeutic drug, are applied to the tumor supplying vascular system and enriched in the tumor region by application of an external magnetic field. We and others have shown successful targeting and enhanced anti-tumor activity of SPION-loaded chemotherapeutic agents when magnetically accumulated in the tumor region [14–16]. Following the principle of magnetic targeting, it is not only possible to transport active ingredients to a target by magnetic control. Cells can also be loaded with SPIONs and moved to the desired region by an external magnetic field. For that, the SPIONs must either bind to the cell surface or be taken up into the cell. Dendritic cells, stem cells, and endothelial cells have been functionalized with SPIONs and were magnetically guided to the target tissue for tumor vaccination, to control tissue injury or application in regenerative medicine [17–19].

Concerning diagnosis, magnetic nanoparticles can be visualized in magnetic resonance imaging (MRI) and have been used for liver imaging previously. Analogously, SPION-loaded cells can be visualized by MRI, which has been used to track the path of T cells for a better understanding of migration and survival of antigen-specific T cells under pathophysiological situations [20–22]. Interestingly, for future translation into clinics, the magnetic field coils inherent to clinical MRI scanners can not only be used for tracking but also steering magnetic nanoparticles or nanoparticle-loaded cells into the wanted region [23,24].

To enable their site-directed targeting, lymphocytes such as NK cells and T cells have been previously functionalized with SPIONs [25–32]. Physicochemical factors of the nanoparticles such as surface charge, size, shape, and coating were reported to influence nanoparticle uptake and biocompatibility [33]. Since lymphocytes are not phagocytic, their nanoparticle uptake is usually low [34], which might reduce magnetic retention. Positively charged coatings of the SPIONs such as 3-aminopropyl-triethoxysilane can enhance interaction with the negatively charged plasma membrane [31] but might also lead to nanoparticle clusters that can detach and spill from the loaded cells. Moreover, NK cells were loaded with positively charged magnetic nanocomplexes and enabled image-based guiding by MRI after intra-arterial infusion [35]. Another employed strategy was the binding of magnetic nanoclusters onto T cells via a PD-1 antibody, which enabled efficient recruiting of T cells to tumor sites, where the PD-1 antibody was released and both, together, exerted a synergistic anti-tumor effect [28]. Zupke et al. showed that the loading efficacy of T cells with nanoparticles strongly depended on nanoparticle concentration, incubation time, and temperature, as well as the presence of serum in the incubation media [36]. The question if SPIONs spill from labeled cells to other cells and the stability of SPIONs in the loaded cells, however, has not been intensively investigated so far.

Besides other challenges, for magnetic guidability the basis is the stable and sufficient loading of SPIONs on or into the T cells without compromising their viability and effector functions. Here, we investigated, with negatively charged citrate-coated SPI-ONs, how experimental conditions influence nanoparticle uptake and viability of primary human T cells.

2. Materials and Methods

2.1. Materials

Cell culture plates were purchased from TPP (Trasadingen, Switzerland). Merck (Darmstadt, Germany) provided the Muse® Count & Viability Assay Kit. Ringer's solution was acquired from Fresenius Kabi (Bad Homburg, Germany). Hoechst 33342 (Hoe), 1,1′dimethyl-3,3,3′,3′-tetramethylindodicarbocyanine iodide (DiIC1(5)), Gibco™ RPMI medium 1640, GlutaMAX supplement, penicillin-streptomycin solution 5000 U/mL, Vybrandt™ DiD, Syto16, Fixable Viability Dye eFluor™ 450, CellTrace™ Violet (CTV), and L-glutamine (200 mM) were purchased from Thermo Fisher Scientific (Waltham, MA, USA). Propidium iodide (PI) and phosphate-buffered saline (PBS) were obtained from Sigma-Aldrich (Taufkirchen, Germany), fetal calf serum (FCS), and amphotericin B from Biochrom (Berlin, Germany).

The Human CD3 Fab-TACS Gravity Kit was obtained from IBA (Goettingen, Germany), and the nitric acid 65 % from Carl Roth (Karlsruhe, Germany). The H_2O used in all experiments was prepared in-house using the Merck Milli-Q® Direct water purification system (Darmstadt, Germany). Falcon® 40 μm and 70 μm cell strainers were purchased from Corning by Life Sciences (Corning, NY, USA) and LabSolute (Renningen, Ger-many). Annexin A5 fluorescein isothiocyanate (FITC) conjugate (AxV-FITC) and recombinant human Interleukin (rhIL)-2 and rhIL-7 were obtained from ImmunoTools (Friesoythe, Germany), the ImmunoCult™ Human CD3/CD28/CD2 T cell Activator was obtained from Stemcell Technologies (Vancouver, BC, Canada), and the S-Monovettes 10 mL 9 NC were obtained from Sarstedt (Nuembrecht, Germany). BD Insyte-W intravenous cannula, 16GA were purchased from BD (Haryana, India). Iron reference standards (1 g/L) were purchased from Bernd Kraft GmbH (Duisburg, Germany).

BioLegend (San Diego, CA, USA) provided APC anti-human CD8a, APC Mouse IgG1, κ Isotype Control, PerCP/Cy5.5 anti-human CD4, PerCP Mouse IgG1, κ Isotype Control, and PE anti-human IL-2. FITC mouse anti-human CD3, Pacific Blue mouse anti-human CD8, Pacific Blue mouse IgG1, κ, Isotype Control and FITC Mouse IgG1, κ, Isotype Control were obtained from BD Pharmingen (San Jose, CA, USA). Inside Stain Kit, PE anti-human Tumor Necrosis Factor (TNF)-α, APC-Vio 770® anti-human Interferon (IFN)-γ were purchased from Miltenyi Biotec (Bergisch Gladbach, Germany). Ibidi (Gräflingen, Germany) provided μ-Slide I Luer with a channel height of 0.4 mm. The peristaltic pump Ismatec® IPC and PharMed® BPT tubes (2.06 mm inner diameter) were obtained from Cole-

Parmer GmbH (Wertheim, Germany). Neodymium disc-shaped magnets (5 mm × 5 mm, approximately 400 mT) were purchased from Webcraft GmbH (Gottmadingen, Germany).

The Gallios flow cytometer and Kaluza analysis Software (1.3, 2.1) were obtained from Beckman Coulter (Brea, CA, USA). The SpectraMax iD3 plate reader was purchased from Molecular Devices (San José, CA, USA).

For the analysis of T cells in transmission electron microscopy (TEM) the following was used: Carl Roth (Karlsruhe, Germany) supplied glutardialdehyde 25 % (for microscopy) in PO_4-buffer, ethanol (absolute), acetone (100 % anhydrous), hardener MNA, DPA, glycidyl ether (100 %), accelerator DMP 30, and NaOH (1 M). Science Service (Munich, Germany) supplied OsO_4, copper grid. Merck (Darmstadt, Germany) supplied $K_3(Fe(Cn)_6$, agarose (low melting point), and lead(II)citrate-3-hydrate. Serva (Heidelberg, Germany) supplied uranyl acetate. The following equipment from these companies was used: Emerson (Saint Louis Missouri, United States) Branson 1200 ultra-sonic machine, Leica (Wetzlar, Germany) Ultracut UCT, Zeiss (Oberkochen, Germany) TEM Leo 906, TRS Tröndle (Moorenweis, Germany) CCD camera, ImageSP SysPROG.

2.2. Synthesis of SPIONs and Physicochemical Characterization

SPIONs were synthesized based on an adjusted protocol of Elbialy et al. [37] in three batches. Particles were sterilized by filtration through syringe filters with 0.2 μm pore size (Sartorius, Goettingen, Germany). Subsequently, SPIONs were analyzed regarding their size, iron content, magnetic susceptibility, and zeta potential according to Mühlberger et al. [26]. The iron content was investigated after a dilution of 1:25 in deionized H_2O and dissolving in 65 % nitric acid with atomic emission spectroscopy (AES), using the Agilent 4200 MP-AES (Agilent Technologies, Santa Clara, CA, USA) with an iron solution of 1000 mg/L as an external standard (Bernd Kraft, Duisburg, Germany). Triplicate measurements were performed at a wavelength of 371,993 nm, which were then averaged.

2.3. Isolation of T Cells from Human Whole Blood

Human cells were isolated from peripheral human blood obtained from healthy volunteers after informed consent (approved by the ethics committee of the Friedrich-Alexander-Universität Erlangen-Nürnberg; reference number 257_14 B). To obtain CD3+ T cells from citrate-anticoagulated human blood, the CD3 Fab-TACS™ Gravity Kit was used according to the manufacturer's instructions for freshly drawn blood. Isolated T cells were counted in a MUSE cell Analyzer using the MUSE® Count and Viability assay kit.

2.4. Determination of T Cell Purity

To compare the T cell frequency in whole blood before isolation with the one after isolation, erythrocytes were lysed in 100 μL whole blood for 20 s using 600 μL 0.12 % formic acid (pH 2.7). Immediately after lysis, the solution was neutralized with 265 μL solution containing 6 g/L sodium carbonate, 14.5 g/L sodium chloride, and 31.1 g/L sodium sulphate (pH 11.2). Then, the cells were washed with 1 mL PBS and centrifuged at 300 rcf for 5 min at room temperature. Lysed whole blood and freshly isolated T cells were then stained with antibodies against CD3, CD4, CD8, and corresponding isotype controls for 30 min at 4 °C. Subsequently, the cells were investigated by flow cytometry and the data was analyzed with the Kaluza software (version 2.1).

2.5. Determination of Cell Viability

Cell viability was determined in flow cytometry by staining 50 μL cell suspension with 250 μL of staining mixture for 20 min at 4 °C. The staining mixture contained 20 nM Hoe, 2 μL/mL AxV-FITC, 4 nM DiIC1(5), and 66.6 ng/mL PI per ml Ringer's solution. Cells were analyzed in a Gallios flow cytometer. Electric compensation was used to eliminate fluorescence bleed through. Data were analyzed with the Kaluza software (version 2.1).

2.6. Colloidal Stability of SPIONs in Various Media and Cellular Nanoparticle Uptake

If not indicated otherwise, T cells were cultured in a humidified 5% CO_2 atmosphere at 37 °C in a standard T cell medium composed of RPMI 1640 medium supplemented with 10% heat-inactivated (HI) FCS, 2% penicillin-streptomycin solution, 1% L-glutamine, and 1% amphotericin B.

To investigate the colloidal stability dependent on the medium composition, nanoparticles (0, 26.7, 53.3, 80.0, and 106.7 µg/mL) were incubated in (1) T cell medium with 10% FCS or (2) T cell medium with only 2% FCS or (3) PBS with 2% FCS overnight in cell culture conditions. The next day, 100 µL supernatant of each sample, as well as 100 µL of remaining fluid with resuspended SPIONs, was transferred to separate wells of a 96-well plate. The optical density of the samples was measured at 320 nm in the SpectraMax iD3. To analyze the formation of SPION agglomerates depending on the medium composition, 50 µL aliquots of each well were added to 250 µL of Ringer's solution and then analyzed by flow cytometry (forward scatter: voltage 500, gain 5, discriminator 20; side scatter: voltage 550, gain 10, discriminator off).

To evaluate the effect of the medium composition on the uptake of SPIONs by T cells, 1×10^6 freshly isolated T cells in 100 µL were incubated with SPIONs for one hour. Then, cells were stained for viability as described above, analyzed by flow cytometry, and their iron content was quantified in AES as detailed in Section 2.2.

2.7. Comparison of Stimulated and Non-Stimulated T Cells

2×10^6 of T cells were stimulated in a concentration of 1×10^6 cells per ml with 30 IU/mL rhIL-2 and 25 µL/mL/1×10^6 cells of Immunocult™ human CD3/CD28/CD2 T cell activator. Unstimulated CD3-positive cells in standard T cell medium served as controls. After 72 h, 2 mL of either the T cell medium only or the T cell medium containing 30 IU/mL rhIL-2 was added to the controls or stimulated T cells, respectively.

To determine the effects of the stimulation on the viability of the T cells, flow cytometry was performed after 0 h and 96 h as described above. Additionally, MUSE cell count and viability analyses were conducted to confirm the flow cytometry results.

2.8. Determination of Nanoparticle Uptake

T cells were stimulated as described in Section 2.7. After the second stimulation at 72 h, 200 µL SPION solution was added to the test samples or the controls, respectively. A control was also established, which received only 200 µL deionized H_2O without SPIONs. After an additional 24 h, the T cells were washed twice with PBS to remove free nanoparticles. The T cell number was determined via the Muse Cell Analyzer. Cells were then sedimented by centrifugation at 300 rcf for 5 min. After discarding the supernatant, the pellets were dried at 95 °C, 300 rpm for 30 min in a 1.5 mL Eppendorf tube, before they were lysed with 100 µL nitric acid at 95 °C, 750 rpm for 15 min. 900 µL of water was added, followed by measurement of their iron concentration by AES.

The long-term stability of the ingested SPIONs was also investigated. T cells were loaded with 80 µg Fe/mL SPIONs overnight, controls received only deionized H_2O. The cells were then seeded at a concentration of 2×10^6 cells in 2 mL of T cell medium. 10 ng/mL IL-7 was added to the T cell medium to increase long-term survival. After 24 h, 48 h, and 72 h, the cells were collected, washed three times to remove any unbound iron, and cell count was determined using the MUSE Cell Analyzer. Samples were dried at 95 °C for 30 min, lysed with 50 µL HNO_3 at 95 °C for 15 min, and dissolved in 450 µL deionized H_2O. The iron content was determined with AES and the pg iron per cell was calculated by dividing the iron amount by the cell number.

2.9. SPION Exchange with Non-Loaded T Cells

2 mL of 1×10^6 CD3+ T cells per 1 mL in T cell medium were seeded in 12-well plates, half of them were incubated with SPIONs overnight; the control group was treated the same way with deionized H_2O. The next day, nanoparticles were washed from the

samples at 300 rcf for 5 min at room temperature. Loaded T cells were stained with FITC anti-human CD3 antibody for 30 min at 4 °C with a working solution of 1:200 in T cell medium, were washed two times and resuspended. Non-loaded cells were resuspended in medium and placed into 96-well plates. Triplicates of 100 μL pure non-loaded T cells, triplicates of 100 μL loaded T cells, and triplicates with mixed 50 μL non-loaded and 50 μL stained loaded T cells were placed into 96-well plates and incubated for three hours in an incubator at 37 °C. All samples were placed in the flow cytometer and the side scatter was analyzed. Populations of non-loaded and loaded CD3+ T cells were differentiated by a CD3 stain.

To investigate the long-term stability of the ingested iron in the cells, T cells were stained with 5 μM of CTV for 30 min at 37 °C and then loaded with 80 μg Fe/mL SPIONs overnight as described above; an unstained control received only deionized H_2O. The cells were then washed to remove excess particles and seeded in a 12-well plate at a concentration of 1×10^6 cells in 2 mL of T cell medium. IL-7 was added to the T cell medium at a concentration of 10 ng/mL, to increase the long-term survival. The exchange between loaded and non-loaded cells was investigated by the seeding of 0.5×10^6 of each loaded and non-loaded T cells in 2 mL of T cell medium. After 24 h, 48 h, and 72 h, cells were collected, stained with 2 μL/mL AxV-FITC and 66.6 ng/mL PI for 20 min at 4 °C, and then analyzed by flow cytometry.

2.10. Transmission Electronic Microscopy

For determination of the incorporated amount of SPIONs, transmission electronic microscopy (TEM) was performed. For that, T cells were isolated and incubated for 72 h in T cell medium. Afterwards, SPIONs were added to 8×10^6 T cells per sample for 24 h to receive 2667 μg/mL or 80 μg/mL of iron. Cells that received only H_2O without SPIONs served as controls. Subsequently, the cells were washed three times with PBS whereupon they were fixated in 2.5 % glutaraldehyde in 0.1 M PO_4 buffer for 4 h, subsequently washed 3 times with 0.1 M PO_4 buffer and left overnight at 4 °C. The staining of the cells for TEM was performed with 1% osmium treta oxide (OsO_4) in 3% potassium ferricyanide ($K_3(Fe(Cn)_6)$) for 2 h at room temperature (RT), and additionally washing the stained cells with 0.1 M PO_4 buffer overnight at 4 °C. Before embedding the cells in Epon resin, the cells were transferred into a matrix of 2% agarose (low melting point) in 0.1 M PO_4 buffer in an Eppendorf reaction tube overnight. For Epon resin embedding, the cells were dehydrated in agarose through an ethanol step and finally transferred to Epon using acetone and an acetone–epon mixture, followed by a polymerization step for 48 h at 60 °C. Ultra-thin sections of approximately 50 nm section thickness of the samples were prepared with the Leica Ultracut UCT and placed on copper grids. These ultra-thin sections were then first contrasted with lead citrate for 10 min and then with uranyl acetate for a further 10 min. The images were taken with the Zeiss TEM 906 LEO (from Zeiss) with a CCD-camera residual light amplifier from A. Tröndle, TRS, and the software ImageSP SYS Prog, TRS. The images were taken at an acceleration voltage of 80 kV and a magnification of 12,930-fold.

2.11. Magnetic Accumulation of SPION-Loaded T Cells under Dynamic Conditions

The magnetic attractability of CD3+ T cells after SPION-loading was evaluated under dynamic conditions. To imitate blood flow, a peristaltic pump was used to move SPION-loaded cells through slides to which Neodymium disc shape magnets (5 mm × 5 mm, approximately 400 mT) were attached. Unloaded cells and slides without magnets served as controls. After 1 h of pumping, magnetically accumulated cells were stained with the Hoechst 33,342 stain. Images were taken on a Zeiss Axio Observer Z1 fluorescence microscope (Carl Zeiss AG, Oberkochen, Germany). Subsequently, pictures were analyzed using the ImageJ software (version 1.52a).

2.12. T Cell Activation and Proliferation after Polyclonal Stimulation

The T cells were investigated for cytokine production after stimulation. T cells were isolated and loaded as described above and stimulated with 25 µL per 1×10^6 cells per 1 mL ImmunoCult human CD3/CD28/CD2 T-cell activator mix, however, no IL-2 was added. After 24 h, the T cells were stained for CD4 and CD8 for 30 min at 4 °C, in order to distinguish T cell subclasses. Afterwards, the cells were fixated and permeabilized with the inside stain kit. The fixation was performed at room temperature for 20 min with 100 µL of Inside Fix. Cells were then permeabilized for 30 min at 4 °C with the addition of antibodies against TNF-α, IF-γ, and IL-2. Subsequently, the cells were analyzed by flow cytometry.

2.13. Data Analysis and Statistics

Data were prcessed in Microsoft Excel (Microsoft, Redmond, WA, USA). Statistical analysis and graph creation were performed with GraphPad PRISM 9.0.2 from Graph-Pad Software, Inc. (San Diego, CA, USA). For statistical significance, p-values ≤ 0.05 were considered.

3. Results

3.1. Physicochemical Characterization of SPIONs

Magnetic susceptibility, hydrodynamic Z-average size, and zeta potential of three independently synthesized batches of SPIONs were investigated (Table 1). The magnetic susceptibility of the particles normalized to an iron concentration of 1 mg/mL for the purpose of comparison was determined to be ranging from 4.08×10^{-3} to 4.12×10^{-3}. The hydrodynamic Z-average size of the nanoparticles was in mean 52 nm to 58 nm. In accordance, the polydispersity index (PDI) in water was 0.143 to 0.152. The zeta potential ranged from -48.5 mV to -53.7 mV in deionized H_2O with a pH of 7.3.

Table 1. Physicochemical characterization of SPIONs.

Physicochemical Feature	Batch 1	Batch 2	Batch 3
Magnetic susceptibility (10^{-3})	4.08 ± 0.00	4.12 ± 0.00	4.10 ± 0.00
Z-average size (nm) in H_2O	58 ± 0.1	52 ± 0.1	53 ± 0.2
Polydispersity index (PDI)	0.143 ± 0.005	0.151 ± 0.08	0.152 ± 0.07
Zeta potential (mV) at pH 7.3	-48.5 ± 0.5	-53.7 ± 0.4	-51.8 ± 0.3

3.2. Colloidal Stability of SPIONs in Cell Culture Medium

Since the protein corona is known to play a crucial role in nanoparticle uptake, we reduced the FCS amount from 10% to 2% in the cell culture medium in order to increase cellular nanoparticles loading. Experiments were additionally performed in 2% FCS in PBS. To determine the colloidal stability of SPIONs in the above-mentioned media, the nanoparticles were incubated overnight to allow them to agglomerate and sediment. On the next day, the nanoparticle suspensions were analyzed for agglomerations. In the medium with only 2% FCS, already macroscopically nanoparticle agglomerations were visible in a dose-dependent manner. In PBS with 2% FCS and medium with 10% FCS, we found no obvious larger nanoparticle sedimentations (Figure 1A). To analyze smaller agglomerations, which were not visible to the naked eye, we measured the absorption of the supernatant of sedimented or resuspended nanoparticle suspensions, respectively. Recording of the spectrum (250–500 nm) of nanoparticle dilutions revealed an optimal wavelength for the detection of SPIONs at 320 nm (Figure 1B). In the case of complete colloidal stability, the absorption of sedimented and resuspended samples should be equal. SPIONs in T cell medium containing 10% FCS and 2% FCS in PBS revealed the best colloidal stability as determined by absorption measurements of the supernatant of the sedimented and resuspended samples. Thus, absorption values of the supernatant of sedimented and resuspended samples increased dose-dependently and were not dramatically different between both. In contrast, the medium containing only 2% FCS induced nanoparticle agglomeration and sedimentation, causing large differences in the optical density values of the supernatants of sedimented and resuspended samples. The resuspended samples induced a dose-dependent absorption, whereas already in the smallest tested concentration the values of the sedimented and resuspended samples were different (Figure 1C). Analyzing the agglomerations by flow cytometry supported this conclusion: only in the T cell medium with 2% FCS nanoparticle agglomerations were detected by their relative size (forward scatter) and granularity (side scatter). In the other solutions, the nanoparticles were colloidally stable and the nanoparticle sizes were below the detection limit (Figure 1D).

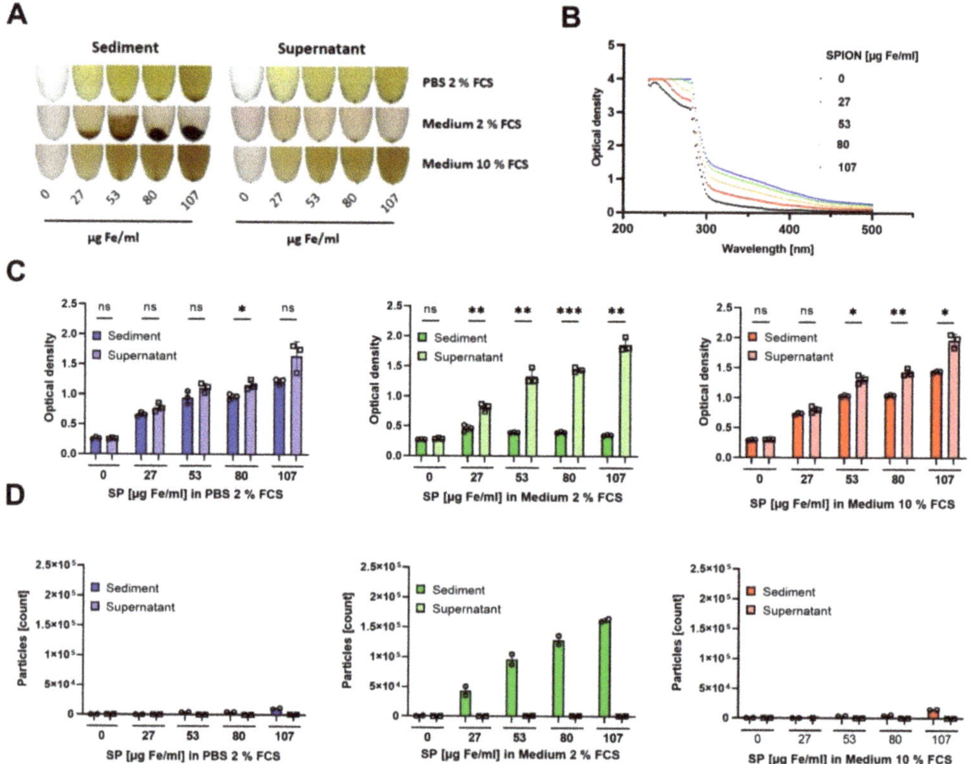

Figure 1. Colloidal stability of SPIONs in various media. SPIONs were incubated overnight in PBS with 2% FCS or RPMI cell culture medium containing 2% or 10% FCS, respectively. (**A**) Macroscopic pictures of SPIONs diluted in PBS supplemented with 2% FCS or T cell medium with 2% or 10% FCS, respectively. (**B**) Spectrum of SPIONs diluted in RPMI with 10% FCS. (**C,D**) T cell medium with 10% FCS is depicted in red, while T cell medium with 2% FCS is shown in green. PBS supplemented with 2% FCS is depicted in blue. (**C**) Optical density at 320 nm of the supernatant of sedimented or resuspended samples. (**D**) Supernatant or sediment was analyzed in flow cytometry for agglomerations. Shown are the mean values of triplicates (**C**) or duplicates (**D**) with standard deviations. Significances were calculated using unpaired t-test with Welch's correction; * $p \leq 0.05$; ** $p \leq 0.01$; *** $p \leq 0.001$. ns: not significant, SP: SPIONs.

3.3. Influence of Medium Composition on SPION-Loading of T Cells

T cell isolation via IBA gravity columns resulted in a CD3+ T cell purity of 91.5% ± 1.1% compared to the portion of CD3+ T cells of 34.9% ± 3.9% in whole blood (Figure S1). To achieve magnetic guidability of the cells by loading with SPIONs, a sufficient amount of iron per cell is crucial. Based on previous data by Mühlberger et al. [27], we selected 80 µg/mL as the optimal iron concentration for the loading of T cells. Similar to the loading strategy of Sanz-Ortega [31], we reduced the medium volume, resulting in a cell count of

1×10^6 cells in 100 µL of medium or PBS in the presence of 80 µg/mL SPIONs for 1 h to enhance contact of cells with the nanoparticles.

Subsequently, cells were washed, counted, and the amount of iron was determined from the cell lysates using AES. Cell count and viability were directly (without washing) determined by flow cytometry after staining for apoptosis and necrosis using AxV-FITC and PI (Figure 2A). AxV-FITC stains phosphatidylserine which is exposed on the outer surface of the cells in apoptosis. Counterstaining with the plasma-membrane impermeable dye PI detects necrotic cells. Thus, AxV-FITC-/PI- cells are considered viable, AxV-FITC+/PI- cells apoptotic, and PI+ cells necrotic.

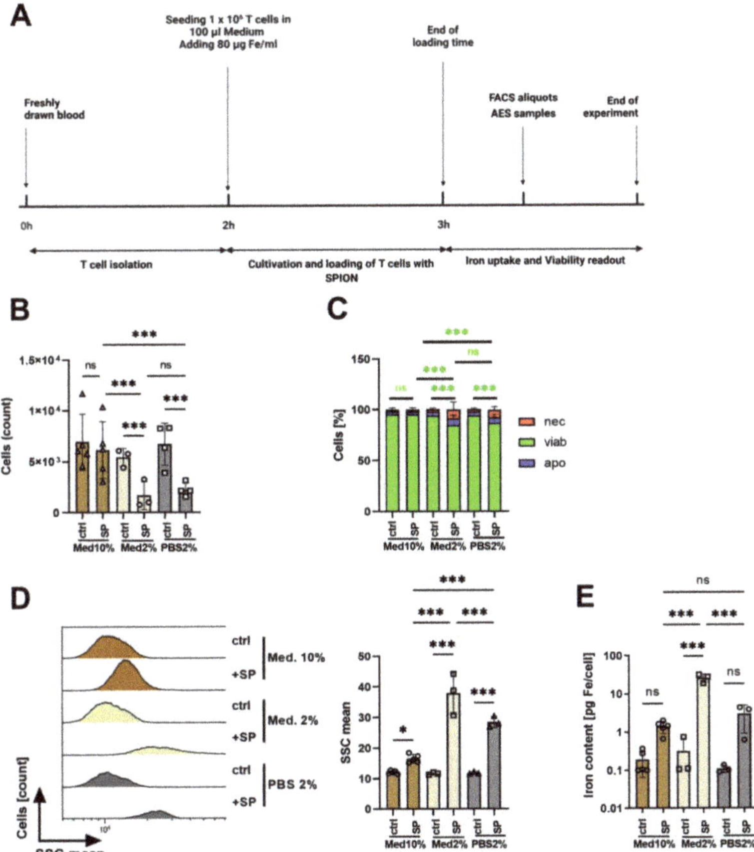

Figure 2. Influence of medium composition on SPION-loading of T cells. (**A**) Experimental setup. (**B**) Determination of the cell count by flow cytometry. (**C**) The viability of the cells was determined by AxV-FITC and PI staining in flow cytometry. AxV-FITC-/PI- cells were considered viable, AxV-FITC+/PI- apoptotic, and PI+ cells necrotic. (**D**) Estimation of cellular nanoparticle loading by analysis of the side scatter increase of viable cells by flow cytometry. (**E**) Cellular iron content [pg Fe/cell] was determined by atomic emission spectroscopy of cell lysates. (**B–E**) Experiments were performed in triplicates of T cells from 3–5 donors. Shown are the mean values with standard deviations. Significances were calculated with a 2-way ANOVA test with a Tukey post hoc test; * $p \leq 0.05$; *** $p \leq 0.001$. ctrl: control; SSC: side scatter; med: medium; ns: not significant; nec: necrotic; apo: apoptotic; viab: viable.

With the medium containing 10% FCS, the cell count was not reduced in the presence of SPIONs. In contrast, SPIONs in both medium or PBS containing 2% FCS reduced the cell count dramatically (Figure 2B), which might be due to iron overload and rupture of cells. The majority of the residual cells, however, were viable (AxV-FITC-/PI-) (Figure 2C).

As shown by Friedrich et al., nanoparticle uptake was accompanied by an increase of side scatter by flow cytometry [38], serving as a marker to estimate SPION-loading of the cells. Since cell death processes also cause the alteration of cell morphology, we gated only on viable cells based on their AxV-FITC and PI negativity. In RPMI medium with 2% FCS, in which the nanoparticles agglomerated, the cells showed the largest side scatter increase. In PBS containing 2% FCS, in which the nanoparticles were colloidally stable, the side scatter increase was smaller, and in medium with 10% FCS hardly detectable (Figure 2D). We analyzed the iron amount in the cell lysates using AES and calculating the iron amount in relation to the cell count. In line with the side scatter data, we achieved the highest iron values per cell for T cells in medium with 2% FCS with 26.03 pg $\pm$ 5.32 pg iron per cell, which might be due to nanoparticle agglomerations in the cell lysates. T cell loading in PBS with 2% FCS resulted in 2.96 $\pm$ 1.67 pg iron per cell. Cells in the medium with 10% FCS contained only 1.40 $\pm$ 0.44 pg iron per cell (Figure 2E).

3.4. Influence of Polyclonal Stimulation on SPION-Loading of T Cells

Previously, it has been shown that uptake of nanoparticles by cells is dependent on the cell cycle state, whereas uptake in G2/M was stronger than in the S and G0/G1 phase, respectively [39]. In order to increase nanoparticle uptake by cells, we aimed to foster the division of the cells by polyclonal primary and costimulatory stimulation using CD3/CD28/CD2 beads in the presence of 30 IU/mL rhIL-2 for 72 h (Figure 3A). Then, cells were re-stimulated with 30 IU/mL rhIL-2 for another 24 h. At this time point, 80 µg/mL SPIONs were added as well. Then, after 24 h incubation with SPIONs and an overall 96 h of stimulation, viability and iron uptake were determined by flow cytometry and AES, respectively. At this time point, the stimulated T cells were at the cell division stage. Unstimulated cells served as controls.

Flow cytometry revealed that with stimulation the cell count increased from 5.038 $\pm$ 1.637 to 12.154 $\pm$ 2.865 cells, whereas the presence of nanoparticles had no influence on the cell count after stimulation (4.932 $\pm$ 1.537 to 13.004 $\pm$ 3.037 cells) (Figure 3B). The cell viability, however, was reduced by the stimulation from 75.8% $\pm$ 9.8% to 62.4% $\pm$ 8.0%. The presence of SPIONs further reduced the cell viability of non-stimulated and stimulated cells to 71.8% $\pm$ 10.8% and 52.9% $\pm$ 8.0%, respectively (Figure 3C).

The side scatter of non-stimulated cells increased from 11.67 $\pm$ 0.53 to 16.17 $\pm$ 1.17 in the presence of nanoparticles. For stimulated cells, the side scatter of the cells was generally higher, possibly because of alterations in morphology due to the proliferation. Nonetheless, incubation with nanoparticles further significantly increased it from 31.58 $\pm$ 0.64 to 32.5 $\pm$ 0.89 (Figure 3D). Determination of the iron content in the cell pellets via AES and calculation of the iron content per cell confirmed the nanoparticle loading of the cells in the presence of SPIONs. Interestingly, however, we saw no significant differences in the cellular iron amount dependent on the polyclonal stimulation of the cells. We observed 0.976 $\pm$ 0.43 pg Fe/cell for unstimulated and 1.03 $\pm$ 0.33 pg Fe/cell for stimulated cells (Figure 3E).

3.5. SPION-Loaded T Cells Do Not Exchange Nanoparticles with Non-Loaded T Cells

After nanoparticle loading, it is mandatory that SPIONs are not released from the cells or exchanged with other cells, to minimize loss of magnetic controllability, spill over to formerly non-loaded cells, and the bias of other cells getting also magnetically attracted.

Transmission electron microscopy of the unstimulated cells revealed both the binding of the SPIONs to the plasma membrane as well as internalization into vesicles (Figure 4A). With 27 µg/mL of SPIONs, most of the nanoparticles were located intracellularly in vesicles. Only some particles were attached to the plasma membrane. With 80 µg/mL, obviously more nanoparticles were associated with the plasma membrane. These nanoparticles seemed to be tightly attached in a uniform layer at one side of the cell. In the magnification (Figure 4A, picture 3), the formation of a membrane invagination with SPIONs is visible.

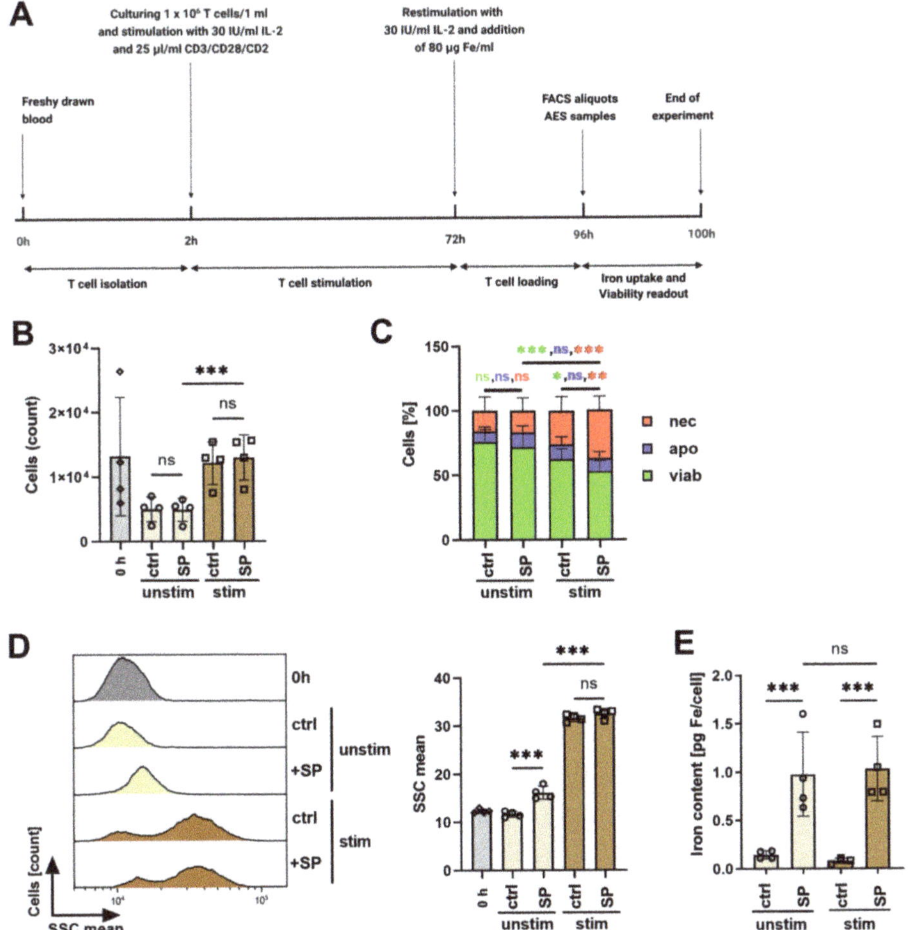

Figure 3. Influence of T cell stimulation on nanoparticle uptake. (**A**) Experimental setup. 1×10^6 CD3+ T cells in 1 mL T cell medium were stimulated for 72 h and then re-stimulated and incubated with 80 µg/mL SPIONs for 24 h. Unstimulated and/or cells without SPIONs served as controls. Subsequently, cells were analyzed by flow cytometry and atomic emission spectroscopy (AES). Determination of cell count (**B**) and viability (**C**) in flow cytometry. AxV-FITC-/PI- cells were considered viable, AxV-FITC+/PI- cells apoptotic, and PI+ cells necrotic. (**D**) Side Scatter (SSC) increase of AxV-FITC-/PI- cells indicates nanoparticle uptake. E) Determination of the iron amount (pg/cell) by AES. (**B–E**) Experiments were performed in triplicates of T cells from four donors. Shown are the mean values with standard deviations. Significance was estimated by a 2-way ANOVA with a Tukey post hoc test; * $p \leq 0.05$; ** $p \leq 0.01$; *** $p \leq 0.001$. ctrl: control; ns: not significant; SP: SPIONs, stim: stimulated; unstim: unstimulated; nec: necrotic; apo: apoptotic; viab: viable.

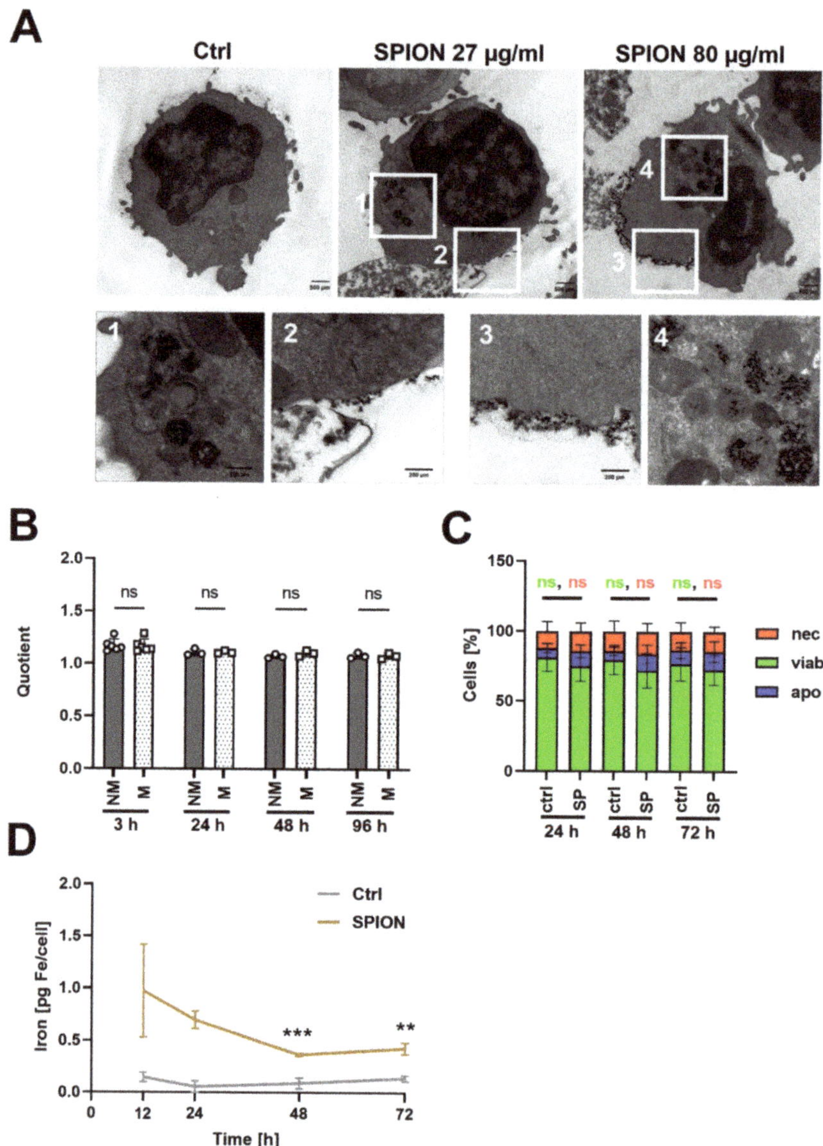

Figure 4. SPIONs were not exchanged between loaded and non-loaded CD3+ T cells. (**A**) Subcellular localization of SPIONs determined by TEM. Scale bars depict 500 μm (upper row) and 250 μm (lower row). Squares indicate areas with SPION-loading in T cells. (**B**) Isolated CD3+ T cells (1×10^6 cells/mL) were loaded with 80 μg Fe/mL SPIONs overnight and stained for CD3 or CTV; non-loaded T cells served as controls. Side scatter values were used for estimation of nanoparticle uptake of SPION-loaded and non-loaded cells. A quotient was calculated by division of the SSC of loaded T cells with that of non-loaded T cells of the same donor at the same time point. (**C**) Cell viability of SPION-loaded cells. Cells were stained with AxV-FITC and PI. Ax-FITC-/PI- cells were considered viable (viab), Ax-FITC+/PI- apoptotic (apo), and PI+ necrotic (nec). (**D**) Iron content per cell as determined by atomic emission spectroscopy and division through cell count. Shown are the mean values with standard deviations. Significances were estimated using an unpaired t test with Welch's correction; ** $p \leq 0.01$, *** $p \leq 0.001$; ctrl: control, SP: SPIONs, NM: not mixed, M: mixed.

To investigate nanoparticle release from SPION-loaded T cells, cells were loaded overnight with 80 µg Fe/mL SPIONs and stained for CD3 or with CTV for identification by flow cytometry. Subsequently, we mixed loaded T cells with non-loaded T cells and analyzed them after 3 h, 24 h, 48 h, and 72 h by flow cytometry regarding SSC increase. Non-loaded T cells and loaded T cells in separate tubes served as controls. For comparison between the time points, a quotient was calculated from the SSC of loaded T cells divided by the SSC of unloaded T cells from the same donor at the same time point. The quotients of non-mixed and mixed samples were not significantly different at each time point (Figure 4B), indicating that nanoparticles were not exchanges between loaded and formerly non-loaded cells. In parallel, we determined the viability after 24 h, 48 h, or 72 h after loading by AxV-FITC and PI staining. As depicted in Figure 4C, after 24 h 81.1% ± 10.1% of unloaded T cells were still viable compared to 74.7% ± 10.5% of loaded T cells. This slightly reduced viability was also detectable after 48 h (non-loaded: 79.5% ± 10.4%, loaded: 71.9% ± 12.4%) and 72 h (non-loaded: 76.7% ± 12.0%, loaded 72.5% ± 10.8%), however was not significant (Figure 4C). Furthermore, the iron content per cell after 24 h, 48 h, and 72 h was determined by AES. Starting with an iron content around 1 pg/cell at 12 h after loading, the iron amount was reduced to 0.5 µg/mL at 48 h until it remained constant (Figure 4D).

3.6. Loading of T Cells with SPIONs Allows Magnetic Enrichment under Dynamic Conditions

The magnetic attractability of CD3+ T cells after SPION-loading was evaluated under dynamic conditions. To imitate blood flow in a physiological vascular system, a peristaltic pump was used to move SPION-loaded cells through slides to which magnets were attached. Non-loaded cells and slides without magnets served as controls (Figure 5A). The catching of SPION-loaded cells (containing roughly 1 pg iron/cell) was visible even macroscopically. After 1 h of pumping, magnetically accumulated cells were analyzed by fluorescence microscopy by Hoechst staining. For cells without SPION-loading, no cell accumulation was detected. With SPION loading, the numbers of cells increased from 62 ± 90 to 2188 ± 186 (stimulation) in the presence of a magnet, proving their magnetic attractability (Figure 5B–D).

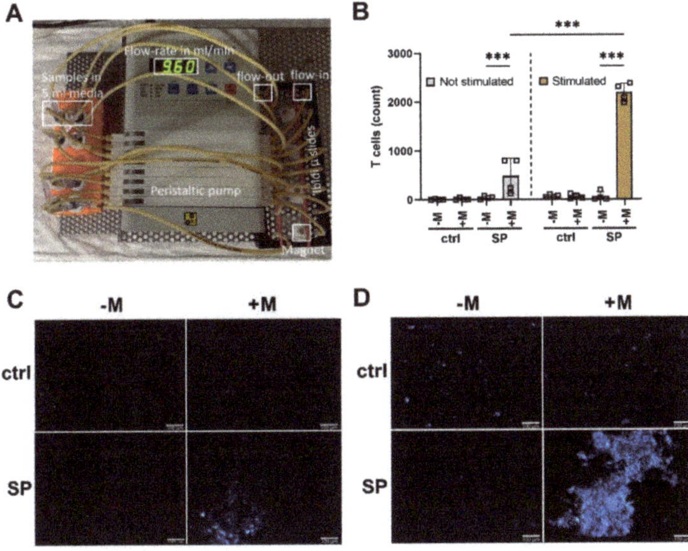

Figure 5. Dynamic magnetic enrichment of T cells. CD3+ T cells were isolated and stimulated for

72 h with a CD3/CD28/CD2 activation mix with 30 IU/mL IL-2. After three days, the T cells received an additional 30 IU/mL IL-2 as well as 80 µg Fe/mL SPIONs for 24 h. T cells without SPIONs or un-stimulated T cells served as controls. Subsequently, the cells were stained with Hoechst 33342. Dynamic enrichment was performed with a peristaltic pump at a flow velocity of 9.6 mL/min for 1 h in µ-slides I Luer. Magnets were added to locally enrich T cells. For the controls, no magnets were added. The slides were then imaged in fluorescent microscopy to analyze cell count. (**A**) Depiction of the experimental setup. (**B**) T cell count after the dynamic enrichment and (**C,D**) corresponding fluorescent microscopy images of the area where the magnet was placed of non-stimulated (**C**) and stimulated cells (**D**). Statistical significances were calculated with an unpaired t test with Welch's correction; *** $p \leq 0.001$. ctrl: control; SP: SPION; -M: without Magnet; +M: with Magnet; the white scale bar indicates 100 µm.

3.7. SPION Loading Does Not Impair Cytokine Release and Differentiation of T Cells after Polyclonal Stimulation

The ability of loaded T cells to perform an immune reaction after stimulation is mandatory. We polyclonally stimulated SPION-loaded and non-loaded T cells with CD3/CD28/CD2 T-cell activator mix overnight. Subsequently, cells were intracellularly stained for cytokine expression such as INFγ, TNFα, and IL-2. We found that neither the loading with SPIONs induced cytokine production nor inhibited cytokine expression for both CD4+ and CD8+ cells after polyclonal stimulation (Figure 6).

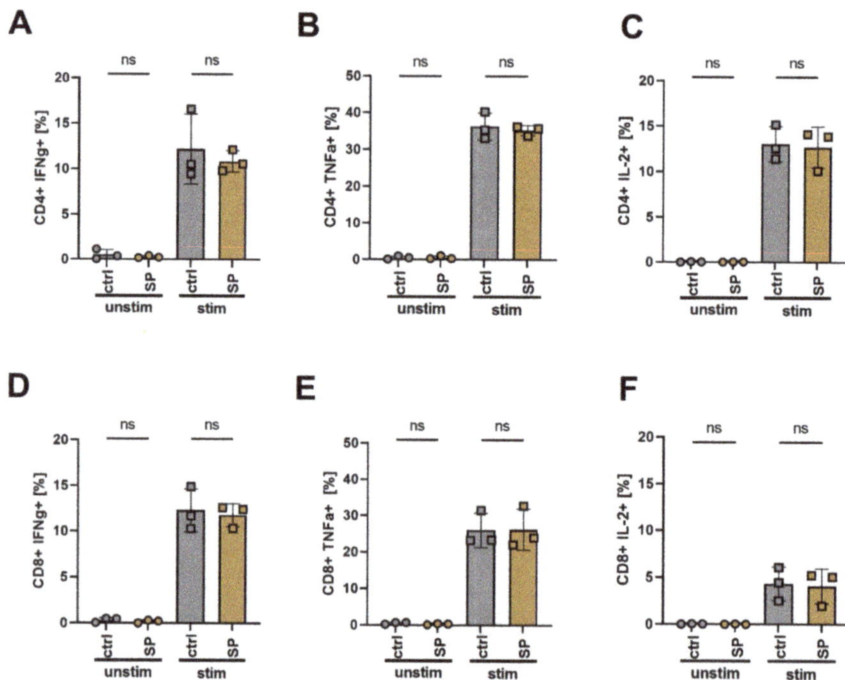

Figure 6. Cytokine expression of T cells after stimulation. Isolated T cells were loaded with 80 µg Fe/mL SPIONs and were stimulated with a CD3/CD28/CD2 activation mix overnight. H_2O treated cells and cells without stimulation mix served as controls. The cells were then stained for CD4 and CD8, viability, and cytokine expression. Depicted are the percentages of CD4 (**A–C**) or CD8 (**D–F**) T cells expressing TNFα (**A,D**), IFNγ (**B,E**) or IL-2 (**C,F**). Statistical significances were estimated using an unpaired t test with Welch's correction; (ns): $p \geq 0.05$. ctrl: control; SP: SPIONs; unstim: unstimulated T cells; stim: stimulated T cells.

4. Discussion

The loading of cells with SPIONs enables their control by external magnetic forces. This strategy has been applied for a long time for magnetic associated cell sorting (MACS) in vitro [40] but also represents a great potential for in vivo applications, e.g., for targeted adoptive T cell therapies. For MACS, cells are usually labeled via antibody-conjugated magnetic microbeads, thus, cell labeling enables the isolation of wanted (touched) or unwanted (untouched) cell populations. In contrast to direct targeting of surface markers by antibody-labeled nanoparticles, we use nanoparticles with a citrate shell without any specific targeting moiety. With this, cells can be labeled independently from specific surface structures. Our aim is to make T cells magnetically attractable to enable site-directed targeting for biomedical applications. For this purpose, T cells must be loaded with sufficient iron and at the same time, the cellular function must not be impaired. Here we analyzed how experimental conditions influence nanoparticle uptake and cell viability.

Our SPIONs have a negative zeta potential of around -50 mV at pH 7.3 and a hydrodynamic diameter of 54 nm in water (Table 1). When exposed to physiological fluids, it is known that the nanoparticle surface is immediately covered by various biomolecules to lower the surface energy [41–43]. Rocker et al. showed that serum proteins such as the abundant albumin, bind nanoparticles [44]. This protein adsorption, referred to as protein corona, essentially determines the "biological identity" of the nanoparticle, which can also explain reduced targeting efficacy of, e.g., antibody-conjugated nanoparticles in vivo [45]. With 10% FCS in the cell culture medium, our SPIONs remained colloidally stable and roughly 1.4 pg iron was associated with the cells after 3 h of incubation (Figures 1 and 2). Reduction of the FCS content in cell culture medium to 2% led to nanoparticle agglomerations and their sedimentation (Figure 1). In parallel, we observed increased cellular levels of iron (25 pg/cell), which was probably caused by insufficient removability by washing due to nanoparticle sedimentation. In PBS with 2% FCS, in fact, no nanoparticle agglomerations were detected, but cellular iron amounts also increased to nearly 3 pg/cell. Thus, in conditions with 2% FCS only, iron uptake was enhanced but also accompanied by a strong reduction in cell counts (Figure 2). These findings are in line with previous reports by Lesniak et al., who showed much more nanoparticle uptake in the absence of proteins. Analyzing the composition of the protein corona, they found that in the absence of proteins nanoparticles tend to cover themselves with proteins from the plasma membranes, which was exhausting for the cells [46]. This, and the tendency of SPION-loaded cells to aggregate, might have led to a loss of cells in conditions with low protein amounts [47]. Zupke et al. also reported that the presence of human serum proteins reduces the uptake of nanoparticles by CD4+ T cells but preserves cell viability [36].

Besides protein amount, the activation status of the cell was reported to also influence nanoparticle uptake. Others have shown increased incorporation of nanoparticles by activated lymphocytes compared to freshly isolated ones [31,36]. This has been ascribed to the increased macropinocytosis by primary mouse and human T cells after polyclonal activation to support T cell growth [48]. Based on these findings, we polyclonally stimulated T cells to foster nanoparticle uptake. As expected, the T cell count was increased after 72 h, but the stimulated cells did not contain more nanoparticles than the resting ones (Figure 3). Contrary to our data, polyclonal activation of the T cells by anti-CD3, anti-CD28, IL-2, and concanavalin A has previously been shown to increase nanoparticle uptake [49]. In this context, the dependence of nanoparticle uptake on the cell cycle has been discussed. Although cells in different phases of the cell cycle internalized nanoparticles at similar rates, cells in the G2/M phase contained more nanoparticles than those in the S or G0/G1 phase. As soon as cells split, cell-associated nanoparticles were divided between daughter cells [39] and, therefore, it was discussed that increased amounts of nanoparticles were rather due to cell size and the number of endosomes than the cell cycle state [50,51].

When we stimulated T cells after SPION uptake, we found that the release of the cytokines IFNγ, TNFα, and IL-2 was not influenced (Figure 6). Thus, these data confirm earlier investigations of our group, that activation after polyclonal stimulation of human primary T cells was not impaired by SPION loading [27]. In addition, others have previously shown the biocompatibility of SPIONs [30–32,36].

To analyze the subcellular location of nanoparticles after incubation with T cells in a cell culture medium containing 10% FCS, we performed TEM. Although it has been described that the internalization of cationic nanoparticles is more efficient in comparison to neutral or anionic nanoparticles due to the interaction with the positively charged plasma membrane [52,53], we found not only a strong association with the plasma membrane but also remarkable amounts of particles intracellularly enriched in vesicles (Figure 4A). The nanoparticles associated with the plasma membranes were rather bound as thin layers, not in clusters. When comparing with the TEM pictures from others using cationic SPIONs, which achieved up to ten-fold higher iron amounts per cell, their nanoparticles were not taken up but attached to the plasma membrane in multiple layers and clusters, with the risk of detachment from the cells [31]. Since lymphoid cells are non-phagocytic, the uptake of nanoparticles has been frequently described to be low and supporting techniques such as functionalization of the nanoparticles by RGD peptides, coatings derived from viruses, or electroporation of the target cells, were employed for better engulfment [54,55]. The difference in the uptake of nanoparticles dependent on incubation temperature, however, indicated an energy-dependent uptake process with the involvement of endocytic processes for T cells as well [36].

When analyzing the SSC as a marker for nanoparticle uptake of co-incubated loaded and non-loaded cells, we found that SSC of the single-cell populations did not change; indicating that SPIONs did not spill from loaded cells to initially non-loaded ones. However, when analyzing the iron amount in the loaded T cells, the AES measurements revealed a reduction in iron during the 72 h observation time, which was not due to cell division, since the analysis was performed with non-proliferating T cells. Others have previously characterized the export of nanoparticles from T cells as an energy-dependent active process, taking part within 24 h, which was reduced by lowering of the temperature or use of inhibitors of cell metabolism [36]. Concerning the cellular persistence of SPIONs, both the retention for several days as well as bisection of the MRI signal within the first 24 h has been reported [56–58]. Whereas SPIONs are superparamagnetic, their degradation products, mainly ferritin and hemosiderin, were anti-ferromagnetic, exhibiting a detectable, but reduced MRI signal [59]. Others previously analyzing the degradation of citrate-coated SPIONs in stem cell spheroids found endosomal degradation and upregulation iron homeostasis genes coding for ferritin light chain (iron loading) and ferroportin (iron export) from day 3 onwards [60]. Moreover, after i.v. injection of radioactively labeled SPIONs in mice, after 7 days, 59Fe from the administered nanoparticles appeared in the hemoglobin of newly formed erythrocytes, indicating the intracellular degradation of the nanoparticles. From studies applying SPIONs as contrast agents (Endorem, Resovist), the phenomenon of Fe incorporation into erythrocytes has been well-known [61]. Concerning T cells, it has been found that endosomal acidification was slower and not as robust as in other cells [62]. If SPIONs, in our case, are excreted from the T cells or degraded intracellularly in the lysosomes remains, so far, elusive and must be further investigated. Nonetheless, we detected no spilling to initially non-labeled cells (Figure 4B).

Finally, we analyzed the magnetic retention of the SPION loaded T cells under flow conditions. In magnetic accumulation, the viscosity of the medium, cell radius, and cell velocity are known to play a role [47]. When a permanent magnet generates a magnetic field of about 10–50 T/m over a distance of 1 cm, with a 10 pg iron load, a cell experiences a corresponding force of 1 pN to a few nN [63]. For the investigation of the magnetic accumulation of SPION loaded cells, we used a flow rate of 9.6 mL/min for 1 h at a channel height of 0.4 mm, and a 5×5 mm sized permanent neodymium magnet with approximately 400 mT. With this experimental setup, we were at least able to show in vitro magnetic retention of the SPION-loaded T cells in the wanted area (Figure 5). As with magnetic drug targeting, we are aware that magnetic targeting of cells to the tumor region in vivo is much more ambitious and faces several challenges and pitfalls [64–67]. Exemplarily, tumor vascularization may be unfavorable, hindering the delivery of the cells or nanoparticles for drug delivery. However, others previously functionalized lymphocytes with magnetic nanoparticles and showed increased therapeutic success with the adoptive transfer of T cells and NK cells [28,35]. Further, contactless magnetic movement by permanent magnets and dynamically programmable magnetic fields is under intense investigation [29,68]. Interestingly, for future translation into clinics, the magnetic field coils inherent to clinical MRI scanners can not only be used for tracking but also for steering magnetic nanoparticles or nanoparticle-loaded cells into the wanted region [23,24], possibly enabling an image-based therapy in the future.

5. Conclusions

Adoptive T cell transfer suffers from poor efficacy in many patients with solid cancers, which is due to low infiltration of the cells into the tumor. The stable and effective loading of T cells with SPIONs enables their magnetic controllability and accumulation. In summary, we showed here that we can load primary human CD3+ T cells with SPIONs. We found that loading efficacy and cell viability was dependent on the amount of serum present in the cell culture medium. The activation of the cells, however, did not affect nanoparticle uptake. With our loading strategy, we achieved 1.4 pg Fe/iron per cell, which was enough to accumulate the cells in a dynamic flow system. Furthermore, we found that SPIONs were located intracellularly in vesicles or tightly attached to the plasma membrane, without spillover to non-loaded cells. With stable and sufficient iron loading, T cells become not only magnetically controllable but can enable tracking of injected cells using MRI.

Supplementary Materials: The following are available online at https://www.mdpi.com/article/10.3390/cancers13164143/s1: Supplementary Figure S1: Efficacy of T cell isolation using IBA CD3 Fab-TACS gravity columns.

Author Contributions: Conceptualization: C.J. and C.A.; methodology, M.M., L.F. and S.L.; software, P.B., F.P. and L.F.; validation, E.S. and J.B.; formal analysis, P.B. and F.P.; investigation, P.B., F.P., R.S., B.F., J.B. and E.S.; resources, C.A. and D.D; data curation, P.B., F.P. and C.J.; writing—original draft preparation, P.B., and F.P.; writing—review and editing, C.A., C.J. and S.L.; visualization, P.B., F.P. and L.F.; supervision, C.J. and C.A.; project administration, C.J.; funding acquisition, C.A., C.J. and D.D. All authors have read and agreed to the published version of the manuscript.

Funding: This research was funded by the Else Kröner Fresenius Stiftung, Bad Homburg v.d.H., Germany (2018_A88) to C.J. and Deutsche Forschungsgemeinschaft (DFG)-SFB TRR 305-B05, DU5485-1 to D.D. as well as intramurally by the Emerging Fields Initiative EFI-BIG-Thera and the Staedler-Stiftung to D.D. and C.A.

Institutional Review Board Statement: The study was conducted according to the guidelines of the Declaration of Helsinki, and approved by the Ethics Committee of the Friedrich-Alexander Universität Erlangen-Nürnberg (protocol code 257_14B; approved on 18 September 2014 and pro-longed on 2 February 2018).

Informed Consent Statement: Informed consent was obtained from all subjects involved in the study.

Data Availability Statement: No new data were created or analyzed in this study. Data sharing is not applicable to this article.

Acknowledgments: The present work was performed in fulfilment of the requirements for ob-taining the degree "Dr. med." (Philipp Boosz) at the Friedrich-Alexander-Universität (FAU), Er-langen-Nürnberg, Germany. We acknowledge Andrea Kerpes for her excellent technical support in the cell culture and Anita Hecht and Andrea Hilpert for operating the TEM.

Conflicts of Interest: The authors declare no conflict of interest.

References

1. Mattiuzzi, C.; Lippi, G. Current Cancer Epidemiology. *J. Epidemiol. Glob. Health* **2019**, *9*, 217–222. [CrossRef]
2. Galon, J.; Costes, A.; Sanchez-Cabo, F.; Kirilovsky, A.; Mlecnik, B.; Lagorce-Pages, C.; Tosolini, M.; Camus, M.; Berger, A.; Wind, P.; et al. Type, density, and location of immune cells within human colorectal tumors predict clinical outcome. *Science* **2006**, *313*, 1960–1964. [CrossRef] [PubMed]
3. Jass, J.R. Lymphocytic infiltration and survival in rectal cancer. *J. Clin. Pathol.* **1986**, *39*, 585–589. [CrossRef] [PubMed]
4. Kim, S.T.; Jeong, H.; Woo, O.H.; Seo, J.H.; Kim, A.; Lee, E.S.; Shin, S.W.; Kim, Y.H.; Kim, J.S.; Park, K.H. Tumor-infiltrating lymphocytes, tumor characteristics, and recurrence in patients with early breast cancer. *Am. J. Clin. Oncol.* **2013**, *36*, 224–231. [CrossRef] [PubMed]
5. Kmiecik, J.; Poli, A.; Brons, N.H.; Waha, A.; Eide, G.E.; Enger, P.O.; Zimmer, J.; Chekenya, M. Elevated CD3+ and CD8+ tumor-infiltrating immune cells correlate with prolonged survival in glioblastoma patients despite integrated immunosuppressive mechanisms in the tumor microenvironment and at the systemic level. *J. Neuroimmunol.* **2013**, *264*, 71–83. [CrossRef]
6. Dudley, M.E.; Rosenberg, S.A. Adoptive-cell-transfer therapy for the treatment of patients with cancer. *Nat. Rev. Cancer* **2003**, *3*, 666–675. [CrossRef]
7. Bernhard, H.; Neudorfer, J.; Gebhard, K.; Conrad, H.; Hermann, C.; Nahrig, J.; Fend, F.; Weber, W.; Busch, D.H.; Peschel, C. Adoptive transfer of autologous, HER2-specific, cytotoxic T lymphocytes for the treatment of HER2-overexpressing breast cancer. *Cancer Immunol. Immunother.* **2008**, *57*, 271–280. [CrossRef]
8. Pockaj, B.A.; Sherry, R.M.; Wei, J.P.; Yannelli, J.R.; Carter, C.S.; Leitman, S.F.; Carasquillo, J.A.; Steinberg, S.M.; Rosenberg, S.A.; Yang, J.C. Localization of 111indium-labeled tumor infiltrating lymphocytes to tumor in patients receiving adoptive immunotherapy. Augmentation with cyclophosphamide and correlation with response. *Cancer* **1994**, *73*, 1731–1737. [CrossRef]
9. Sanz-Ortega, L.; Rojas, J.M.; Barber, D.F. Improving Tumor Retention of Effector Cells in Adoptive Cell Transfer Therapies by Magnetic Targeting. *Pharmaceutics* **2020**, *12*, 812. [CrossRef]
10. Bargou, R.; Leo, E.; Zugmaier, G.; Klinger, M.; Goebeler, M.; Knop, S.; Noppeney, R.; Viardot, A.; Hess, G.; Schuler, M.; et al. Tumor regression in cancer patients by very low doses of a T cell-engaging antibody. *Science* **2008**, *321*, 974–977. [CrossRef]
11. Schlegel, P.; Lang, P.; Zugmaier, G.; Ebinger, M.; Kreyenberg, H.; Witte, K.E.; Feucht, J.; Pfeiffer, M.; Teltschik, H.M.; Kyzirakos, C.; et al. Pediatric posttransplant relapsed/refractory B-precursor acute lymphoblastic leukemia shows durable remission by therapy with the T-cell engaging bispecific antibody blinatumomab. *Haematologica* **2014**, *99*, 1212–1219. [CrossRef] [PubMed]
12. Topp, M.S.; Kufer, P.; Gokbuget, N.; Goebeler, M.; Klinger, M.; Neumann, S.; Horst, H.A.; Raff, T.; Viardot, A.; Schmid, M.; et al. Targeted therapy with the T-cell-engaging antibody blinatumomab of chemotherapy-refractory minimal residual disease in B-lineage acute lymphoblastic leukemia patients results in high response rate and prolonged leukemia-free survival. *J. Clin. Oncol.* **2011**, *29*, 2493–2498. [CrossRef] [PubMed]
13. Klinger, M.; Brandl, C.; Zugmaier, G.; Hijazi, Y.; Bargou, R.C.; Topp, M.S.; Gökbuget, N.; Neumann, S.; Goebeler, M.; Viardot, A.; et al. Immunopharmacologic response of patients with B-lineage acute lymphoblastic leukemia to continuous infusion of T cell-engaging CD19/CD3-bispecific BiTE antibody blinatumomab. *Blood* **2012**, *119*, 6226–6233. [CrossRef] [PubMed]
14. Tietze, R.; Lyer, S.; Durr, S.; Struffert, T.; Engelhorn, T.; Schwarz, M.; Eckert, E.; Goen, T.; Vasylyev, S.; Peukert, W.; et al. Efficient drug-delivery using magnetic nanoparticles—biodistribution and therapeutic effects in tumour bearing rabbits. *Nanomedicine* **2013**, *9*, 961–971. [CrossRef]
15. Liang, P.C.; Chen, Y.C.; Chiang, C.F.; Mo, L.R.; Wei, S.Y.; Hsieh, W.Y.; Lin, W.L. Doxorubicin-modified magnetic nanoparticles as a drug delivery system for magnetic resonance imaging-monitoring magnet-enhancing tumor chemotherapy. *Int. J. Nanomed.* **2016**, *11*, 2021–2037.
16. Pan, C.; Liu, Y.; Zhou, M.; Wang, W.; Shi, M.; Xing, M.; Liao, W. Theranostic pH-sensitive nanoparticles for highly efficient targeted delivery of doxorubicin for breast tumor treatment. *Int. J. Nanomed.* **2018**, *13*, 1119–1137. [CrossRef]
17. Jin, H.; Qian, Y.; Dai, Y.; Qiao, S.; Huang, C.; Lu, L.; Luo, Q.; Chen, J.; Zhang, Z. Magnetic Enrichment of Dendritic Cell Vaccine in Lymph Node with Fluorescent-Magnetic Nanoparticles Enhanced Cancer Immunotherapy. *Theranostics* **2016**, *6*, 2000–2014. [CrossRef]
18. Polyak, B.; Fishbein, I.; Chorny, M.; Alferiev, I.; Williams, D.; Yellen, B.; Friedman, G.; Levy, R.J. High field gradient targeting of magnetic nanoparticle-loaded endothelial cells to the surfaces of steel stents. *Proc. Natl. Acad. Sci. USA* **2008**, *105*, 698–703. [CrossRef]
19. Tukmachev, D.; Lunov, O.; Zablotskii, V.; Dejneka, A.; Babic, M.; Sykova, E.; Kubinova, S. An effective strategy of magnetic stem cell delivery for spinal cord injury therapy. *Nanoscale* **2015**, *7*, 3954–3958. [CrossRef]
20. Baeten, K.; Adriaensens, P.; Hendriks, J.; Theunissen, E.; Gelan, J.; Hellings, N.; Stinissen, P. Tracking of myelin-reactive T cells in experimental autoimmune encephalomyelitis (EAE) animals using small particles of iron oxide and MRI. *NMR Biomed.* **2010**, *23*, 601–609. [CrossRef]
21. Liu, L.; Ye, Q.; Wu, Y.; Hsieh, W.Y.; Chen, C.L.; Shen, H.H.; Wang, S.J.; Zhang, H.; Hitchens, T.K.; Ho, C. Tracking T-cells in vivo with a new nano-sized MRI contrast agent. *Nanomedicine* **2012**, *8*, 1345–1354. [CrossRef]
22. Beer, A.J.; Holzapfel, K.; Neudorfer, J.; Piontek, G.; Settles, M.; Kronig, H.; Peschel, C.; Schlegel, J.; Rummeny, E.J.; Bernhard, H. Visualization of antigen-specific human cytotoxic T lymphocytes labeled with superparamagnetic iron-oxide particles. *Eur. Radiol.* **2008**, *18*, 1087–1095. [CrossRef] [PubMed]

23. Muthana, M.; Kennerley, A.J.; Hughes, R.; Fagnano, E.; Richardson, J.; Paul, M.; Murdoch, C.; Wright, F.; Payne, C.; Lythgoe, M.F.; et al. Directing cell therapy to anatomic target sites in vivo with magnetic resonance targeting. *Nat. Commun.* **2015**, *6*, 8009. [CrossRef] [PubMed]

24. Riegler, J.; Wells, J.A.; Kyrtatos, P.G.; Price, A.N.; Pankhurst, Q.A.; Lythgoe, M.F. Targeted magnetic delivery and tracking of cells using a magnetic resonance imaging system. *Biomaterials* **2010**, *31*, 5366–5371. [CrossRef] [PubMed]

25. Mühlberger, M.; Janko, C.; Unterweger, H.; Friedrich, R.P.; Friedrich, B.; Band, J.; Cebulla, N.; Alexiou, C.; Dudziak, D.; Lee, G.; et al. Functionalization of T Lymphocytes with Citrate-Coated Superpara-magnetic Iron Oxide Nanoparticles for Magnetically Controlled Immune Therapy. *Int. J. Nanomed.* **2019**, *14*, 8421–8432. [CrossRef] [PubMed]

26. Mühlberger, M.; Janko, C.; Unterweger, H.; Schreiber, E.; Band, J.; Lehmann, C.; Dudziak, D.; Lee, G.; Alexiou, C.; Tietze, R. Functionalization of T lymphocytes for magnetically controlled immune therapy: Selection of suitable superparamagnetic iron oxide nanoparticles. *J. Magn. Magn. Mater.* **2019**, *473*, 61–67. [CrossRef]

27. Mühlberger, M.; Unterweger, H.; Band, J.; Lehmann, C.; Heger, L.; Dudziak, D.; Alexiou, C.; Lee, G.; Janko, C. Loading of Primary Human T Lymphocytes with Citrate-Coated Superparamagnetic Iron Oxide Nanoparticles Does Not Impair Their Activation after Polyclonal Stimulation. *Cells* **2020**, *9*, 342. [CrossRef]

28. Nie, W.; Wei, W.; Zuo, L.; Lv, C.; Zhang, F.; Lu, G.H.; Li, F.; Wu, G.; Huang, L.L.; Xi, X.; et al. Magnetic Nanoclusters Armed with Responsive PD-1 Antibody Synergistically Improved Adoptive T-Cell Therapy for Solid Tumors. *ACS Nano* **2019**, *13*, 1469–1478. [CrossRef]

29. Pai, A.; Cao, P.; White, E.E.; Hong, B.; Pailevanian, T.; Wang, M.; Badie, B.; Hajimiri, A.; Berlin, J.M. Dynamically Programmable Magnetic Fields for Controlled Movement of Cells Loaded with Iron Oxide Nanoparticles. *ACS Appl. Bio Mater.* **2020**, *3*, 4139–4147. [CrossRef]

30. Sanz-Ortega, L.; Portilla, Y.; Pérez-Yagüe, S.; Barber, D.F. Magnetic targeting of adoptively transferred tumour-specific nanoparticle-loaded CD8(+) T cells does not improve their tumour infiltration in a mouse model of cancer but promotes the retention of these cells in tumour-draining lymph nodes. *J. Nanobiotechnol.* **2019**, *17*, 87. [CrossRef]

31. Sanz-Ortega, L.; Rojas, J.M.; Marcos, A.; Portilla, Y.; Stein, J.V.; Barber, D.F. T cells loaded with magnetic nanoparticles are retained in peripheral lymph nodes by the application of a magnetic field. *J. Nanobiotechnol.* **2019**, *17*, 14. [CrossRef]

32. Sanz-Ortega, L.; Rojas, J.M.; Portilla, Y.; Pérez-Yagüe, S.; Barber, D.F. Magnetic Nanoparticles Attached to the NK Cell Surface for Tumor Targeting in Adoptive Transfer Therapies Does Not Affect Cellular Effector Functions. *Front. Immunol.* **2019**, *10*, 2073. [CrossRef] [PubMed]

33. Foroozandeh, P.; Aziz, A.A. Insight into Cellular Uptake and Intracellular Trafficking of Nanoparticles. *Nanoscale Res. Lett.* **2018**, *13*, 339. [CrossRef] [PubMed]

34. Waiczies, S.; Niendorf, T.; Lombardi, G. Labeling of cell therapies: How can we get it right? *Oncoimmunology* **2017**, *6*, e1345403. [CrossRef] [PubMed]

35. Sim, T.; Choi, B.; Kwon, S.W.; Kim, K.S.; Choi, H.; Ross, A.; Kim, D.H. Magneto-Activation and Magnetic Resonance Imaging of Natural Killer Cells Labeled with Magnetic Nanocomplexes for the Treatment of Solid Tumors. *ACS Nano* **2021**. [CrossRef] [PubMed]

36. Zupke, O.; Distler, E.; Jürchott, A.; Paiphansiri, U.; Dass, M.; Thomas, S.; Hartwig, U.F.; Theobald, M.; Landfester, K.; Mailänder, V.; et al. Nanoparticles and antigen-specific T-cell therapeutics: A comprehensive study on uptake and release. *Nanomedicine* **2015**, *10*, 1063–1076. [CrossRef] [PubMed]

37. Elbialy, N.S.; Fathy, M.M.; Khalil, W.M. Doxorubicin loaded magnetic gold nanoparticles for in vivo targeted drug delivery. *Int. J. Pharm.* **2015**, *490*, 190–199. [CrossRef] [PubMed]

38. Friedrich, R.P.; Janko, C.; Poettler, M.; Tripal, P.; Zaloga, J.; Cicha, I.; Dürr, S.; Nowak, J.; Odenbach, S.; Slabu, I.; et al. Flow cytometry for intracellular SPI-ON quantification: Specificity and sensitivity in comparison with spectroscopic methods. *Int. J. Nanomed.* **2015**, *10*, 4185–4201. [CrossRef]

39. Kim, J.A.; Aberg, C.; Salvati, A.; Dawson, K.A. Role of cell cycle on the cellular uptake and dilution of nanoparticles in a cell population. *Nat. Nanotechnol.* **2011**, *7*, 62–68. [CrossRef]

40. Miltenyi, S.; Muller, W.; Weichel, W.; Radbruch, A. High gradient magnetic cell separation with MACS. *Cytometry* **1990**, *11*, 231–238. [CrossRef]

41. Lynch, I.; Salvati, A.; Dawson, K.A. Proteinnanoparticle interactions: What does the cell see? *Nat. Nanotechnol.* **2009**, *4*, 546–547. [CrossRef]

42. Mahmoudi, M.; Lynch, I.; Ejtehadi, M.R.; Monopoli, M.P.; Bombelli, F.B.; Laurent, S. Protein-nanoparticle interactions: Opportunities and challenges. *Chem. Rev.* **2011**, *111*, 5610–5637. [CrossRef]

43. Nel, A.E.; Mädler, L.; Velegol, D.; Xia, T.; Hoek, E.M.V.; Somasundaran, P.; Klaessig, F.; Castranova, V.; Thompson, M. Understanding biophysicochemical interactions at the nano–bio interface. *Nat. Mater.* **2009**, *8*, 543–557. [CrossRef] [PubMed]

44. Rocker, C.; Potzl, M.; Zhang, F.; Parak, W.J.; Nienhaus, G.U. A quantitative fluorescence study of protein monolayer formation on colloidal nanoparticles. *Nat. Nanotechnol.* **2009**, *4*, 577–580. [CrossRef] [PubMed]

45. Hadjidemetriou, M.; Al-Ahmady, Z.; Mazza, M.; Collins, R.F.; Dawson, K.; Kostarelos, K. In Vivo Biomolecule Corona around Blood-Circulating, Clinically Used and Antibody-Targeted Lipid Bilayer Nanoscale Vesicles. *ACS Nano* **2015**, *9*, 8142–8156. [CrossRef] [PubMed]

46. Lesniak, A.; Fenaroli, F.; Monopoli, M.P.; Aberg, C.; Dawson, K.A.; Salvati, A. Effects of the presence or absence of a protein corona on silica nanoparticle uptake and impact on cells. *ACS Nano* **2012**, *6*, 5845–5857. [CrossRef] [PubMed]
47. Kolosnjaj-Tabi, J.; Wilhelm, C.; Clement, O.; Gazeau, F. Cell labeling with magnetic nanoparticles: Opportunity for magnetic cell imaging and cell manipulation. *J. Nanobiotechnol.* **2013**, *11* (Suppl. 1), S7. [CrossRef]
48. Charpentier, J.C.; Chen, D.; Lapinski, P.E.; Turner, J.; Grigorova, I.; Swanson, J.A.; King, P.D. Macropinocytosis drives T cell growth by sustaining the activation of mTORC1. *Nat. Commun.* **2020**, *11*, 180. [CrossRef] [PubMed]
49. Dodd, C.H.; Hsu, H.C.; Chu, W.J.; Yang, P.; Zhang, H.G.; Mountz, J.D., Jr.; Zinn, K.; Forder, J.; Josephson, L.; Weissleder, R.; et al. Normal T-cell response and in vivo magnetic resonance imaging of T cells loaded with HIV transactivator-peptide-derived superparamagnetic nanoparticles. *J. Immunol. Methods* **2001**, *256*, 89–105. [CrossRef]
50. Panet, E.; Mashriki, T.; Lahmi, R.; Jacob, A.; Ozer, E.; Vecsler, M.; Lazar, I.; Tzur, A. The interface of nanoparticles with proliferating mammalian cells. *Nat. Nanotechnol.* **2017**, *12*, 598–600. [CrossRef]
51. Rees, P.; Wills, J.W.; Brown, M.R.; Barnes, C.M.; Summers, H.D. The origin of heterogeneous nanoparticle uptake by cells. *Nat. Commun.* **2019**, *10*, 2341. [CrossRef] [PubMed]
52. Frohlich, E. The role of surface charge in cellular uptake and cytotoxicity of medical nanoparticles. *Int. J. Nanomed.* **2012**, *7*, 5577–5591. [CrossRef]
53. Jambhrunkar, S.; Qu, Z.; Popat, A.; Yang, J.; Noonan, O.; Acauan, L.; Ahmad Nor, Y.; Yu, C.; Karmakar, S. Effect of surface functionality of silica nanoparticles on cellular uptake and cytotoxicity. *Mol. Pharm.* **2014**, *11*, 3642–3655. [CrossRef] [PubMed]
54. Garden, O.A.; Reynolds, P.R.; Yates, J.; Larkman, D.J.; Marelli-Berg, F.M.; Haskard, D.O.; Edwards, A.D.; George, A.J. A rapid method for labelling CD4+ T cells with ultrasmall paramagnetic iron oxide nanoparticles for magnetic resonance imaging that preserves proliferative, regulatory and migratory behaviour in vitro. *J. Immunol. Methods* **2006**, *314*, 123–133. [CrossRef] [PubMed]
55. Lankoff, A.; Arabski, M.; Wegierek-Ciuk, A.; Kruszewski, M.; Lisowska, H.; Banasik-Nowak, A.; Rozga-Wijas, K.; Wojewodzka, M.; Slomkowski, S. Effect of surface modification of silica nanoparticles on toxicity and cellular uptake by human peripheral blood lymphocytes in vitro. *Nanotoxicology* **2013**, *7*, 235–250. [CrossRef] [PubMed]
56. Kaufman, C.L.; Williams, M.; Ryle, L.M.; Smith, T.L.; Tanner, M.; Ho, C. Superparamagnetic iron oxide particles transactivator protein-fluorescein isothiocyanate particle labeling for in vivo magnetic resonance imaging detection of cell migration: Uptake and durability. *Transplantation* **2003**, *76*, 1043–1046. [CrossRef]
57. Kircher, M.F.; Allport, J.R.; Graves, E.E.; Love, V.; Josephson, L.; Lichtman, A.H.; Weissleder, R. In vivo high resolution three-dimensional imaging of antigen-specific cytotoxic T-lymphocyte trafficking to tumors. *Cancer Res.* **2003**, *63*, 6838–6846.
58. Smirnov, P.; Lavergne, E.; Gazeau, F.; Lewin, M.; Boissonnas, A.; Doan, B.-T.; Gillet, B.; Combadière, C.; Combadière, B.; Clément, O. In vivo cellular imaging of lymphocyte trafficking by MRI: A tumor model approach to cell-based anticancer therapy. *Magn. Reson. Med.* **2006**, *56*, 498–508. [CrossRef]
59. Briley-Saebo, K.; Bjornerud, A.; Grant, D.; Ahlstrom, H.; Berg, T.; Kindberg, G.M. Hepatic cellular distribution and degradation of iron oxide nanoparticles following single intravenous injection in rats: Implications for magnetic resonance imaging. *Cell Tissue Res.* **2004**, *316*, 315–323. [CrossRef]
60. Mazuel, F.; Espinosa, A.; Luciani, N.; Reffay, M.; Le Borgne, R.; Motte, L.; Desboeufs, K.; Michel, A.; Pellegrino, T.; Lalatonne, Y.; et al. Massive Intracellular Biodegradation of Iron Oxide Nanoparticles Evidenced Magnetically at Single-Endosome and Tissue Levels. *ACS Nano* **2016**, *10*, 7627–7638. [CrossRef]
61. Feliu, N.; Docter, D.; Heine, M.; Del Pino, P.; Ashraf, S.; Kolosnjaj-Tabi, J.; Macchiarini, P.; Nielsen, P.; Alloyeau, D.; Gazeau, F.; et al. In vivo degeneration and the fate of inorganic nanoparticles. *Chem. Soc. Rev.* **2016**, *45*, 2440–2457. [CrossRef] [PubMed]
62. Olden, B.R.; Cheng, E.; Cheng, Y.; Pun, S.H. Identifying key barriers in cationic polymer gene delivery to human T cells. *Biomater. Sci.* **2019**, *7*, 789–797. [CrossRef] [PubMed]
63. Wilhelm, C.; Riviere, C.; Biais, N. Magnetic control of Dictyostelium aggregation. *Phys. Rev. E Stat. Nonlin. Soft Matter Phys.* **2007**, *75 Pt 1*, 041906. [CrossRef]
64. Park, K. Facing the truth about nanotechnology in drug delivery. *ACS Nano* **2013**, *7*, 7442–7447. [CrossRef] [PubMed]
65. Shi, J.; Kantoff, P.W.; Wooster, R.; Farokhzad, O.C. Cancer nanomedicine: Progress, challenges and opportunities. *Nat. Rev. Cancer* **2017**, *17*, 20–37. [CrossRef] [PubMed]
66. Torrice, M. Does Nanomedicine Have a Delivery Problem? *ACS Cent. Sci.* **2016**, *2*, 434–437. [CrossRef] [PubMed]
67. Wilhelm, C.; Gazeau, F. Universal cell labelling with anionic magnetic nanoparticles. *Biomaterials* **2008**, *29*, 3161–3174. [CrossRef] [PubMed]
68. Blumler, P.; Friedrich, R.P.; Pereira, J.; Baun, O.; Alexiou, C.; Mailander, V. Contactless Nanoparticle-Based Guiding of Cells by Controllable Magnetic Fields. *Nanotechnol. Sci. Appl.* **2021**, *14*, 91–100. [CrossRef]

cancers

Article

Magnetic-Guided Axillary UltraSound (MagUS) Sentinel Lymph Node Biopsy and Mapping in Patients with Early Breast Cancer. A Phase 2, Single-Arm Prospective Clinical Trial

Allan Jazrawi [1,2], Eirini Pantiora [3,4], Shahin Abdsaleh [3,5], Daniel Vasiliu Bacovia [6], Staffan Eriksson [1,2], Henrik Leonhardt [7], Fredrik Wärnberg [3,4,8] and Andreas Karakatsanis [3,4,*]

[1]   Centre for Clinical Research, County Västmanland, Uppsala University, 72189 Västerås, Sweden; allan.jazrawi@regionvastmanland.se (A.J.); staffan.eriksson@regionvastmanland.se (S.E.)
[2]   Department of Surgery, Västmanlands County Hospital, 72189 Västerås, Sweden
[3]   Department of Surgical Sciences, Uppsala University, 75185 Uppsala, Sweden; eirini.pantiora@akademiska.se (E.P.); shahin.abdsaleh@akademiska.se (S.A.); fredrik.warnberg@vgregion.se (F.W.)
[4]   Department of Surgery, Section for Endocrine and Breast Surgery, Uppsala University Hospital, 75185 Uppsala, Sweden
[5]   Aleris Mammography Unit, 75320 Uppsala, Sweden
[6]   Department Immunology, Genetics and Pathology, Uppsala University, 75185 Uppsala, Sweden; daniel.vasiliu-bacovia@igp.uu.se
[7]   Department of Radiology, Institute of Clinical Sciences, Sahlgrenska Academy, University of Gothenburg, 41343 Gothenburg, Sweden; Henrik.leonhardt@vgregion.se
[8]   Department of Surgery, Institute of Clinical Sciences, Sahlgrenska Academy, University of Gothenburg, 41345 Gothenburg, Sweden
*   Correspondence: andreas.karakatsanis@surgsci.uu.se

Citation: Jazrawi, A.; Pantiora, E.; Abdsaleh, S.; Bacovia, D.V.; Eriksson, S.; Leonhardt, H.; Wärnberg, F.; Karakatsanis, A. Magnetic-Guided Axillary UltraSound (MagUS) Sentinel Lymph Node Biopsy and Mapping in Patients with Early Breast Cancer. A Phase 2, Single-Arm Prospective Clinical Trial. *Cancers* **2021**, *13*, 4285. https://doi.org/10.3390/cancers13174285

Academic Editors: Moriaki Kusakabe, Akihiro Kuwahata and Katja Pinker-Domenig

Received: 22 June 2021
Accepted: 20 August 2021
Published: 25 August 2021

Publisher's Note: MDPI stays neutral with regard to jurisdictional claims in published maps and institutional affiliations.

Simple Summary: Superparamagnetic iron oxide nanoparticles (SPIO) have been shown to identify sentinel lymph nodes (SLNs) in patients with breast cancer. This study investigated whether a minimally invasive approach with MRI-LG after SPIO injection in the breast followed by a magnetic guided axillary ultrasound and core biopsy of the SLN (MagUS) could accurately stage the axilla. The study included not only patients planned for primary surgery but also patients with recurrent cancer after previous surgery, but also patients scheduled for neoadjuvant treatment (NAT). The latter underwent minimally invasive SLNB prior to treatment and had their SLN clipped; surgery in the axilla was performed after NAT. In 79 included patients, MagUS detected all patients with macrometastasis and performed comparably with surgical sentinel lymph node dissection (SLND). It also allowed for marking of the SLN in patients planned for PST and enabled tailored decision making in breast cancer recurrence.

Abstract: Lymph Node Dissection (SLND) is standard of care for diagnosing sentinel lymph node (SLN) status in patients with early breast cancer. Study aim was to determine whether the combination of Superparamagnetic iron oxide nanoparticles (SPIO) MRI-lymphography (MRI-LG) and a Magnetic-guided Axillary UltraSound (MagUS) with biopsy can allow for minimally invasive, axillary evaluation to de-escalate surgery. Patients were injected with 2 mL of SPIO and underwent MRI-LG for SN mapping. Thereafter MagUS and core needle biopsy (CNB) were performed. Patients planned for neoadjuvant treatment, the SLN was clipped and SLND was performed after neoadjuvant with the addition of isotope. During surgery, SLNs were controlled for signs of previous biopsy or clip. The primary endpoint was MagUS SLN detection rate, defined as successful SLN detection of at least one SLN of those retrieved in SLND. In 79 patients, 48 underwent upfront surgery, 12 received neoadjuvant and 19 had recurrent cancer. MagUS traced the SLN in all upfront and neoadjuvant cases, detecting all patients with macrometastases ($n = 10$). MagUS missed only one micrometastasis, outperforming baseline axillary ultrasound AUS (AUC: 0.950 vs. 0.508, $p < 0.001$) and showing no discordance to SLND ($p = 1.000$). MagUS provides the niche for minimally invasive axillary mapping that can reduce diagnostic surgery.

Keywords: sentinel lymph node biopsy; breast cancer; superparamagnetic iron oxide; magnetic tracer; sentinel lymph node

1. Introduction

Primary tumor biology and axillary status guide therapeutic decisions in breast cancer treatment [1,2]. Sentinel Lymph Node Dissection (SLND) is considered the standard method of axillary staging, both in upfront surgery as well as after neoadjuvant treatment (NAT) [3–8].

Preoperative identification of patients with a negative SLN, or low-volume axillary disease that does not warrant further surgery, but guides therapeutic decisions, may allow for tailored approaches avoiding upfront SLND [6,9,10]. In patients scheduled for NAT, identifying those with a true negative axilla, but also those with low-volume disease, as de-escalation of axillary surgery after conversion from cN1 to cN0, could be safely attempted. [7,11,12].

At the same time, SLND is not an indolent procedure and is related to complications and considerable short- and long-term morbidity [13–16]. Therefore, non- or minimally invasive modalities have been proposed in order to address this problem. All of them are based on the principle of injecting a contrast interstitially in the breast in the same manner as when SLND is performed. The contrast will then be taken up by the lymphatics and reach the SLNs and will subsequently be visualized by a radiological modality. Previously, several methods such as single-photon emission computed tomography (SPECT), tridimensional computed tomography lymphography (3D-CTLG) or contrast enhanced ultrasound with microbubbles (CEUS) have been evaluated as alternatives to surgery [17–19]. Most of these have shown promising results, but larger studies are missing and, complicated logistics, need for access to nuclear medicine facilities and demanding learning curves are restricting their introduction into clinical practice.

Superparamagnetic iron oxide nanoparticles (SPIO) are used as a SLND tracer with comparable detection to the combination of radioisotope and blue dye, as shown in previous studies [20,21]. Additionally, when SPIO is injected in the breast, it can identify SLNs in axillary magnetic resonance imaging lymphography (MRI-LG) [22]. At the same time, SPIO yields the benefit that it resides in the tissue for a prolonged period of time without migrating to higher lymph node echelons and, thus, allows for the identification for SLNs during a much wider timeframe [23]. In this manner the SLNs that are identified during surgery should be visible in an MRI and, at the same time, transcutaneous signal detected by a magnetic probe, as in surgery, should be able to guide the axillary ultrasound to allow for transcutaneous identification and biopsy of the SLNs. Such a concept would have the perceived advantages of combining and tailoring modalities and at the same time, allowing for preoperative work up in a timeframe wider than the short halftime of Tc^{99} used for SPECT or that in the case of CEUS [19,24].

The development of an integrated technique bridging non-invasive and minimally invasive procedures for enhancement of the standard, axillary ultrasound-based diagnostic work-up is highly relevant [23–25]. The aim of this study was to determine whether the preoperative work-up with SPIO MRI-LG and Magnetic-guided Axillary UltraSound (MagUS), can accurately localize SLNs and predict SLN status and whether such a technique has the potential of replacing SLN surgery in the future.

2. Methods

2.1. Patients

Adult patients with clinically and ultrasound node-negative early breast cancer (cN0) planned for SLND at Uppsala University Hospital, from September 2017 to December 2020, were enrolled in the study after written informed consent. Patients with hypersensitivity to dextran compounds or SPIO, iron overload disease or planned for NAT and monitored

with breast MRI for tumor response, were excluded. If a diagnostic breast MRI was needed, it was performed separately, before SPIO injection and axillary MRI-LG. The study was approved by the Regional Ethics Board in Uppsala (DNR 2016/385).

2.2. MRI-LG

Patients were injected peritumorally in the breast with 2 mL of SPIO (Magtrace®, Endomag., Cambridge, UK) and underwent MRI-LG one to 14 days after the injection. MRI-LG was performed with the patient in a supine position and adduction of the ipsilateral arm. The examination was performed without iv-contrast and took ca 8 min to complete. In cases of previous breast and axillary surgery or parasternal cancers, the contralateral axilla was also included in the MRI-LG to identify aberrant lymphatic outflow [26]. The MRI images were obtained using a 1,5-T and 3-T system (Philips®, Amsterdam, The Netherlands) with T2W cor, T2* tra and T2* cor sequences. Any lymph node with SPIO uptake in a T1 sequence or SPIO related void artifact on T2 sequence was considered a SLN, as previously described imaging was reviewed and the number of identified SLNs was documented [22]. SLN localization was described according to the classification proposed by Clough et al. [27], in relation to the lateral thoracic vein and the second intercostobrachial nerve. SLN metastatic status was assessed according to criteria previously proposed by Motomura et al. [22]; a lymph node was considered non-metastatic if there was a homogenous low intensity signal uptake of SPIO and metastatic if the entire node or a focal area did not show low signal intensity uptake.

2.3. Magnetic Guided Axillary UltraSound (MagUS) and Core Needle Biopsy (CNB)

After reviewing of MRI-LG, the radiologist performed a second look axillary ultrasound in another session. The examination was focused to the area where the SLNs were identified on MRI (Figure 1). After a primary assessment for lymph nodes, a handheld magnetometer (Sentimag®, Endomag, Cambridge, UK) was used to identify the "pre-incision hotspot" which is the area with the highest magnetic uptake on the skin, and concordance with the MRI localization was registered.

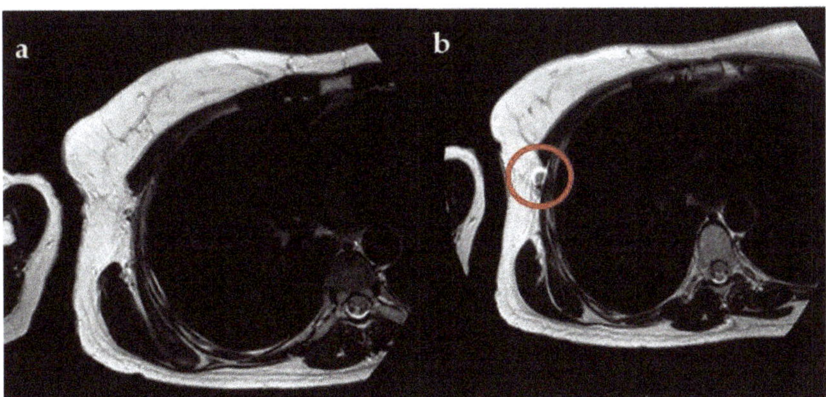

Figure 1. (a,b). Visualization of SLN with MRI before and after SPIO. In an enhancement of the SLN is visualized after injection of SPIO. The red circle visualizes the enhanced SLN after the injection of SPIO.

Subsequently, the identified lymph node(s) were assessed, and the percutaneous CNB of the SLN was performed with ultrasound guidance under local anesthesia (Figure 2). The CNB was evaluated for the presence of brown staining and magnetic uptake with the SentiMag probe (Figure 3). If more than one pathological lymph nodes were identified at this stage, the protocol stated that multiple efforts could be performed only after patient consent; otherwise, if the bioptic material obtained was considered representative and adequate, only the most prominent node was biopsied. Standard histopathologic analyses to assess metastasis was also performed, including verification of SPIO presence in the SLN. In patients undergoing NAT, the SLN was clipped simultaneously after the CNB, at the same session. When CNB was completed, the area was scanned for bleeding.

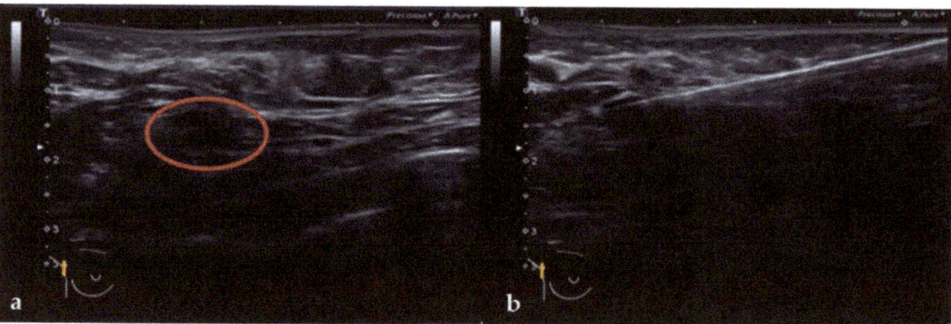

Figure 2. (**a,b**). MagUS with the SLN visualized in the red circle (**left**). Magnetic probe localizes the magnetic "hotspot" and after that CNB is performed (**right**). Monitor width 3.9 cm.

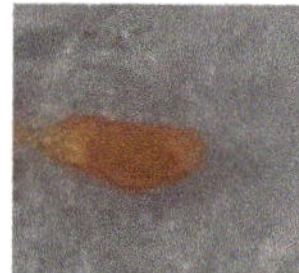

Figure 3. MagUS SLN-biopsy specimen (size 1 cm).

The study protocol ruled that the first five patients would undergo axillary MRI-LG before and after SPIO administration, and that MagUS and CNB was performed in the operation theatre, after the induction of anesthesia and right before surgery. In cases of recurrent breast cancer with aberrant SLN localization on MRI-LG and MagUS, a decision to attempt SLND was made at the multidisciplinary conference and after discussion with the patient. In patients undergoing NAT, a new axillary MRI-LG was performed after NAT, with no subsequent SPIO injection to see whether SPIO uptake in the SLNs was still visible. The number and localization of SLNs on MRI images was documented and axillary transcutaneous SentiMag signal was recorded. During subsequent SLND, concomitant radioisotope injection was administered and during surgery we registered which SLNs were magnetic, radioactive or both as well as the signal of the clipped node with both tracers.

2.4. Surgery and Specimen Pathology

During surgery, SLND was performed and the retrieved SLNDs were controlled macroscopically and microscopically for signs of previous biopsy, hematoma or the presence of clip, if placed. Standard pathology of the SLN specimen served as a reference to the microscopical examination of the CNB.

The entire MagUs flowchart is summarized in Figure 4.

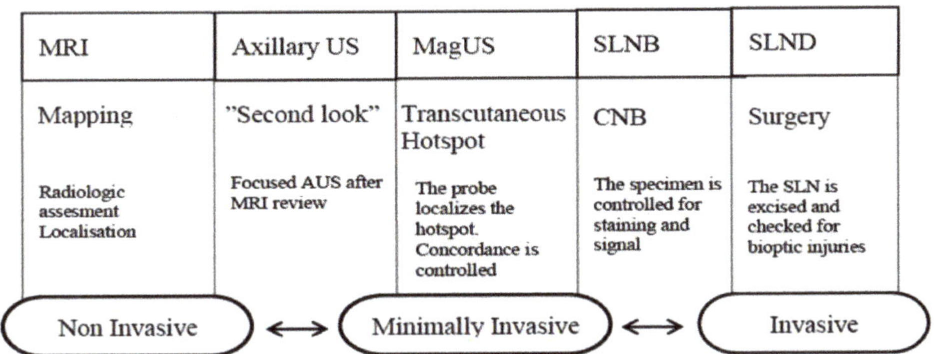

Figure 4. Flowchart showing the MagUS process.

2.5. Trial Design and Study Endpoints

To assess whether the MagUS concept has the niche to replace surgical axillary evaluation (SLND), it was necessary to ensure concordance and agreement across the different modalities. With other words, it was necessary to verify that the SLNs identified and retrieved during surgery, were the same lymph nodes visualized on the MRI and the same that were detected by the magnetic probe, identified by the ultrasound and subsequently biopsied with a core needle. The common denominator was the presence of SPIO in the node and how this is demonstrated throughout the different modalities (MRI, MagUS, Surgery). Therefore, the outcome of interest was a minimum agreement in the assessment obtained by the MRI/MagUS with the standard of care, that is surgery. For this, it was clinically relevant to assess if the technique at hand is feasible, before venturing on a large clinical trial. Subsequently, the MagUS trial was conceived as a single stage phase 2 trial following the A'Hern's design [28]. For a one-sided test a type one error a = 0.025 and 80% power, a sample size of 75 or more was required between a maximum futility proportion of 95% (corresponding to the proportion of successful detection above which the method can be further considered) and a minimum efficacy of proportion of 85% (corresponding to the proportion of successful detection under which, the method should not warrant further investigation).

The primary endpoint was determination of the MagUS SLN detection rate, defined as successful SLN detection of at least one SLN of those retrieved in the following SLND. Secondary endpoints were false-negative rate (FNR) of the MagUS technique, defined as no diagnosis of SLN metastasis (index test = negative) but presence of metastases by histopathology in any of the retrieved SLNs (reference test = positive) and overall accuracy, sensitivity, specificity and positive and negative predictive value (PPV, NPV).

Another aim of the study was to determine whether the MagUS technique could improve preoperative workup accuracy. For this, discordance in axillary evaluation from baseline clinical and ultrasonographical assessment was assessed.

Subgroup analyses were carried out to review the role of each component of the MagUS technique (MRI-LG. MagUS and MagUS core biopsy) and their potential role in tailored axillary mapping and inform on a future phase 3 trial.

The manuscript was prepared according to the Standards for the Reporting of Diagnostic Accuracy Studies (STARD) statement [29]. Descriptive statistics were performed by means of median (range) for continuous variables. Subsequently, non-parametric tests were used for comparisons. The McNemar's test was used for the assessment of discordance in paired observations. For diagnostic accuracy statistics, Receiver Operating Characteristics (ROC) curves were constructed and the area under the curve (AUC) is provided. Effect sizes are provided with 95% confidence intervals (95% CI). Data analyses were performed using SPSS (V 26.0. IBM Corp, Armonk, NY, USA) and Stata®, version 16 (StataCorp LP, College Station, TX, USA).

3. Results

The study is summarized in (Figure 5) and patient characteristics are presented in Table 1. In a total of 79 patients, 48 had early breast cancer and underwent upfront surgery, 12 underwent NAT and 19 had recurrent breast cancer after previous breast and axillary surgery.

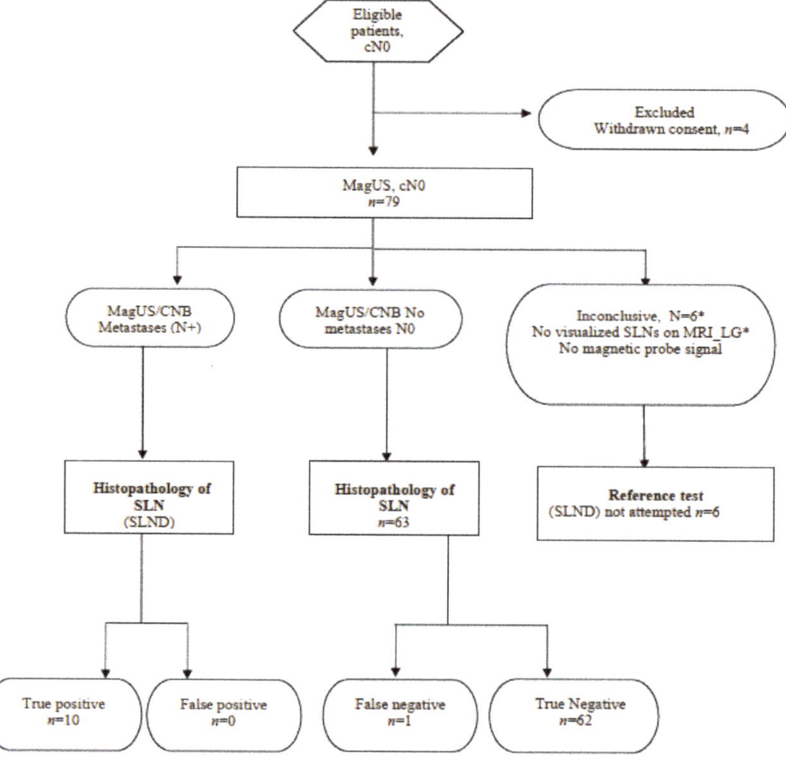

Figure 5. STARD flow diagram. * MRI_LG: Magnetic resonance imaging Lymphography. SLND: Sentinel Lymph Node Dissection.

Table 1. Patient characteristics.

Patient Characteristics.	
Patient age at operation (median, range)	64 (38–87)
Body mass index (median, range)	24.8 (19.1–43.8)
Preoperative tumor extent mm (median, range)	20 (5–120)
Days between injection and Surgery (median, range)	12 (0–140)
Laterality, number, % Right Left	 41 (51.9) 38 (48.1)
Previous breast surgery Right Left	 21 (26.6) 58 (73.4)
Previous axillary surgery Right Left	 19 (24.4) 59 (75.6)

Table 1. *Cont.*

Patient Characteristics.	
Neo adjuvant treatment	
Right	12 (15.2)
Left	67 (84.8)
Localization in the breast, number, %	
Upper outer	31 (39.2)
Upper inner	12 (15.2)
Lower outer	9 (11.4)
Lower inner	7 (8.9)
Central	7 (8.9)
Multicentric	11 (13.9)
Chest wall	2 (2.5)
Histological type ($n = 79$)	
Invasive ductal (n, (%))	66 (83.5)
Invasive lobular (n, (%))	11 (13.9)
Other Histology (n, (%))	2 (2.5)
Intrinsic Subtype ($n = 79$)	
Luminal A (n, (%))	36
Luminal B, erbb2− (n, (%))	20
Luminal B, erbb2+ (n, (%))	10
Non luminal erbb2+ (n, (%))	3
Triple negative (n, (%))	9
Type of surgery ($n = 79$)	
Wide local excision (n, (%))	28 (35.4)
Mastectomy (n, (%))	23 (29.1)
Oncoplastic breast conservation (n, (%))	28 (34.4)

MRI-LG was performed a median of 3 days after SPIO injection (range 1–12) and the MagUS with transcutaneous SLNB ± SLN clipping a median of 3 days (range 1–5) after MRI-LG. In all 73 patients where MagUS SLNB was performed, transcutaneous detection was successful and the SLN was located. Minimally invasive SLNB (MagUS CNB) retrieved lymphatic tissue with magnetic signal on the SentiMag® probe, and the presence of SPIO was confirmed on post-operative histopathology. At surgery, the node with signs of previous biopsy and/or clip was always retrieved. In one case, the lymph node that was biopsied was a non-sentinel node (i.e., ex vivo signal less than 10% of the signal of the SLN with the maximal signal), but the true SLN was just behind it and recovered during SLND.

Metastases on specimen pathology was found in 11 patients (11/73, 15.1%, 95% confidence intervals: 7.8; 25.4). MagUS identified all patients with SLN macrometastases ($n = 10$) and missed only one SLN with a micrometastasis, resulting in a FNR of 8.3% and an overall accuracy of 98.6% (Tables 2 and 3). In terms of diagnostic performance, when compared to the results of surgical pathology, MagUS performed very accurately (AUC: 0.955; 0.865, 1.000, $p < 0.001$) whereas AUS was not predictive at all (AUC: 0.505; 0.410, 0.601, $p = 0.916$).

Table 2. Comparison between MagUS and final pathology.

		Preoperative MagUS Assessment for Metastases		
		No n, (%)	Yes n, (%)	Total n, (%)
Metastases at	No	62 (98.4)	0 (0)	62 (84.9)
histopathology	Yes	1 (1.6)	10 (100)	11 (15.1)
Total		63 (100)	10 (100)	73 (100)

Mc Nemar's test, $p = 1.000$.

Table 3. Diagnostic performance of the MagUS technique.

	Rate	Lower 95% CI	Upper 95% CI
Sensitivity	90.9%	58.7%	99.8%
Specificity	100%	94.2%	100%
PPV	100%	69.1%	100%
NPV	98.4%	91.5%	99.9%
Accuracy	98.6%	92.6%	99.9%

The number of SLNs identified on MRI-LG (median 4, range 1–6) did not differ from the number of SLNs retrieved (median 3, range 1–6) (Wilcoxon signed rank test, $p = 0.331$) with high correlation (Cronbach's Alpha = 0.719; 0.481, 0.848, $p < 0.001$). Additionally, topographic concordance between MRI-LG, MagUS and SLND was 100%. In 63 patients (86%), the nodes were located medial to the lateral thoracic vein and caudal to the intercostobrachial nerve.

In patients receiving NAT, the MagUS allowed for accurate axillary mapping, identification and clipping of the true SLN prior to the initiation of NAT. After the completion of NAT, a median of 130 days (range 86–140) after SPIO injection, the SLNs were still visualized in MRI-LG and were detectable during surgery in all patients. There was excellent correlation between the number of SLNs identified on MRI (median 4, range 2–6) and the magnetic SLNs retrieved (median 3.5, range 1–6) with Cronbach's Alpha = 0.919; 0.699, 0.978, $p < 0.001$.

In patients with local recurrence after previous breast and axillary surgery ($n = 19$), MagUS showed either aberrant lymphatic outflow or no outflow in 9 patients (47.3%), preventing unnecessary ipsilateral axillary exploration. In the remaining 10 patients, both MagUS SLNB and subsequent surgery were successful.

4. Discussion

In this phase 2 trial, the MagUS technique (MRI-LG and MagUS) provided comparable results in accuracy and FNR with the standard of SLND. It was more accurate than the standard b-mode AUS in preoperatively detecting low-volume axillary disease. In this trial, it was demonstrated that accurate minimally invasive axillary staging can be achieved with a multimodal platform that can be modified to meet tailored patient needs.

SLND is not an indolent procedure and is related to short- and long-term morbidity such as postoperative pain, restricted shoulder range of motion, axillary web syndrome and lymphedema, as suggested in recent meta-analysis [13,14,30]. These findings indicate the need of establishing techniques for less invasive axillary staging that might result in less surgery, less subsequent postoperative complications and a reduction of costs and resources related with surgery [31,32]. Additionally, this MagUS workup can be performed in a wide timeframe and in an outpatient basis, as SPIO resides in the tissue a long period of time.

Recently, the necessity of surgical axillary mapping has been challenged in particular clinical scenarios. Observational data suggest that SLND may be safely omitted in older patients with primary tumors with small size and favorable biology [33–35]. The SOUND randomized trial examines whether a negative AUS can allow for the omission of SLND in patients with unifocal tumors < 2 cm planned for breast conservation and radiotherapy [36]. However, this approach does not take in consideration recent data that suggest that, in women with small tumors that are SLN negative, radiotherapy may be safely omitted nor that diagnosis of low-volume axillary disease, may allow for tailoring of radiotherapy or systemic treatment [6,9,37–39]. The results of the MagUS trial suggest that this technique may be used instead of SLND in selected cases.

It has been shown that 25% of patients considered as cN0 by AUS+/−FNAC will have a positive SLN in surgery. MagUS has the potential to correctly identify this low-volume axillary disease group, so that further treatment decisions may be tailored but without further axillary surgery, as it has been shown in landmark trials such as AMAROS,

ACOSOG Z0011 or, more recently, the RxPonder trial [6,9,40]. Reversely, in women with one positive lymph node on standard AUS, MagUS could assess the volume of axillary disease in a more accurate manner. This is a group that often harbors a higher nodal disease burden [41]. However, other studies show that this is explained by the fact that the sensitivity of AUS + FNAC increases significantly in patients with higher risk for nodal metastasis [42]. At the same time, up to 43.2% of this patient group, will be found to have two or less metastatic nodes, meaning that ALND will have been overtreatment [10]. If MagUS shows that there is only low-volume axillary disease, then the patient may have the possibility to avoid overtreatment and tailor treatment decisions may be made after discussion in the multidisciplinary meeting [43].

Subsequently, MagUS may also address issues regarding axillary staging in the setting of NAT, as it yields the potential of differentiating patients that are clinically node negative from those who are also SLN negative prior to NAT. In this manner, therapeutic decisions regarding the axilla, such as axillary radiotherapy may be better tailored, while its definitive role in this setting remains still to be elucidated [44,45]. At the same time, it may answer whether, in cN positive patients, the metastatic node is a sentinel or if, at presentation, there are non-sentinel metastases, which is suggestive of a higher axillary nodal burden. In this manner, it becomes safer to identify more appropriate potential candidates for axillary conservation post-NAT as recently suggested in the Lucerne toolbox [12]. Moreover, MRI-LG before and after NAT allows for an estimate of the number of SLNs in the axilla. This may address the problem of FNR after NAT, that has been discussed in landmark trials, such as Sentina and ACOSOG Z1071 [46–50]. In these trials FNR was shown to decrease with the removal of ≥3 nodes, including clipped nodes, if such, whereas double tracer was shown to increase detection rate [7,46–50]. In the present study, post-NAT MRI-LG showed uptake in the same SLNs, suggesting that SPIO did not migrate in higher nodal echelons during NAT. Intraoperatively, there was transcutaneous magnetic signal and SLNs were detected in all cases. It may be so that, a MagUS could be repeated after NAT to allow for more focused axillary evaluation, as standard AUS has not shown promising results in this setting [51]. As omission of axillary surgery post neoadjuvant is discussed in several breast cancer subtypes, provided that there is pathologic complete response (PCR) in the breast, MagUS could provide a safer manner to discuss omission of surgery, rather than, in case of non-PCR, performing SLND that will be subject to the risk of false negatives post NAT and after a previous excision in the breast [52,53]. A given restriction is that SPIO injection in the breast impairs the diagnostic accuracy of the MRI, suggesting that the tumor response should be performed with other modalities. Reassuringly, modalities, such as ultrasound and PET-CT have shown comparable accuracy in this setting, without the known risk of false positive findings from the MRI [54–57].

Evaluating nodal status for breast cancer after previous breast and axillary surgery is a challenge. SLN detection rate is lower and aberrant, extra-axillary lymphatic drainage is not unusual [26,58,59]. For this reason, the use of preoperative mapping by means of scintigraphy is recommended in this setting. However, whilst accurate, scintigraphy complicates logistics and this is why it recent data suggest that it is no longer necessary for patients without previous breast or axillary surgery undergoing upfront SLND [60]. MagUS has, in this setting, allowed for tailored patient treatment with flexibility, as the MRI-LG performed preoperatively, allowed in good time to know whether SLND would be attempted on the day of surgery. In this manner, logistics were facilitated, and treatment decisions could be tailored with more precision and accuracy.

The strictly controlled study design allowed for safe results, despite the absence of a control arm. However, this is a phase 2 trial and these results need to be refined and reproduced in a larger scale. Consequently, a phase 3 randomized controlled trial is needed prior to standardization and routine adaptation of the technique instead of surgical SLND. The results suggest that MagUS has the potential to provide a substantial niche to avoid axillary surgery. The cost of surgery is the most substantial, especially if one takes the expenses related with leave of absence, morbidity and complication risks into consideration.

Moreover, it is currently unclear whether the technique will always be implemented with the combination of an MRI and MagUS, something which might complicate and prolong the preoperative assessment of the patient. Finally if clinical MRI of the breast is intended, it should be performed first, to be followed by MagUs in another, different session. However, study results suggest that in women without risk factors for decreased ultrasound accuracy and transcutaneous magnetic probe detection (obesity, previous axillary surgery, etc.), MagUS and CNB were sufficient to accurately stage the axilla, suggesting that MRI is probably necessary in a small subgroup of patients (obesity, previous axillary surgery, etc.). This means that tailoring the technique to the specific patient will result in different routines and probably costs. Another substantial benefit is that this can be performed during the period between diagnosis a breast surgery, so that axillary mapping can be performed preoperatively and on an outpatient basis.

MagUS seems to be a method that can allow for alternatives to surgical axillary mapping. It comes to add to the armamentarium of other minimally invasive techniques that have previously been proposed [17,19,22,61] allowing for tailored axillary mapping in breast cancer. Its presumed advantages are the combination of different imaging modalities, together with that SPIO remains in the node a longer period, so as to allow for delayed SLND. Technique refinement and larger studies will allow for elucidation of the possibilities and its role in breast cancer diagnosis and treatment.

5. Conclusions

MagUS provides the niche for minimally invasive axillary mapping that can meet tailored patient needs and reduce diagnostic surgery. A phase 3 RCT is planned to further evaluate the technique.

Author Contributions: Conceptualization, A.K.; methodology, A.K.; software, A.K.; validation, A.K.; formal analysis, A.K.; investigation, A.J., E.P., S.A., D.V.B., F.W., S.E. and A.K.; resources, A.J., E.P., S.A. and D.V.B.; data curation, F.W. and A.K.; writing—original draft preparation, A.J.; writing—review and editing, A.J., E.P., D.V.B., S.A., F.W., H.L., S.E. and A.K.; visualization, A.J.; supervision, F.W., S.E. and A.K.; project administration, F.W. and A.K.; funding acquisition, A.K. All authors have read and agreed to the published version of the manuscript.

Funding: Institutional grants by Uppsala University and the Swedish Association for Breast Cancer ("Bröstcancerförbundet", www.brostcancerforbundet.se, ref. KDB1621/17) are acknowledged. Sponsors and funding bodies had no role in study design, data collection, analysis, or interpretation.

Institutional Review Board Statement: The study was conducted according to the guidelines of the Helsinki Declaration of ethical principles involving human subjects and was approved by Uppsala University regional ethical committee (decision number 2017/063).

Informed Consent Statement: Informed consent was obtained from all subjects involved in the study.

Data Availability Statement: The data presented in this study are available on request from the corresponding author. The data are not publicly available due to ethical considerations and data regulations.

Conflicts of Interest: The authors declare no conflict of interest.

References

1. Kim, M.K.; Park, H.S.; Kim, J.Y.; Kim, S.; Nam, S.; Park, S.; Kim, S.I. The clinical implication of the number of lymph nodes harvested during sentinel lymph node biopsy and its effects on survival outcome in patients with node-negative breast cancer. *Am. J. Surg.* **2016**, *214*, 726–732. [CrossRef] [PubMed]
2. Stenmark Tullberg, A.; Lundstedt, D.; Olofsson Bagge, R.; Karlsson, P. Positive sentinel node in luminal A-like breast cancer patients-implications for adjuvant chemotherapy? *Acta Oncol.* **2019**, *58*, 162–167. [CrossRef]
3. Veronesi, U.; Paganelli, G.; Viale, G.; Luini, A.; Zurrida, S.; Galimberti, V.; Intra, M.; Veronesi, P.; Robertson, C.; Maisonneuve, P.; et al. A Randomized Comparison of Sentinel-Node Biopsy with Routine Axillary Dissection in Breast Cancer. *N. Engl. J. Med.* **2003**, *349*, 546–553. [CrossRef] [PubMed]

4. Krag, D.N.; Anderson, S.; Julian, T.B.; Brown, A.M.; Harlow, S.P.; Costantino, J.P.; Ashikaga, T.; Weaver, D.L.; Mamounas, E.P.; Jalovec, L.M.; et al. Sentinel-lymph-node resection compared with conventional axillary-lymph-node dissection in clinically node-negative patients with breast cancer: Overall survival findings from the NSABP B-32 randomised phase 3 trial. *Lancet Oncol.* **2010**, *11*, 927–933. [CrossRef]

5. Morrow, M.; Jagsi, R.; McLeod, M.C.; Shumway, D.; Katz, S.J. Surgeon Attitudes Toward the Omission of Axillary Dissection in Early Breast Cancer. *JAMA Oncol.* **2018**, *4*, 1511–1516. [CrossRef]

6. Donker, M.; van Tienhoven, G.; Straver, M.E.; Meijnen, P.; van de Velde, C.J.H.; Mansel, R.E.; Cataliotti, L.; Westenberg, A.H.; Klinkenbijl, J.H.G.; Orzalesi, L.; et al. Radiotherapy or surgery of the axilla after a positive sentinel node in breast cancer (EORTC 10981-22023 AMAROS): A randomised, multicentre, open-label, phase 3 non-inferiority trial. *Lancet Oncol.* **2014**, *15*, 1303–1310. [CrossRef]

7. Boughey, J.C.; Suman, V.J.; Mittendorf, E.A.; Ahrendt, G.M.; Wilke, L.G.; Taback, B.; Leitch, A.M.; Kuerer, H.M.; Bowling, M.; Flippo-Morton, T.S.; et al. Sentinel lymph node surgery after neoadjuvant chemotherapy in patients with node-positive breast cancer: The ACOSOG Z1071 (Alliance) clinical trial. *JAMA* **2013**, *310*, 1455–1461. [CrossRef] [PubMed]

8. Stearns, V.; Ewing, C.A.; Slack, R.; Penannen, M.F.; Hayes, D.F.; Tsangaris, T.N. Sentinel lymphadenectomy after neoadjuvant chemotherapy for breast cancer may reliably represent the axilla except for inflammatory breast cancer. *Ann. Surg Oncol.* **2002**, *9*, 235–242. [CrossRef] [PubMed]

9. Giuliano, A.E.; Ballman, K.V.; McCall, L.; Beitsch, P.D.; Brennan, M.B.; Kelemen, P.R.; Ollila, D.W.; Hansen, N.M.; Whitworth, P.W.; Blumencranz, P.W.; et al. Effect of Axillary Dissection vs. No Axillary Dissection on 10-Year Overall Survival Among Women With Invasive Breast Cancer and Sentinel Node Metastasis: The ACOSOG Z0011 (Alliance) Randomized Clinical Trial. *JAMA* **2017**, *318*, 918–926. [CrossRef]

10. Ahmed, M.; Jozsa, F.; Baker, R.; Rubio, I.T.; Benson, J.; Douek, M. Meta-analysis of tumour burden in pre-operative axillary ultrasound positive and negative breast cancer patients. *Breast Cancer Res. Treat.* **2017**, *166*, 329–336; Erratum in **2017**, *166*, 337. [CrossRef]

11. Caudle, A.S.; Yang, W.T.; Krishnamurthy, S.; Mittendorf, E.A.; Black, D.M.; Gilcrease, M.Z.; Bedrosian, I.; Hobbs, B.P.; DeSnyder, S.M.; Hwang, R.F.; et al. Improved Axillary Evaluation Following Neoadjuvant Therapy for Patients With Node-Positive Breast Cancer Using Selective Evaluation of Clipped Nodes: Implementation of Targeted Axillary Dissection. *J. Clin. Oncol.* **2016**, *34*, 1072–1078. [CrossRef]

12. Dubsky, P.; Pinker, K.; Cardoso, F.; Montagna, G.; Ritter, M.; Denkert, C.; Rubio, I.T.; de Azambuja, E.; Curigliano, G.; Gentilini, O.; et al. Breast conservation and axillary management after primary systemic therapy in patients with early-stage breast cancer: The Lucerne toolbox. *Lancet Oncol.* **2021**, *22*, e18–e28. [CrossRef]

13. Lucci, A.; McCall, L.M.; Beitsch, P.D.; Whitworth, P.W.; Reintgen, D.S.; Blumencranz, P.W.; Leitch, A.M.; Saha, S.; Hunt, K.K.; Giuliano, A.E. Surgical Complications Associated With Sentinel Lymph Node Dissection (SLND) Plus Axillary Lymph Node Dissection Compared With SLND Alone in the American College of Surgeons Oncology Group Trial Z0011. *J. Clin. Oncol.* **2007**, *25*, 3657–3663. [CrossRef] [PubMed]

14. Mansel, R.E.; Fallowfield, L.; Kissin, M.; Goyal, A.; Newcombe, R.G.; Dixon, J.M.; Yiangou, C.; Horgan, K.; Bundred, N.; Monypenny, I.; et al. Randomized Multicenter Trial of Sentinel Node Biopsy Versus Standard Axillary Treatment in Operable Breast Cancer: The ALMANAC Trial. *J. Natl. Cancer Inst.* **2006**, *98*, 599–609. [CrossRef] [PubMed]

15. Verbelen, H.; Tjalma, W.; Meirte, J.; Gebruers, N. Long-term morbidity after a negative sentinel node in breast cancer patients. *Eur. J. Cancer Care* **2019**, *28*, e13077. [CrossRef] [PubMed]

16. Verbelen, H.; Gebruers, N.; Eeckhout, F.-M.; Verlinden, K.; Tjalma, W. Shoulder and arm morbidity in sentinel node-negative breast cancer patients: A systematic review. *Breast Cancer Res. Treat.* **2014**, *144*, 21–31. [CrossRef]

17. Navalkissoor, S.; Wagner, T.; Gnanasegaran, G.; Buscombe, J. SPECT/CT in imaging sentinel nodes. *Clin. Transl. Imaging* **2015**, *3*, 203–215. [CrossRef]

18. Nakagawa, M.; Morimoto, M.; Takechi, H.; Tadokoro, Y.; Tangoku, A. Preoperative diagnosis of sentinel lymph node (SLN) metastasis using 3D CT lymphography (CTLG). *Breast Cancer* **2015**, *23*, 519–524. [CrossRef]

19. Cox, K.; Taylor-Phillips, S.; Sharma, N.; Weeks, J.; Mills, P.; Sever, A.; Lim, A.; Haigh, I.; Hashem, M.; De Silva, T.; et al. Enhanced pre-operative axillary staging using intradermal microbubbles and contrast-enhanced ultrasound to detect and biopsy sentinel lymph nodes in breast cancer: A potential replacement for axillary surgery. *Br. J. Radiol.* **2017**, *91*, 20170626. [CrossRef] [PubMed]

20. Teshome, M.; Wei, C.; Hunt, K.K.; Thompson, A.; Rodriguez, K.; Mittendorf, E.A. Use of a Magnetic Tracer for Sentinel Lymph Node Detection in Early-Stage Breast Cancer Patients: A Meta-analysis. *Ann. Surg. Oncol.* **2016**, *23*, 1508–1514. [CrossRef]

21. Karakatsanis, A.; Christiansen, P.M.; Fischer, L.; Hedin, C.; Pistioli, L.; Sund, M.; Rasmussen, N.R.; Jørnsgård, H.; Tegnelius, D.; Eriksson, S.; et al. The Nordic SentiMag trial: A comparison of super paramagnetic iron oxide (SPIO) nanoparticles versus Tc(99) and patent blue in the detection of sentinel node (SN) in patients with breast cancer and a meta-analysis of earlier studies. *Breast Cancer Res. Treat.* **2016**, *157*, 281–294. [CrossRef]

22. Motomura, K.; Izumi, T.; Tateishi, S.; Tamaki, Y.; Ito, Y.; Horinouchi, T.; Nakanishi, K. Superparamagnetic iron oxide-enhanced MRI at 3 T for accurate axillary staging in breast cancer. *Br. J. Surg.* **2016**, *103*, 60–69. [CrossRef] [PubMed]

23. Karakatsanis, A.; Hersi, A.; Pistiolis, L.; Bagge, R.O.; Lykoudis, P.M.; Eriksson, S.; Wärnberg, F.; Nagy, G.; Mohammed, I.; Sundqvist, M.; et al. Effect of preoperative injection of superparamagnetic iron oxide particles on rates of sentinel lymph node

dissection in women undergoing surgery for ductal carcinoma in situ (SentiNot study). *J. Br. Surg.* **2019**, *106*, 720–728. [CrossRef] [PubMed]

24. Ahmed, M.; Purushotham, A.D.; Douek, M. Novel techniques for sentinel lymph node biopsy in breast cancer: A systematic review. *Lancet Oncol.* **2014**, *15*, e351–e362. [CrossRef]

25. Karakatsanis, A.; Daskalakis, K.; Stålberg, P.; Olofsson, H.; Andersson, Y.; Eriksson, S.; Bergkvist, L.; Wärnberg, F. Superparamagnetic iron oxide nanoparticles as the sole method for sentinel node biopsy detection in patients with breast cancer. *J. Br. Surg.* **2017**, *104*, 1675–1685. [CrossRef] [PubMed]

26. Ahmed, M.; Baker, R.; Rubio, I.T. Meta-analysis of aberrant lymphatic drainage in recurrent breast cancer. *BJS* **2016**, *103*, 1579–1588. [CrossRef]

27. Clough, K.B.; Nasr, R.; Nos, C.; Vieira, M.; Inguenault, C.; Poulet, B. New anatomical classification of the axilla with implications for sentinel node biopsy. *Br. J. Surg.* **2010**, *97*, 1659–1665. [CrossRef]

28. Neven, A.; Mauer, M.; Hasan, B.; Sylvester, R.; Collette, L. Sample size computation in phase II designs combining the A'Hern design and the Sargent and Goldberg design. *J. Biopharm. Stat.* **2020**, *30*, 305–321. [CrossRef]

29. Bossuyt, P.M.; Reitsma, J.B.; Bruns, D.E.; Gatsonis, C.A.; Glasziou, P.P.; Irwig, L.; Lijmer, J.G.; Moher, D.; Rennie, D.; de Vet, H.C.W.; et al. For the STARD Group. STARD 2015: An Updated List of Essential Items for Reporting Diagnostic Accuracy Studies. *BMJ* **2015**, *351*, h5527. [CrossRef] [PubMed]

30. Pilger, T.L.; Francisco, D.F.; Candido Dos Reis, F.J. Effect of sentinel lymph node biopsy on upper limb function in women with early breast cancer: A systematic review of clinical trials. *Eur. J. Surg. Oncol.* **2021**, *30*, 1497–1506. [CrossRef]

31. Boughey, J.C.; Moriarty, J.P.; Degnim, A.C.; Gregg, M.S.; Egginton, J.S.; Long, K.H. Cost modeling of preoperative axillary ultrasound and fine-needle aspiration to guide surgery for invasive breast cancer. *Ann. Surg. Oncol.* **2010**, *17*, 953–958. [CrossRef] [PubMed]

32. Turaga, K.K.; Chau, A.; Eatrides, J.M.; Kiluk, J.V.; Khakpour, N.; Laronga, C.; Lee, M.C. Selective application of routine preoperative axillary ultrasonography reduces costs for invasive breast cancers. *Oncologist* **2011**, *16*, 942–948. [CrossRef] [PubMed]

33. Chung, A.; Gangi, A.; Amersi, F.; Zhang, X.; Guiliano, A. Not Performing a Sentinel Node Biopsy for Older Patients with Early-Stage Invasive Breast Cancer. *JAMA Surg.* **2015**, *150*, 683–684. [CrossRef]

34. Ingvar, C.; Ahlgren, J.; Emdin, S.; Lofgren, L.; Nordander, M.; Nimeus, E.; Arnesson, L.-G. Long-term outcome of pT1a-b, cN0 breast cancer without axillary dissection or staging: A prospective observational study of 1543 women. *Br. J. Surg.* **2020**, *107*, 1299–1306. [CrossRef]

35. O'Connell, R.L.; Rusby, J.E.; Stamp, G.F.W.; Conway, A.; Roche, N.; Barry, P.; Khabra, K.; Bonomi, R.; Rapisarda, I.F.; Della Rovere, G.Q. Long term results of treatment of breast cancer without axillary surgery–predicting a sound approach? *Eur. J. Surg. Oncol.* **2016**, *42*, 942–948. [CrossRef]

36. Sentinel Node Vs Observation after Axillary Ultra-souND (SOUND). Available online: https://clinicaltrials.gov/ct2/show/NCT02167490 (accessed on 18 May 2021).

37. Hughes, K.S.; Schnaper, L.A.; Bellon, J.R.; Cirrincione, C.T.; Berry, D.A.; McCormick, B.; Muss, H.B.; Smith, B.L.; Hudis, C.A.; Winer, E.P.; et al. Lumpectomy plus tamoxifen with or without irradiation in women age 70 years or older with early breast cancer: Long-term follow-up of CALGB 9343. *J. Clin. Oncol.* **2013**, *31*, 2382–2387. [CrossRef]

38. Kunkler, I.H.; Williams, L.J.; Jack, W.J.; Cameron, D.A.; Dixon, J.M.; PRIME II Investigators. Breast-conserving surgery with or without irradiation in women aged 65 years or older with early breast cancer (PRIME II): A randomised controlled trial. *Lancet Oncol.* **2015**, *16*, 266–273. [CrossRef]

39. Wickberg, Å.; Liljegren, G.; Killander, F.; Lindman, H.; Björhle, J.; Carlberg, M.; Blomqvist, C.; Ahlgren, J.; Villman, K. Omitting radiotherapy in women ≥ 65 years with low-risk early breast cancer after breast-conserving surgery and adjuvant endocrine therapy is safe. *Eur. J. Surg. Oncol.* **2018**, *44*, 951–956. [CrossRef] [PubMed]

40. Kalinsky, K.; Barlow, W.E.; Meric-Bernstam, F.; Gralow, J.R.; Albain, K.S.; Hayes, D.; Lin, N.; Perez, E.A.; Goldstein, L.J.; Chia, S.; et al. First results from a phase III randomized clinical trial of standard adjuvant endocrine therapy (ET) +/- chemotherapy (CT) in patients (pts) with 1-3 positive nodes, hormone receptor-positive (HR+) and HER2-negative (HER2-) breast cancer (BC) with recurrence score (RS) < 25: SWOG S1007 (RxPonder). In Proceedings of the 2020 San Antonio Breast Cancer Virtual Symposium, San Antonio, TX, USA, 8–11 December 2020. Abstract nr GS3-00.

41. Van Wely, B.J.; de Wilt, J.H.; Francissen, C.; Teerenstra, S.; Strobbe, L.J. Meta-analysis of ultrasound-guided biopsy of suspicious axillary lymph nodes in the selection of patients with extensive axillary tumour burden in breast cancer. *Br. J. Surg.* **2015**, *102*, 159–168. [CrossRef] [PubMed]

42. Houssami, N.; Ciatto, S.; Turner, R.M.; Cody, H.S., 3rd; Macaskill, P. Preoperative ultrasound-guided needle biopsy of axillary nodes in invasive breast cancer: Meta-analysis of its accuracy and utility in staging the axilla. *Ann. Surg.* **2011**, *254*, 243–251. [CrossRef]

43. Caudle, A.; Hunt, K.K.; Kuerer, H.M.; Meric-Berstein, F.; Lucci, A.; Bedrosian, I.; Babiera, G.V.; Hwang, R.F.; Ross, M.I.; Feig, B.W.; et al. Multidisciplinary considerations in the implementation of the findings from the American College of Surgeons Oncology Group (ACOSOG) Z0011 study: A practice-changing trial. *Ann. Surg. Oncol.* **2011**, *18*, 2407–2412. [CrossRef]

44. Available online: https://clinicaltrials.gov/ct2/show/NCT01901094 (accessed on 18 May 2021).

45. Standard or Comprehensive Radiation Therapy in Treating Patients With Early-Stage Breast Cancer Previously Treated with Chemotherapy and Surgery. Available online: https://clinicaltrials.gov/ct2/show/NCT01872975 (accessed on 18 May 2021).

46. Boughey, J.C.; Ballman, K.V.; Le-Petross, H.T.; McCall, L.M.; Mittendorf, E.A.; Ahrendt, G.M.; Wilke, L.G.; Taback, B.; Feliberti, E.C.; Hunt, K.K. Identification and Resection of Clipped Node Decreases the False-negative Rate of Sentinel Lymph Node Surgery in Patients Presenting With Node-positive Breast Cancer (T0–T4, N1–N2) Who Receive Neoadjuvant Chemotherapy: Results From ACOSOG Z1071 (Alliance). *Ann. Surg.* **2016**, *263*, 802–807. [CrossRef]

47. Haffty, B.G.; McCall, L.M.; Ballman, K.V.; Buchholz, T.A.; Hunt, K.K.; Boughey, J.C. Impact of Radiation on Locoregional Control in Women with Node-Positive Breast Cancer Treated with Neoadjuvant Chemotherapy and Axillary Lymph Node Dissection: Results from ACOSOG Z1071 Clinical Trial. *Int. J. Radiat. Oncol. Biol. Phys.* **2019**, *105*, 174–182. [CrossRef] [PubMed]

48. Haffty, B.G.; McCall, L.M.; Ballman, K.V.; McLaughlin, S.; Jagsi, R.; Ollila, D.W.; Hunt, K.K.; Buchholz, T.A.; Boughey, J.C. Patterns of Local-Regional Management Following Neoadjuvant Chemotherapy in Breast Cancer: Results from ACOSOG Z1071 (Alliance). *Int. J. Radiat. Oncol. Biol. Phys.* **2016**, *94*, 493–502. [CrossRef]

49. Kuemmel, S.; Heil, J.; Rueland, A.; Seiberling, C.; Harrach, H.; Schindowski, D.; Lubitz, J.; Hellerhoff, K.; Ankel, C.; Graßhoff, S.T.; et al. A Prospective, Multicenter Registry Study to Evaluate the Clinical Feasibility of Targeted Axillary Dissection (TAD) in Node-Positive Breast Cancer Patients. *Ann. Surg.* **2020**. [Epub ahead of print]. [CrossRef] [PubMed]

50. Kuehn, T.; Bauerfeind, I.; Fehm, T.; Fleige, B.; Hausschild, M.; Helms, G.; Lebeau, A.; Liedtke, C.; von Minckwitz, G.; Nekljudova, V.; et al. Sentinel-lymph-node biopsy in patients with breast cancer before and after neoadjuvant chemotherapy (SENTINA): A prospective, multicentre cohort study. *Lancet Oncol.* **2013**, *14*, 609–618. [CrossRef]

51. Schwentner, L.; Helms, G.; Nekljudova, V.; Ataseven, B.; Bauerfeind, I.; Ditsch, N.; Fehm, T.; Fleige, B.; Hauschild, M.; Heil, J.; et al. Using ultrasound and palpation for predicting axillary lymph node status following neoadjuvant chemotherapy-Results from the multi-center SENTINA trial. *Breast* **2017**, *31*, 202–207. [CrossRef] [PubMed]

52. van der Noordaa, M.E.M.; van Duijnhoven, F.H.; Cuijpers, F.N.E.; van Werkhoven, E.; Wiersma, T.G.; Elkhuizen, P.H.M.; Winter-Warnars, G.; Dezentje, V.; Sonke, G.S.; Groen, E.J.; et al. Toward omitting sentinel lymph node biopsy after neoadjuvant chemotherapy in patients with clinically node-negative breast cancer. *Br. J. Surg.* **2020**, *108*, 667–674. [CrossRef] [PubMed]

53. Reimer, T.; Glass, A.; Botteri, E.; Loibl, S.; DGentilini, O. Avoiding Axillary Sentinel Lymph Node Biopsy after Neoadjuvant Systemic Therapy in Breast Cancer: Rationale for the Prospective, Multicentric EUBREAST-01 Trial. *Cancers* **2020**, *12*, 3698. [CrossRef] [PubMed]

54. Lee, M.C.; Gonzalez, S.J.; Lin, H.; Zhao, X.; Kiluk, J.V.; Laronga, C.; Mooney, B. Prospective trial of breast MRI versus 2D and 3D ultrasound for evaluation of response to neoadjuvant chemotherapy. *Ann. Surg. Oncol.* **2015**, *22*, 2888–2894. [CrossRef] [PubMed]

55. Marinovich, M.L.; Macaskill, P.; Irwig, L.; Sardanelli, F.; von Minckwitz, G.; Mamounas, E.; Brennan, M.; Ciatto, S.; Houssami, N. Meta-analysis of agreement between MRI and pathologic breast tumour size after neoadjuvant chemotherapy. *Br. J. Cancer* **2013**, *109*, 1528–1536. [CrossRef] [PubMed]

56. Sheikhbahaei, S.; Trahan, T.J.; Xiao, J.; Taghipour, M.; Mena, E.; Connolly, R.M.; Subramaniam, R.M. FDG-PET/CT and MRI for Evaluation of Pathologic Response to Neoadjuvant Chemotherapy in Patients With Breast Cancer: A Meta-Analysis of Diagnostic Accuracy Studies. *Oncologist* **2016**, *21*, 931–939. [CrossRef] [PubMed]

57. Chen, L.; Yang, Q.; Bao, J.; Liu, D.; Huang, X.; Wang, J. Direct comparison of PET/CT and MRI to predict the pathological response to neoadjuvant chemotherapy in breast cancer: A meta-analysis. *Sci. Rep.* **2017**, *7*, 8479. [CrossRef]

58. Ugras, S.; Matsen, C.; Eaton, A.; Stempel, M.; Morrow, M.; Cody, H.S., 3rd. Reoperative Sentinel Lymph Node Biopsy is Feasible for Locally Recurrent Breast Cancer, But is it Worthwhile? *Ann. Surg. Oncol.* **2016**, *23*, 744–748. [CrossRef] [PubMed]

59. Vugts, G.; Maaskant-Braat, A.J.; Voogd, A.C.; van Riet, Y.E.; Luiten, E.J.; Rutgers, E.J.; Rutten, H.J.; Roumen, R.M.; Nieuwenhuijzen, G.A. Repeat sentinel node biopsy should be considered in patients with locally recurrent breast cancer. *Breast Cancer Res. Treat.* **2015**, *153*, 549–556. [CrossRef]

60. Kuemmel, S.; Holtschmidt, J.; Gerber, B.; Von der Assen, A.; Heil, J.; Thill, M.; Krug, D.; Schem, C.; Denkert, C.; Lubitz, J.; et al. Prospective, Multicenter, Randomized Phase III Trial Evaluating the Impact of Lymphoscintigraphy as Part of Sentinel Node Biopsy in Early Breast Cancer: SenSzi (GBG80) Trial. *J. Clin. Oncol.* **2019**, *37*, 1490–1498. [CrossRef] [PubMed]

61. Motomura, K.; Izumi, T.; Tateishi, S.; Sumino, H.; Noguchi, A.; Horinouchi, T.; Nakanishi, K. Correlation between the area of high-signal intensity on SPIO-enhanced MR imaging and the pathologic size of sentinel node metastases in breast cancer patients with positive sentinel nodes. *BMC Med. Imaging* **2013**, *13*, 32. [CrossRef] [PubMed]

Review

Application of Magnetic Nanoparticles for Rapid Detection and In Situ Diagnosis in Clinical Oncology

Tatsuya Onishi [1,†], Kisyo Mihara [2,†], Sachiko Matsuda [3,*], Satoshi Sakamoto [4], Akihiro Kuwahata [5], Masaki Sekino [6], Moriaki Kusakabe [7,8], Hiroshi Handa [9] and Yuko Kitagawa [3]

[1] Department of Breast Surgery, National Cancer Center Hospital East, 6-5-1, Kashiwanoha, Kashiwa 277-8577, Chiba, Japan; t.onishi77@gmail.com
[2] Department of Surgery, Kawasaki Municipal Kawasaki Hospital, Kawasaki-ku, Kawasaki 210-0013, Kanagawa, Japan; kisyomihara@gmail.com
[3] Department of Surgery, School of Medicine, Keio University, 35 Shinanomachi, Shinjuku-ku, Tokyo 160-8582, Japan; kitagawa@a3.keio.jp
[4] School of Life Science and Technology, Tokyo Institute of Technology, 4259 Nagatsuta-cho, Midori-ku, Yokohama 226-8501, Kanagawa, Japan; ssakamoto@bio.titech.ac.jp
[5] Graduate School of Engineering, Tohoku University, 6-6-05 Aoba, Aramaki-aza, Aoba-ku, Sendai 980-8579, Miyagi, Japan; akihiro.kuwahata.b1@tohoku.ac.jp
[6] Graduate School of Engineering, The University of Tokyo, 7-3-1 Hongo, Bunkyo-ku, Tokyo 113-8656, Japan; sekino@g.ecc.u-tokyo.ac.jp
[7] Graduate School of Agricultural and Life Sciences, Research Center for Food Safety, The University of Tokyo, 1-1-1 Yayoi, Bunkyo-ku, Tokyo 113-8657, Japan; kusabmrl@gmail.com
[8] Matrix Cell Research Institute Inc., 1-35-3 Kamikashiwada, Ushiku 300-1232, Ibaraki, Japan
[9] Department of Nanoparticle Translational Research, Tokyo Medical University, 6-1-1 Shinjuku, Shinjuku-ku, Tokyo 160-8402, Japan; hhanda@tokyo-med.ac.jp
[*] Correspondence: matsudasachio@keio.jp; Tel.: +81-3-3353-1211
[†] These authors contributed equally to this work.

Citation: Onishi, T.; Mihara, K.; Matsuda, S.; Sakamoto, S.; Kuwahata, A.; Sekino, M.; Kusakabe, M.; Handa, H.; Kitagawa, Y. Application of Magnetic Nanoparticles for Rapid Detection and In Situ Diagnosis in Clinical Oncology. *Cancers* **2022**, *14*, 364. https://doi.org/10.3390/cancers14020364

Academic Editor: Manfred Ogris

Received: 14 December 2021
Accepted: 10 January 2022
Published: 12 January 2022

Publisher's Note: MDPI stays neutral with regard to jurisdictional claims in published maps and institutional affiliations.

Simple Summary: Screening, monitoring, and diagnostic methods in oncology are a critical part of treatment. The currently used clinical methods have limitations, most notably the time, cost, and special facilities required for radioisotope-based techniques. The use of magnetic nanoparticles is an alternative approach that offers faster analyses with safer materials over a wide range of oncological applications, such as the detection of cancer biomarkers and immunostaining. Furthermore, magnetic nanoparticles, such as superparamagnetic iron oxide nanoparticles, can detect sentinel lymph nodes for breast cancer in a clinical setting, as well as those for gallbladder cancer in animal models within a timeframe that would enable them to be used during surgery with a magnetic probe.

Abstract: Screening, monitoring, and diagnosis are critical in oncology treatment. However, there are limitations with the current clinical methods, notably the time, cost, and special facilities required for radioisotope-based methods. An alternative approach, which uses magnetic beads, offers faster analyses with safer materials over a wide range of oncological applications. Magnetic beads have been used to detect extracellular vesicles (EVs) in the serum of pancreatic cancer patients with statistically different EV levels in preoperative, postoperative, and negative control samples. By incorporating fluorescence, magnetic beads have been used to quantitatively measure prostate-specific antigen (PSA), a prostate cancer biomarker, which is sensitive enough even at levels found in healthy patients. Immunostaining has also been incorporated with magnetic beads and compared with conventional immunohistochemical methods to detect lesions; the results suggest that immunostained magnetic beads could be used for pathological diagnosis during surgery. Furthermore, magnetic nanoparticles, such as superparamagnetic iron oxide nanoparticles (SPIONs), can detect sentinel lymph nodes in breast cancer in a clinical setting, as well as those in gallbladder cancer in animal models, in a surgery-applicable timeframe. Ultimately, recent research into the applications of magnetic beads in oncology suggests that the screening, monitoring, and diagnosis of cancers could be improved and made more accessible through the adoption of this technology.

Keywords: magnetic nanoparticles; in situ diagnosis; rapid detection; extracellular vesical quantification; presurgical screening; pathological diagnosis; sentinel node mapping

1. Introduction

Magnetic nanoparticles (MNPs) have recently been applied to life sciences as well as clinical settings. MNPs comprise aggregates of iron oxide (FeO, Fe_2O_3, and Fe_3O_4) or ferrite particles (which contain iron oxide as the main component) in the nanometer order, which are dispersed or embedded in polymers, such as polysaccharide, polystyrene, silica, and agarose [1]. Their application to life science research stems from the ability to separate, guide, and detect MNPs using magnetic fields. Additionally, MNPs can be processed to furnish their surface with a variety of functions. Recognition sites, such as functional groups and biomolecules, are immobilized on the surface of the beads and are used to recognize targets for separation or detection [1]. The physical size and magnetization strength of the beads are roughly proportional to the number of iron oxide particles in the polymer. Protein purification and cell separation applications require strong magnetic particles, whereby micro-sized magnetic particles are used with a magnetic field [2]. For stem cell differentiation experiments and gene transfer applications, small magnetic particles (<100 nm) are generally used [3]. Furthermore, some nanosized magnetic particles, such as superparamagnetic iron oxide nanoparticles (SPIONs), are biocompatible and can be used internally in magnetic resonance imaging (MRI) contrast media for the liver [4].

We focused on the applications of MNPs in oncology from a surgeon's perspective when monitoring biomarkers before and after surgery, and for intraoperative diagnosis during surgery (Figure 1). In this review, we provide an overview of the application of MNPs in oncology.

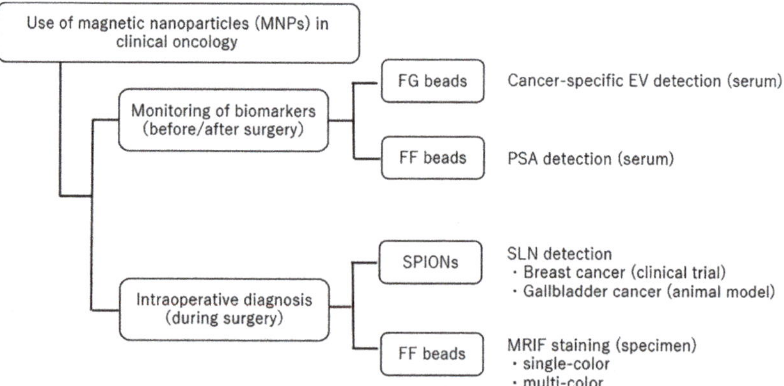

Figure 1. The concept of this review. The usage of magnetic nanoparticles (MNPs) was divided into two objectives: monitoring of biomarkers (before/after surgery) and intraoperative diagnosis (during surgery). The types of MNPs and examples of their use are indicated. FG, ferrite and glycidyl methacrylate; FF, fluorescent FG; SPIONs, superparamagnetic iron oxide nanoparticles; EVs, extracellular vesicles; PSA, prostate-specific antigen; SLN, sentinel lymph node; MRIF, magnetically promoted rapid immunofluorescence.

2. Monitoring Biomarkers before and after Surgery

In oncological clinical settings, early detection and accurate diagnosis are important for cancer treatment, both before and after surgery. Enzyme-linked immuno-sorbent assay (ELISA) [5–7], which uses antigen–antibody reactions as its detection mechanism, is widely used to detect cancer biomarkers in serum for screening or monitoring before surgery, but the enzymatic method is time consuming. However, the MNP method accelerates the antigen–antibody reaction. This is a different mechanism to magnetic separation, in which antibody-immobilized MNPs can be attracted to immobilized antigen via a magnetic field. In this section, we describe two examples of the MNP method for biomarker detection: pancreas cancer-specific extracellular vesicles (EVs) using ferrite and glycidyl methacrylate (FG) beads and prostate-specific antigen (PSA) using fluorescent FG (FF) beads.

2.1. Measuring a Biomarker in Serum Using FG Beads

2.1.1. FG Beads

Handa's group initially developed affinity latex beads, styrene–GMA (SG) beads, which have a polystyrene core and glycidyl methacrylate (GMA) on their surface, known as poly GMA beads (Figure 2A left) [8]. Poly GMA beads have epoxy groups that can immobilize proteins, nucleic acids, and low-molecular-weight compounds. Additionally, the group found that carboxyl and thiol groups bind to the ferrite surface [9,10]. On the basis of these findings, 35–40 nm ferrite was coupled with the adaptor molecule and then coated with a copolymer of styrene and GMA, followed by coating with GMA [10] to generate the FG beads (Figure 2A, middle). FG beads have a 200 nm diameter with several encapsulated ferrite nanoparticles. Similar to the SG beads, specific ligands can be bound to the GMA surface to enable it to bind target molecules (Figure 2A, right). Because of the ferrite core, it can then be attracted or separated using magnetic forces.

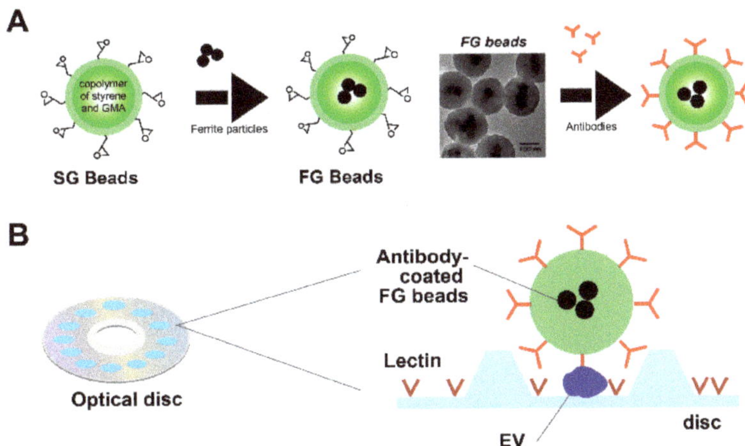

Figure 2. (**A**): Construction of SG and ferrite and glycidyl methacrylate (FG) beads. SG beads are composed of styrene and glycidyl methacrylate (GMA) (left). FG beads are prepared with surface-modified ferrite particles, styrene, and GMA (middle). Transmission electron microscopy image is shown (middle). Antibodies can be immobilized on the surface of FG beads (right). Modified from Inomata et al. and Nishino et al. (**B**): Schematic image of the quantification of extracellular vesicles (EVs). Candidate lectins were coated on the optical disc of the ExoCounter system. Lectin-binding EVs in the sera of pancreatic cancer patient or cell lines were captured on the disc and labeled with anti-CD9 Ab-conjugated nanoparticles. The absolute numbers of labeled EVs were quantified using the optical disc drive of the ExoCounter. Modified from Yokose et al.

2.1.2. Screening or Monitoring of EVs with FG Beads

EVs are granular substances with a diameter of 50–150 nm, and they are secreted by cells [11,12]. Lipids and proteins derived from cell membranes are contained on the surface of EVs, and inside the EVs are intracellular substances, such as nucleic acids [13] (including microRNA, messenger RNA, and DNA) and proteins [14]. Recently, it has been suggested that EVs are involved in cancer development. EVs released from cancer cells are known to function in ways that favor cancer cells, such as cell survival, malignant transformation, and metastasis.

The surface proteins on EVs reflect parental cells, such as CD147 from colorectal cancer cells [15,16], human epidermal growth factor receptor 2 (HER2) from breast cancer cells [17], and CD91 from lung cancer cells [18]. Therefore, measuring specific EVs released from cancer cells has potential in cancer screening and monitoring. The methods used to count EVs are mainly conventional particle-counting methods, such as nanoparticle tracking analysis [19,20] and tunable resistive pulse sensing [21,22], or labeling-detection methods, such as ELISA [18,23] and flow cytometry [14,24].

The ExoCounter system is a unique assay system that uses FG beads to count the absolute number of EVs and analyze surface proteins simultaneously. The system uses an optical disc with periodic grooves that are 160 nm wide at the bottom and 260 nm wide at the top. Individual EVs can be bound at the bottom of the groove and FG beads at the top (Figure 2B). The basic reaction mechanism is a magnetically prompted rapid sandwich immunoassay. Using an optical head based on Blu-ray disc technology, EVs modified with nanoparticles are detected one by one. The immunoassay uses antibody-coated detection FG beads and samples placed on a capture antibody- or ligand-coated optical disc (Figure 2B). A magnet is attached under the disc for 1–2 min to concentrate the FG beads onto the immobilized capture antibody or ligand, and then unbound FG beads are washed out. The captured FG beads are counted by an optical pickup composed of a laser diode and a photodetector.

The ExoCounter system has been used to analyze pancreatic cancer patient serum, in which EVs with glycoprotein are bound to Agaricus bisporus agglutinin (ABA) or Amaranthus caudatus agglutinin (ACA) using CD9 antibody-coated FG beads to detect EVs on an ABA- or ACA-coated disc [25]. Using this method, EVs that have a carbohydrate chain that binds to ABA or ACA can be detected. EV quantification was performed on 90 samples from pancreatic cancer patients (68 preoperative and 22 postoperative samples) and 77 negative control serum samples [25]. The ABA-binding and ACA-binding EVs were significantly higher in the preoperative pancreatic cancer patients than in the negative controls ($p < 0.001$ and $p < 0.001$, respectively) (Figure 3) [25]. Furthermore, the number of labeled EVs was significantly reduced in the post-pancreatectomy sera, almost to the same level as that of the negative controls ($p < 0.001$ and $p < 0.001$, respectively) (Figure 3) [25]. The measurement that captures the characteristics of EVs is quite unique.

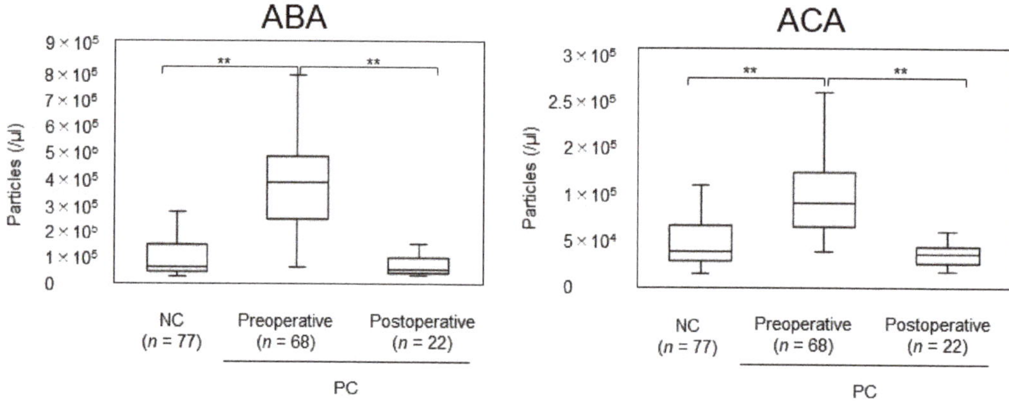

Figure 3. Quantification of ABA- and ACA-positive EVs from the sera of preoperative and postoperative pancreatic cancer patients and negative controls. Patient sera were analyzed using ABA- or ACA-coated discs and anti-CD9 Ab-conjugated beads with ExoCounter. Adapted from Yokose et al. ** $p < 0.01$.

2.2. Measuring a Biomarker in Serum Using FF Beads

2.2.1. FF Beads

The next generation of FG beads is fluorescent FG beads (FF beads). Generally, fluorescent substances are immobilized on the polymer surface by covalency or affinity. However, a unique feature of FF beads is that fluorescent substances, such as europium complexes (Eu (TTA)$_3$ (TOPO)$_2$), can be encapsulated. Europium complexes emit fluorescence at 618 nm under light excitation at 340 nm. FG beads are tolerant to several organic solvents and expand or shrink depending on the type of solvent. When acetone is used, the surface polymer of FF beads swells along with the encapsulated fluorescent substance, and then returns to its original configuration in water (Figure 4). The fluorescence can be directly observed with a fluorescence detector or microscope. In addition to their magnetic attraction function, signal amplification is not necessary, which enables fast and highly sensitive disease diagnosis [26–28].

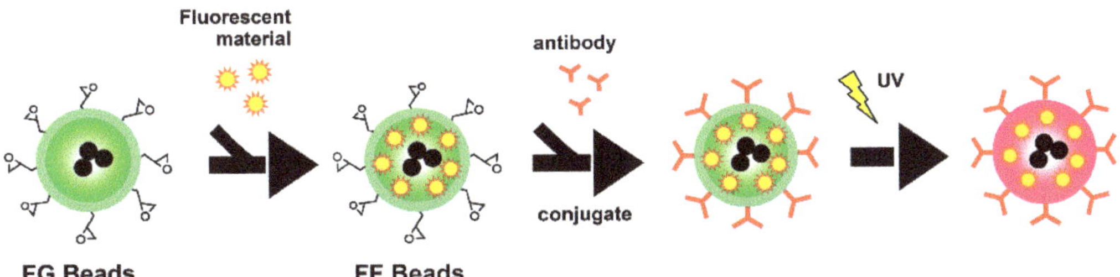

Figure 4. Scheme of FF beads. FF beads were prepared by encapsulating fluorescent materials in FG beads. Antibodies were immobilized on FF beads. FF beads emit fluorescence upon exposure to UV excitation. Modified from Kabe et al.

2.2.2. Screening or Monitoring of Cancer Biomarkers with FF Beads

FF beads were used to measure PSA, a widely used biomarker in patients with prostate cancer, using a magnetically prompted rapid sandwich immunoassay [26]. Detection was undertaken by measuring the fluorescence intensity. The detected antibody-coated FF beads and samples were placed on an antibody-coated capture microplate, and a magnet was attached under the plate for 1–2 min to concentrate the FF beads onto the immobilized antibody. The unbound FF beads were washed out, and the fluorescence of the remaining FF beads was held on the plate through the antigen–antibody reaction, which was then measured directly. When the limit of quantification (LOQ) was defined as the lowest concentration measurable intraassay (CV < 20%) in the sandwich immunoassay with FF beads, then the LOQ of this method was estimated to be 0.02 ng/mL for PSA in serum [26].

Clinical examination of prostate cancer requires the detection of PSA in serum over a range of 0.1 to 10 ng/mL [29,30]. Magnetically prompted rapid sandwich immunoassay is therefore sufficient to analyze a healthy donor who would generally have low concentrations of PSA (<0.1 ng/mL) and patients with prostate cancer who would have concentrations >4.0 ng/mL [26].

3. Intraoperative Diagnosis during Surgery

Cancerous areas are surgically removed and diagnosed pathologically during surgery, often with lymph nodes. The powerful application of MNPs in intraoperative situations includes sentinel lymph node (SLN) mapping and the rapid diagnosis of metastasis in SLNs. Currently, radioisotope (RI) tracers and blue dye are used as the gold standard for SLN mapping during surgery [31,32]; however, the RI method risks radiation exposure to both patients and medical personnel. Furthermore, the locations at which it can be used are limited because RI methods require nuclear medicine facilities. Using biocompatible MNPs, such as SPIONs, SLN detection can be performed without a special RI facility. Moreover, this MRI contrast media can drain into SLNs faster than RI and can be detected using a magnetometer.

The resected lymph nodes can be examined pathologically during surgery. Rapid diagnosis of cancer or metastasis in SLNs is necessary for surgical decision making. To visualize cancer or metastasis, immunostaining can increase the accuracy of diagnosis, but it is usually time consuming.

In this section, magnetic methods for SLN detection and rapid immunostaining are described.

3.1. Detecting Sentinel Node during Surgery Using SPIONs

Lymph nodes are responsible for trapping foreign substances, such as pathogens, before they can spread throughout the body, and eliminating them through an immune response [33]. Metastasis to regional lymph nodes is the most important prognostic indicator of outcome in patients with solid tumors. Tumor cells that have invaded the stroma can reach regional lymph nodes through the lymphatic capillaries and trunks around the tumor, forming lymph node metastases [33]. In melanoma [34] and breast cancer [35], the SLN theory has been established, whereby tumor cells that invade the lymphatic vessels first metastasize to specific lymph nodes, the so-called SLNs [36], and then to regional lymph nodes and organs throughout the body.

Pathologic examination of SLNs during surgery could provide information about the staging of regional lymph nodes. If the SLN is demonstrated to be cancer negative, then radical lymph node dissection would not be necessary. Recently, the applications of SLN theory were reported to be beneficial for many cancers, such as skin [36], breast [37], gastrointestinal [38], and gynecological cancers [39]. There could even be benefits during laparoscopic surgery [40].

The standard approach for the detection of SLNs is the dual-tracer method using an RI tracer (radiolabeled tin colloid) and blue dye [41]. However, the use of RIs requires a nuclear medicine facility. Furthermore, the RI tracer must be injected 2–24 h prior to surgery for accurate SLN detection [32]. These issues indicate the need for non-radioactive, rapid-assessment tracers with an ability to reliably detect SLNs. The RI method could therefore be replaced by a magnetic method.

3.1.1. SPIONs

SPIONs can be categorized as MNPs. SPIONs, such as Sienna+ and Resovist, are hydrophilic colloidal solutions of γ-Fe_2O_3 coated with carboxydextran. The diameter of the iron oxide particles is 4–10 nm, and the total size of SPIONs is approximately 60 nm. SPIONs are biocompatible and are specifically taken up by reticuloendothelial tissues (Kupffer cells), mainly in the liver. MRI is a diagnostic approach that uses a receiving coil to acquire the radio waves generated when a high-frequency magnetic field is applied to hydrogen atoms in a living body, causing a resonance phenomenon, and creates an image on the basis of the signal data. SPIONs are used as a negative contrast agent because they have a strong transverse relaxation time (T2) shortening effect and decrease the MR signal. After administration to the human body, SPIONs are rapidly taken up by Kupffer cells in the liver. Kupffer cells are not present in cancerous tissues and, thus, exert a contrast effect in MRI [42].

SPIONs have also been used as tracers for SLN biopsy. Following injection around the tumor, SPIONs are taken up by the SLNs and detected by a dedicated probe [43]. In this section, we focus on SLN detection by SPIONs.

3.1.2. Magnetic Probes
Magnetic Probes for Breast Cancer

Magnetic field detectors are necessary to detect SPIONs in SLNs for SLN mapping. A number of magnetic probes have been developed. For example, Sentimag is based on the mechanism of an AC pickup coil that is commercially available and is one of the most widely used in clinical settings [43–49]. Other magnetic probes that are based on the fundamental mode of orthogonal fluxgate (FM-OFG) [50–53], such as a magnetic tunneling junction (MTJ) sensor [54] and negatively charged nitrogen-vacancy centers in diamonds, have been developed. DiffMag is based on a pickup coil with AC and DC differential magnetometry [55,56]. These magnetic probes have demonstrated the ability to detect between 280 ng and 500 µg SPIONs from a distance of 1 mm to 2.5 cm.

Sekino et al. [57] showed that the amount of iron uptake in SLNs in breast cancer patients was approximately 140 ± 80 µg [57], which was 0.3% of the injection dose (1.6 mL of Resovist) that contained 44.6 mg of iron. Therefore, the magnetic probe is required to have a detection ability in the order of 100 µg at a typical distance of 2–3 cm for breast cancer to be applicable in the clinic.

A magnetic probe developed by Sekino and Kusakabe's group employed a permanent magnet and a Hall-effect magnetic sensor with a code-less handheld shape [57]. This probe is also commercially available as a medical device (Matrix Cell Research Institute Inc., Ibaraki, Japan, CE mark 93/42EEC; NB:0344, EC certificate No.4201663CE01). The major feature of this probe is that it allows precise positioning of the sensor with respect to the magnetic null point (where the magnetic flux density is zero) to remove environmental effects, such as any ambient magnetic fields and temperature effects. Other features of this probe are its easy handling for surgeons during operations because of its compact shape and low weight (108 g), and its code-less appearance. This probe can detect 56, 140, 280, and 560 µg SPIONs at a distance of 7, 9, 11, and 15 mm, respectively.

Magnetic Probe for Laparoscopic Study

Laparoscopic surgery is a less intensive method of surgery in which an endoscope and forceps are manipulated in four to five small incisions with ports (trocars) [58]. Usually, two sizes of ports are used, and the inner diameter of the larger port is 12 mm. Therefore, there is the need for a magnetic probe of a suitable shape for laparoscopic surgery. The differences between magnetic probes used for breast cancer and laparoscopic surgery are shown in Table 1 [57,59].

Table 1. Comparison of the probes for breast cancer and laparoscopic surgery.

Measure	Probe for Breast Cancer	Probe for Laparoscopic Surgery
Appearance		
Total length	24.5 cm	58.5 cm
Handle length	16 cm	17 cm
Shaft length	7 cm	37 cm
Head size (diameter)	18 mm	10 mm
Weight	100 g	150 g
Detection range (140 µg Resovist)	8 mm	6mm

The benefit of using magnetic nanoparticles, such as SPIONs, for SLN mapping in laparoscopic surgery is not just to avoid RI exposure, but because of the speed at which SPIONs can drain to SLNs from the injection site. SPIONs drain quicker than RI tracers [60], meaning that SPIONs could be used as an SLN detection tracer during surgery. Another benefit is the detection distance, which is shorter than that of RI. Furthermore, the strength of the RI tracer signal means that signals from the injection site can interfere with the detection signal from SLNs [41,61,62]. This so-called shine-through effect is especially pronounced in the narrow intraperitoneal space and is not an issue with magnetic nanoparticles.

Kuwahata et al. [63] developed an AC/DC probe magnetic sensor for laparoscopic surgery. This probe employs a nonlinear response from the magnetic nanoparticles magnetized by an alternating magnetic field with a static magnetic field to achieve sensitive detection. The probe showed a longitudinal detection length of 10 mm for 140 μg iron; the detection limit is approximately 280 ng from a 1 mm distance. The suitability of the probe was demonstrated using a porcine model.

3.1.3. SLN Detection during Surgery

Breast Cancer

Magnetic tracers are taken up by macrophages in the lymph nodes and detected by a handheld magnetometer [43]. In a previous study, it was shown that SPIONs reach the axillary lymph nodes within minutes after injection into the breast [60]. To detect SPIONs, several magnetometers have been developed [43,56,64].

In the EU, Sienna+ (a suspension of SPIONs) and Sentimag (a specialized probe) are used for SLN biopsy of breast cancer. Sienna+ is injected into the tumor periphery to reach the SLNs and can be identified by Sentimag. Sienna+ is a suspension of dark grains and can be recognized as a dye. A meta-analysis of clinical trials of SLN biopsies using magnetic detection systems showed that the identification rate of SLNs was not inferior to that of simultaneous administration of radiocolloid ± dye (conventional method vs. magnetic method: 96.8% vs. 97.1%).

Clinical tests using SPIONs and blue dye tracers in patients with breast cancer have shown that handheld magnetic probes are useful for detecting SLNs containing magnetic nanoparticles [65]. A multicenter study of breast cancer SLN biopsies using TAKUMI and Resovist (ferucarbotran) as a tracer showed that the identification rate of SLNs was not inferior to that of the RI method (RI method vs. magnetic method: 98.1% vs. 94.8%) [66].

Gallbladder Cancer

SLN mapping is challenging for cancers of difficult-to-access visceral organs, such as the gallbladder. This is because the standard method of RI use requires preoperative tracer injection. Indocyanine green (ICG) fluorescence imaging is a promising tool for SLN detection in patients with breast, gastric [67], and colorectal cancers [68]. Lymph flow and SLNs are detected soon after injection with a fluorescence imaging system, even in dense adipose tissue. However, because the ICG tracer is small, it passes through downstream lymph nodes, making it difficult to quantitatively analyze SLNs [69]. Magnetic methods to detect intra-abdominal SLNs can be used to overcome these challenges and have been effectively applied.

In a gallbladder cancer feasibility study using an animal model, the TAKUMI probe, which includes a Hall sensor, was modified for laparoscopic use [59]. Its feasibility for detecting SLNs of the gallbladder was evaluated using a laparoscopic dual-tracer method by injecting ICG and SPIONs into five wild-type pigs without cancer and one immunodeficient (RAG2-knockout) cancer-bearing pig. The laparoscopic probe identified the SPIONs in the lymph nodes of four out of the five wild-type pigs during surgery (Figure 5). The magnetic field counts were 2.5–15.9 μT, and fluorescence was detected in SLNs in all five pigs.

ICG shows a visual lymph-flow map, and SPIONs more accurately identify each SLN with a measurable magnetic field, which is similar to the RI method. It was confirmed using a RAG2-knockout porcine gallbladder cancer model with lymph node metastases that SLN mapping is effective under tumor-burden circumstances. We identified an SLN in the laparoscopic investigation, and the magnetic field count was 3.5 μT. The SLN was histologically determined to be one of two metastatic lymph nodes [59]. This result suggested the possibility of identifying SLNs in the intra-abdominal cavity organs.

3.2. Magnetically Promoted Rapid Immunofluorescence (MRIF) Staining Using FF Beads

Resected SLNs are examined pathologically. Here, we describe the rapid immunostaining of SLNs with positive images observed by fluorescence microscopy.

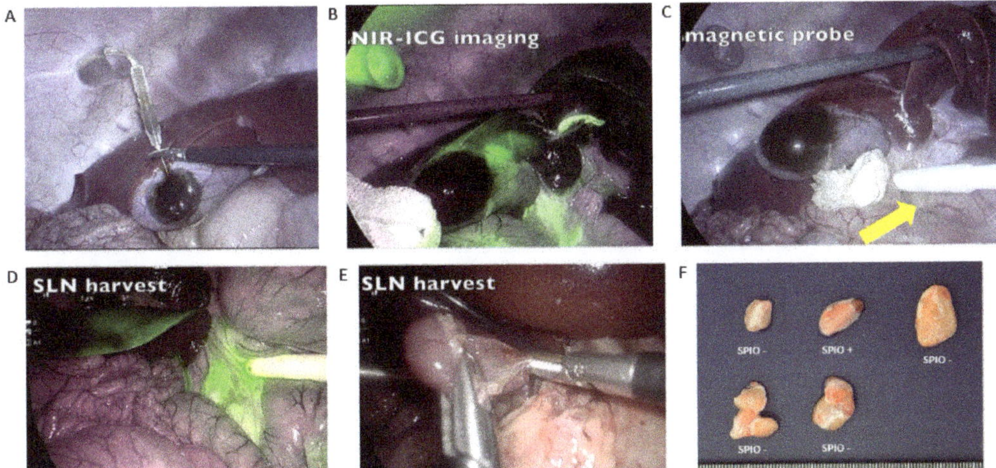

Figure 5. Laparoscopic sentinel lymph node (SLN) detection with a mixed tracer in porcine surgery. (**A**): Injection of the mixed indocyanine green (ICG) dye and magnetic tracer into the gallbladder wall. (**B**): ICG fluorescence signals detected by near-infrared laparoscopy. (**C**): Magnetic field evaluation of lymph nodes with the laparoscopic magnetic probe (yellow arrow). (**D**): Fluorescence signal-oriented identification of SLNs by the magnetic method. (**E**): Resection of the detected SLNs. (**F**): Brown pigmentation with the magnetic tracer in one resected regional lymph node among five. Modified from Mihara et al.

3.2.1. Europium Single Staining

Accurate identification of the extent of a lesion allows the surgeon to minimize removal during minimally invasive surgery of solid tumors. Thus, there is a need for the rapid diagnosis of lesion characteristics and progression during surgery [70,71]. Generally, snap-frozen sections are prepared during surgery and stained with hematoxylin–eosin (HE) for examination by a pathologist. Although HE staining can provide rapid diagnosis, diagnosis can be difficult, such as in cases with small lesions. Immunostaining is one approach to increase the diagnostic accuracy. The avidin–biotin complex method is a commonly used immunostaining system that involves four sequential steps: (1) primary antibody staining; (2) biotin-labeled secondary antibody staining; (3) avidin–biotin–peroxidase complex formation; and (4) development by diaminobenzidine (DAB) staining. Antigen–antibody reaction steps by primary and secondary antibodies are particularly time consuming, and the method is not suitable for rapid intraoperative diagnosis. Thus, there have been attempts to shorten the time of the procedure using ultrasound [72] and microwaves [72,73] that accelerate the antigen–antibody reaction with a stirring effect in addition to Brownian motion. Alternatively, Onishi et al. used FF beads to develop MRIF staining, which shortens reaction and washing times using a magnet [26,74]. MRIF can be performed in two steps without secondary antibody, signal amplification, or DAB staining: (1) incubation with antibody-coated FF beads and (2) washing, because the antigen–antibody complex can be directly observed using a fluorescence microscope to observe the fluorescent material encapsulated in the FF beads (Figure 6). This procedure reduces the time to a 1 min reaction and 1 min wash step with a magnet when applied to frozen sections of xenografted samples of A431 human epidermoid cancer cells that express high levels of epidermal growth factor receptor (EGFR) and anti-EGFR antibody-europium encapsulated FF beads (Figure 7A) [74].

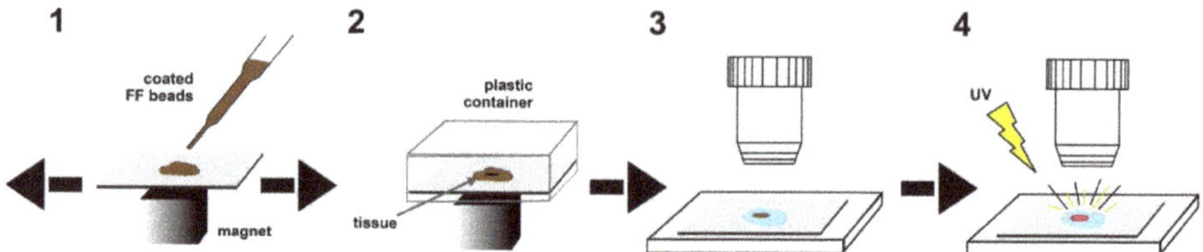

Figure 6. Scheme of magnetically promoted rapid immunofluorescence. 1: Diluted FF beads are dripped onto tumor cells, and the slide is vigorously agitated on the magnet; 2: the slide is inverted into a plastic container and washed with a magnet; 3, 4: FF beads bound to tumor cells can be observed directly by fluorescence microscopy. Modified from Onishi et al.

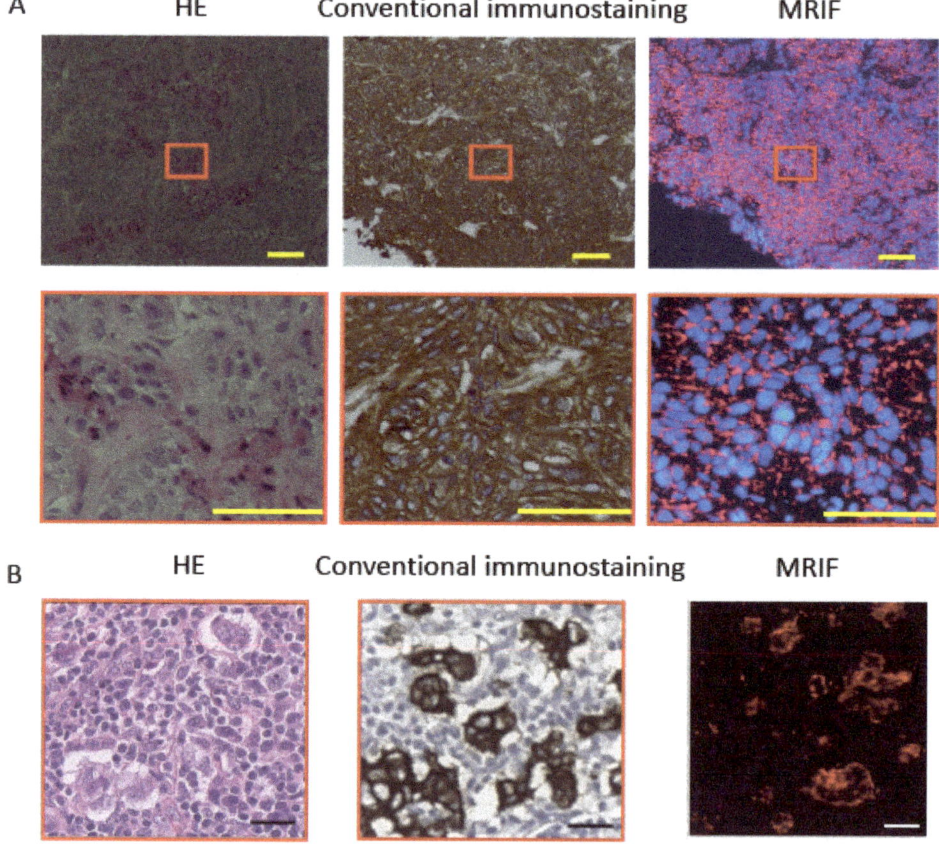

Figure 7. (**A**): Staining of A431 cells by hematoxylin–eosin (HE) (**left**), conventional immunostaining (**middle**), and MRIF (**right**). Images of an A431 (human epidermoid cancer cells with high expression of epidermal growth factor receptor (EGFR)) xenograft in pigs. (**B**): Staining image of a human breast cancer metastatic lymph node by HE (**left**), conventional immunostaining (**middle**), and MRIF (**right**) incubated with anti-pan-cytokeratin antibody-coated FF beads. Image of a paraffin-embedded tissue array of a stained human breast cancer metastatic lymph node. Scale bar = 250 μm and 25 μm for high magnification. Adapted from Onishi et al.

The strength of the magnetic force is critical for obtaining maximum results; therefore, a jig was prepared, and the relationship between the magnetic force and the distance from a 10 mm diameter and 24 mm length cylindrical magnet was examined. The magnetic force (F) acting on an FF bead was calculated as $F = -\nabla(-m_b \cdot B)$, where m_b is the magnetic moment of the FF beads, and B is the magnetic field strength of the magnet [75]. The distribution of the magnetic force was stronger at the margins than at the center of the magnet; therefore, we decided to agitate the magnet to obtain uniform staining [74]. The optimal distance between the A431 xenograft samples and the magnet using anti-EGFR antibody-coated FF beads for a 1 min incubation was within 2–5 mm, whereby the magnetic force $=7.79 \times 10^{-15}$ N to 3.35×10^{-15} N. A distance shorter than 2.0 mm showed unwanted background staining, and a distance greater than 5 mm showed insufficient staining. We also examined the optimal distance for washing. A distance from the samples to the magnet of 11 mm with a magnetic force of 4.78×10^{-16} N showed the best result for anti-EGFR antibody-coated FF beads. A distance >11 mm showed unwanted background staining. The staining efficiency was confirmed by the staining of breast cancer clinical samples for cytokeratin (CK), which is present in all epithelial cells, even in tumorigenesis, and is a widely used epithelial marker. Anti-pan-cytokeratin antibody (AE1/AE3) was used in this study. Figure 7B shows similar patterns of staining by conventional immunostaining and MRIF, which is consistent with the cancer region observed in the HE-stained section. The positive rates of conventional immunostaining were compared with MRIF staining using anti-pan-cytokeratin antibody-coated FF beads and clinical tissue array samples. The positive rate of conventional immunostaining was 96.5% (276/286) and that of MRIF was 92.7% (265/286). The coincidence rate was 94.8% (271/286) [74]. Normal tissue (i.e., breast tissue, tonsil, and lymph nodes) was analyzed. The positive rate of conventional immunostaining was 26.3% (25/95) and that of MRIF was 32.6% (31/95) [74]. The coincidence rate was 91.6% (87/95) (Table 2). Under optimal conditions, this ultrarapid immunostaining approach may be an ancillary method for pathological diagnosis during surgery.

Table 2. Coincidence ratio between conventional IHC and MRIF staining.

Immunostaining Method		MRIF		
	Result	+	−	Total
	+	263	13	276
Conventional Immunostaining	−	2	8	10
	Total	265	21	286

MRIF, magnetically promoted rapid immunofluorescence.

3.2.2. Multi-Colored Staining

Using several hydrophobic fluorophores that can be embedded into the polymer layers of the beads, the construction of multi-colored FF beads becomes possible. We applied a series of compounds, such as 3-dimesityl boryl-2,2′-bithiophene and 5,5′-dimesityl-3-dimesityl boryl-2,2′-bithiophene, which contain boron, to create multi-colored FF beads (patent: JP 6409173). Through the fluorescent labeling of target markers, multi-MRIF would be achieved. Figure 8 shows HE staining, conventional IHC staining, europium single staining, and europium double staining of human lymph nodes with metastasis by multi-MRIF. We designed antibody-coated FF beads to emit fluorescence independently. FF beads were coated with antibodies against CK19, which is expressed in epithelial cells, and tenascin C (TNC), which is a glycoprotein that is expressed in the extracellular matrix around cancer cells. Because some triple-negative breast cancers do not express CK19, tenascin C is a good candidate to compensate for CK19 to increase the detection rate of triple-negative breast cancer. Anti-CK19 antibody-coated FF beads show green fluorescence, and anti-TNC antibody-coated FF beads show red fluorescence. Conventional immunostaining with pan-CK was well correlated with single MRIF staining with pan-CK antibody-coated FF beads, which showed magenta fluorescence derived from europium complexes. The blue color was nuclear with DAPI staining. For CK19 and TNC double staining, both sets of FF beads were equally mixed and stained under the same magnetic conditions as EGFR for a 1 min reaction and a 1 min wash. CK19 (Figure 8D) and tenascin C (Figure 8E) were mostly stained in cancerous regions. Figure 8G shows merged images from D, E, and F. There is still a need to optimize the conditions because the antibody affinity is varied; however, this result demonstrates the possibility of double staining in one step. Furthermore, when frozen sections of six human metastatic lymph node samples from breast cancer were stained with IHC and MRIF, all lymph nodes were positive with a 100% concordance rate. In short, we successfully performed fluorescence multiplex staining of human breast cancer metastatic lymph nodes by binding antibodies against CK19 and TNC to FF beads containing different fluorophores. Because the system is applicable to frozen sections, it enables rapid diagnosis and meets clinical needs.

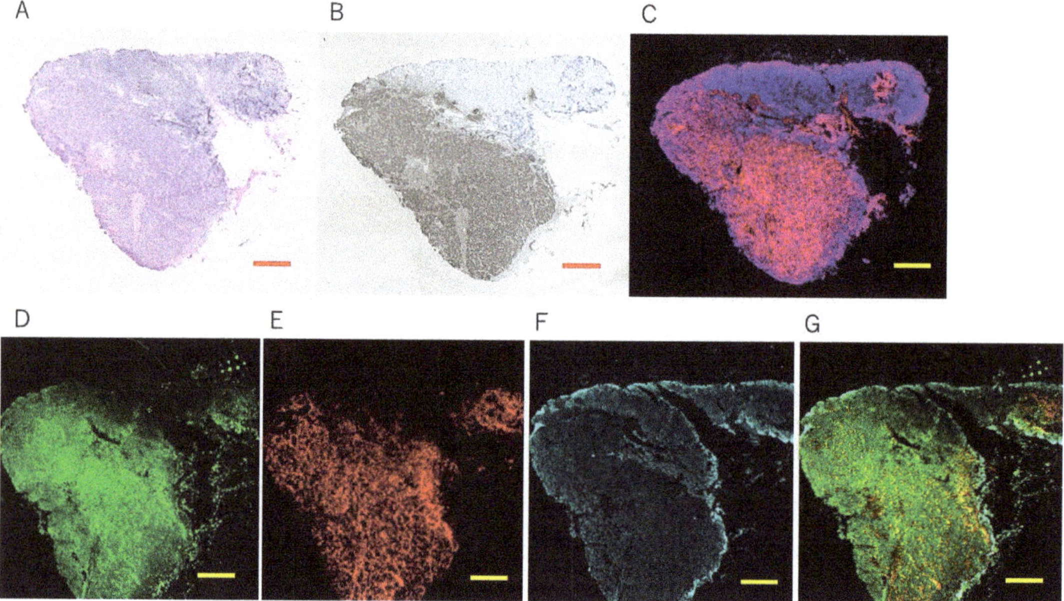

Figure 8. Staining of a frozen tissue section of a human breast cancer metastatic lymph node. (**A**): Hematoxylin and eosin (HE) staining. (**B**): Conventional immunostaining (diaminobenzidine). (**C**): Single staining of magnetically promoted rapid immunofluorescence (MRIF) with anti-pan-cytokeratin antibody-coated FF beads that emitted magenta fluorescence. Multi-colored MRIF using anti-CK19 antibody-coated F beads that emitted green fluorescence (**D**), anti-TNC antibody-coated FF beads that emitted red fluorescence (**E**), DAPI staining (**F**), and merged images (**G**). Scale bar = 1000 μm Adapted from Onishi et al. and new data.

4. Discussion and Future Perspectives

In this review, we described the applications of MNPs in oncology from a surgeon's perspective of monitoring biomarkers before and after surgery, and for intraoperative diagnosis during surgery. Pancreatic cancer-specific EVs and a cancer-specific antigen, PSA, were measured by the magnetic method, which could be used for monitoring cancer development before and after surgery. SLN detection can be performed during surgery by the magnetic method, and immunostaining can even be completed during surgery. Laboratory techniques related to surgical procedures can be undertaken magnetically.

Notably, EV detection and immunostaining are quite unique. Most EV methods require an extraction step of EVs from serum or plasma, but this magnetic method uses serum directly and involves counting the absolute number of EVs that express characteristic cell surface proteins. Pancreatic cancer is one of the most aggressive cancer types because it is difficult to identify in the early stages [76]. This method has potential as an early detection tool. Immunostaining is a powerful tool to increase the accuracy of diagnosis, but to contribute to decisions on surgical procedure, staining must be completed within 20 min [77]. MRIF requires only a 1 min reaction and a 1 min wash, and, thus, this method has the potential for practical application in the clinic. Moreover, because it is easy to define MRIF as positive and negative, it can be automated, reducing the requirement for a pathologist.

The magnetic SLN method is a promising alternative to the RI method. Moreover, it has the potential for clinical application to the laparoscopic method for detecting SLN metastasis from cancers of visceral organs, which are difficult to examine via the surface of the body or by endoscopy. These procedures can enable the identification of SLNs for almost all intra-abdominal organs that are laparoscopically accessible. Moreover, the long shelf life and easy handling of SPIONs and their detector permit the accurate diagnosis of metastatic cancers in mid- to small-scale medical facilities and developing countries.

Because europium is toxic, FG beads also have the potential for magnetic sensing with magnetic probes. Magnetic sensing activities strongly depend on magnetic characteristics, such as the magnetic moment. Compared with the magnetic moment of Resovist (approximately 50 emu/g) [78], the magnetic moment of the beads (20 emu/g) [10] is relatively small. Considering the detectable distance of Resovist of 9 mm with a magnetic probe as demonstrated by Sekino et al. [57], the detectable distance of the FF beads could be several millimeters. This expected magnetic sensing activity potentially enables the intra-abdominal detection of cancer and lymph nodes at a proximal distance.

The problems relating to the rate of false positives and false negatives that this type of methodology generates in each of its applications should be addressed, for example, which test confirms that the biological matrix has correctly come into contact with the analytical system in the presence of a negative result. However, regarding EV measurement, a lectin array [25] could be used to confirm the result; however, the results of lectin arrays are relative and are not quantitative. PSA measurements should be confirmed by conventional methods, such as ELISA, but the authors did not examine the associated rate of false positives and false negatives. The sentinel node is defined as the first lymph node that cancer cells reach, and the number of nodes may vary depending on the detection method. There are usually one or two for the RI method and more for the dye method. It is therefore difficult to discuss false positives and false negatives. In this review, we described the ICG dye method and the SPION method. Regarding MRIF staining, Onishi et al. [74] used conventional immunostaining to confirm that the antibody had correctly come into contact with the antigen and described the concordance rate because tissue array samples are not always serial sections.

5. Conclusions

Screening, monitoring, and diagnosis are critical in oncology treatment. However, current clinical methods are time consuming. The use of magnetic nanoparticles is an alternative approach that offers faster analyses over a wide range of oncological applications, such as the detection of cancer biomarkers and immunostaining. Radioisotope tracers are used for SLN mapping during cancer surgery; however, the RI method risks radiation exposure to both patients and medical personnel and requires nuclear medicine facilities. Using biocompatible MNPs, such as SPIONs, SLN detection can be performed safely without a special RI facility. The magnetic method is an interesting approach and its use is expected in more applications. It is hoped that large-scale clinical trials will be undertaken to demonstrate its usefulness and to validate it for clinical diagnosis.

Author Contributions: T.O., K.M. and S.M. designed and drafted the manuscript. S.S., A.K., M.S., M.K., H.H. and Y.K. drafted a part of the manuscript. All authors have read and agreed to the published version of the manuscript.

Funding: This review received no external funding.

Acknowledgments: The authors thank Y. Nakamura and S. Matsuda for their continued support. We thank Shaun Galbraith and H. Nikki March for editing a draft of this manuscript and for helping to draft the abstract.

Conflicts of Interest: M.K. is a member of Matrix Cell Research Institute Inc., which is TAKUMI's manufacturer. The other authors have no competing interests for this review.

References

1. Dadfar, S.M.; Roemhild, K.; Drude, N.; Von Stillfried, S.; Knüchel, R.; Kiessling, F.; Lammers, T. Iron oxide nanoparticles: Diagnostic, therapeutic and theranostic applications. *Adv. Drug Deliv. Rev.* **2019**, *138*, 302–325. [CrossRef] [PubMed]
2. Plouffe, B.; Murthy, S.K.; Lewis, L.H. Fundamentals and application of magnetic particles in cell isolation and enrichment: A review. *Rep. Prog. Phys.* **2015**, *78*, 016601. [CrossRef] [PubMed]
3. Schwerdt, J.I.; Goya, G.F.; Calatayud, M.P.; Herenu, C.B.; Reggiani, P.C.; Goya, R.G. Magnetic field-assisted gene delivery: Achievements and therapeutic potential. *Curr. Gene Ther.* **2012**, *12*, 116–126. [CrossRef] [PubMed]
4. Reimer, P.; Tombach, B. Hepatic MRI with SPIO: Detection and characterization of focal liver lesions. *Eur. Radiol.* **1998**, *8*, 1198–1204. [CrossRef]
5. Engvall, E.; Jonsson, K.; Perlmann, P. Enzyme-linked immunosorbent assay. II. Quantitative assay of protein antigen, immunoglobulin g, by means of enzyme-labelled antigen and antibody-coated tubes. *Biochim. Biophys. Acta (BBA)-Protein Struct.* **1971**, *251*, 427–434. [CrossRef]
6. Engvall, E.; Perlmann, P. Enzyme-linked immunosorbent assay (ELISA) quantitative assay of immunoglobulin G. *Immunochemistry* **1971**, *8*, 871–874. [CrossRef]
7. Van Weemen, B.K.; Schuurs, A.H. The influence of heterologous combinations of antiserum and enzyme-labeled estrogen on the characteristics of estrogen enzyme-immunoassays. *Immunochemistry* **1975**, *12*, 667–670. [CrossRef]
8. Inomata, Y.; Wada, T.; Handa, H.; Fujimoto, K.; Kawaguchi, H. Preparation of DNA-carrying affinity latex and purification of transcription factors with the latex. *J. Biomater. Sci. Polym. Ed.* **1994**, *5*, 293–302. [CrossRef]
9. Nishio, K.; Gokon, N.; Hasegawa, M.; Ogura, Y.; Ikeda, M.; Narimatsu, H.; Tada, M.; Yamaguchi, Y.; Sakamoto, S.; Abe, M.; et al. Identification of a chemical substructure that is immobilized to ferrite nanoparticles (FP). *Colloids Surf. B Biointerfaces* **2007**, *54*, 249–253. [CrossRef]
10. Nishio, K.; Masaike, Y.; Ikeda, M.; Narimatsu, H.; Gokon, N.; Tsubouchi, S.; Hatakeyama, M.; Sakamoto, S.; Hanyu, N.; Sandhu, A.; et al. Development of novel magnetic nano-carriers for high-performance affinity purification. *Colloids Surf. B Biointerfaces* **2008**, *64*, 162–169. [CrossRef]
11. Colombo, M.; Raposo, G.; Théry, C. Biogenesis, secretion, and intercellular interactions of exosomes and other extracellular vesicles. *Annu. Rev. Cell Dev. Biol.* **2014**, *30*, 255–289. [CrossRef]
12. Kalra, H.; Drummen, G.P.C.; Mathivanan, S. Focus on Extracellular Vesicles: Introducing the Next Small Big Thing. *Int. J. Mol. Sci.* **2016**, *17*, 170. [CrossRef]
13. Belov, L.; Matic, K.J.; Hallal, S.; Best, G.; Mulligan, S.P.; Christopherson, R.I. Extensive surface protein profiles of extracellular vesicles from cancer cells may provide diagnostic signatures from blood samples. *J. Extracell. Vesicles* **2016**, *5*, 25355. [CrossRef]
14. Witwer, K.W.; Buzás, E.I.; Bemis, L.T.; Bora, A.; Lässer, C.; Lötvall, J.; Nolte-'t Hoen, E.N.; Piper, M.G.; Sivaraman, S.; Skog, J.; et al. Standardization of sample collection, isolation and analysis methods in extracellular vesicle research. *J. Extracell. Vesicles* **2013**, *2*, 20360. [CrossRef]
15. Millimaggi, D.; Mari, M.; D'Ascenzo, S.; Carosa, E.; Jannini, E.A.; Zucker, S.; Carta, G.; Pavan, A.; Dolo, V. Tumor Vesicle—Associated CD147 Modulates the Angiogenic Capability of Endothelial Cells. *Neoplasia* **2007**, *9*, 349–357. [CrossRef]
16. Yoshioka, Y.; Kosaka, N.; Konishi, Y.; Ohta, H.; Okamoto, H.; Sonoda, H.; Nonaka, R.; Yamamoto, H.; Ishii, H.; Mori, M.; et al. Ultra-sensitive liquid biopsy of circulating extracellular vesicles using ExoScreen. *Nat. Commun.* **2014**, *5*, 3591. [CrossRef]
17. Marleau, A.M.; Chen, C.-S.; Joyce, J.A.; Tullis, R.H. Exosome removal as a therapeutic adjuvant in cancer. *J. Transl. Med.* **2012**, *10*, 134. [CrossRef]
18. Ueda, K.; Ishikawa, N.; Tatsuguchi, A.; Saichi, N.; Fujii, R.; Nakagawa, H. Antibody-coupled monolithic silica microtips for highthroughput molecular profiling of circulating exosomes. *Sci. Rep.* **2014**, *4*, srep06232. [CrossRef]
19. Gardiner, C.; Ferreira, Y.J.; Dragovic, R.A.; Redman, C.W.G.; Sargent, I.L. Extracellular vesicle sizing and enumeration by nanoparticle tracking analysis. *J. Extracell. Vesicles* **2013**, *2*, 19671. [CrossRef]
20. Tatischeff, I.; Larquet, E.; Falcon-Perez, J.M.; Turpin, P.-Y.; Kruglik, S.G. Fast characterisation of cell-derived extracellular vesicles by nanoparticles tracking analysis, cryo-electron microscopy, and Raman tweezers microspectroscopy. *J. Extracell. Vesicles* **2012**, *1*, 19179. [CrossRef]
21. Coumans, F.A.W.; van der Pol, E.; Böing, A.N.; Hajji, N.; Sturk, G.; van Leeuwen, T.; Nieuwland, R. Reproducible extracellular vesicle size and concentration determination with tunable resistive pulse sensing. *J. Extracell. Vesicles* **2014**, *3*, 25922. [CrossRef]
22. Maas, S.L.N.; De Vrij, J.; Broekman, M.L.D. Quantification and Size-profiling of Extracellular Vesicles Using Tunable Resistive Pulse Sensing. *J. Vis. Exp.* **2014**, *8*, 1443–1458. [CrossRef]

23. Logozzi, M.; De Milito, A.; Lugini, L.; Borghi, M.; Calabro', L.; Spada, M.; Perdicchio, M.; Marino, M.L.; Federici, C.; Iessi, E.; et al. High Levels of Exosomes Expressing CD63 and Caveolin-1 in Plasma of Melanoma Patients. *PLoS ONE* **2009**, *4*, e5219. [CrossRef]

24. Théry, C.; Amigorena, S.; Raposo, G.; Clayton, A. Isolation and Characterization of Exosomes from Cell Culture Supernatants and Biological Fluids. In *Current Protocols in Cell Biology*; John Wiley & Sons: Hoboken, NJ, USA, 2006.

25. Yokose, T.; Kabe, Y.; Matsuda, A.; Kitago, M.; Matsuda, S.; Hirai, M.; Nakagawa, T.; Masugi, Y.; Hishiki, T.; Nakamura, Y.; et al. O-Glycan-Altered Extracellular Vesicles: A Specific Serum Marker Elevated in Pancreatic Cancer. *Cancers* **2020**, *12*, 2469. [CrossRef]

26. Sakamoto, S.; Omagari, K.; Kita, Y.; Mochizuki, Y.; Naito, Y.; Kawata, S.; Matsuda, S.; Itano, O.; Jinno, H.; Takeuchi, H.; et al. Magnetically Promoted Rapid Immunoreactions Using Functionalized Fluorescent Magnetic Beads: A Proof of Principle. *Clin. Chem.* **2014**, *60*, 610–620. [CrossRef]

27. Hatakeyama, M.; Mochizuki, Y.; Kita, Y.; Kishi, H.; Nishio, K.; Sakamoto, S.; Abe, M.; Handa, H. Characterization of a magnetic carrier encapsulating europium and ferrite nanoparticles for biomolecular recognition and imaging. *J. Magn. Magn. Mater.* **2009**, *321*, 1364–1367. [CrossRef]

28. Kabe, Y.; Sakamoto, S.; Hatakeyama, M.; Yamaguchi, Y.; Suematsu, M.; Itonaga, M.; Handa, H. Application of high-performance magnetic nanobeads to biological sensing devices. *Anal. Bioanal. Chem.* **2019**, *411*, 1825–1837. [CrossRef]

29. Chatterjee, S.K.; Zetter, B.R. Cancer biomarkers: Knowing the present and predicting the future. *Future Oncol.* **2005**, *1*, 37–50. [CrossRef]

30. Finne, P.; Auvinen, A.; Määttänen, L.; Tammela, T.L.; Ruutu, M.; Juusela, H.; Martikainen, P.; Hakama, M.; Stenman, U.-H. Diagnostic value of free prostate-specific antigen among men with a prostate-specific antigen level of <3.0 microg per liter. *Eur Urol.* **2008**, *54*, 362–370. [CrossRef]

31. Krag, D.N.; Weaver, D.L.; Alex, J.C.; Fairbank, J.T. Surgical resection and radiolocalization of the sentinel lymph node in breast cancer using a gamma probe. *Surg. Oncol.* **1993**, *2*, 335–340. [CrossRef]

32. Schneebaum, S.; Stadler, J.; Cohen, M.; Yaniv, D.; Baron, J.; Skornick, Y. Gamma probe-guided sentinel node biopsy—optimal timing for injection. *Eur. J. Surg. Oncol.* **1998**, *24*, 515–519. [CrossRef]

33. Vander, A.J.; Sherman, J.H.; Luciano, D.S. *Human Physiology: The Mechanisms of Body Function*; Mcgraw-Hill: New York, NY, USA, 1994.

34. Morton, D.L.; Thompson, J.F.; Cochran, A.J.; Mozzillo, N.; Elashoff, R.; Essner, R.; Nieweg, O.E.; Roses, D.F.; Hoekstra, H.J.; Karakousis, C.P.; et al. Sentinel-Node Biopsy or Nodal Observation in Melanoma. *N. Engl. J. Med.* **2006**, *355*, 1307–1317. [CrossRef] [PubMed]

35. Cox, C.E.; Kiluk, J.V.; Riker, A.I.; Cox, J.M.; Allred, N.; Ramos, D.C.; Dupont, E.L.; Vrcel, V.; Diaz, N.; Boulware, D. Significance of Sentinel Lymph Node Micrometastases in Human Breast Cancer. *J. Am. Coll. Surg.* **2008**, *206*, 261–268. [CrossRef] [PubMed]

36. Morton, D.L.; Wen, D.-R.; Wong, J.H.; Economou, J.S.; Cagle, L.A.; Storm, F.K.; Foshag, L.J.; Cochran, A.J. Technical Details of Intraoperative Lymphatic Mapping for Early Stage Melanoma. *Arch. Surg.* **1992**, *127*, 392–399. [CrossRef]

37. Giuliano, A.E.; Kirgan, D.M.; Guenther, J.M.; Morton, D.L. Lymphatic Mapping and Sentinel Lymphadenectomy for Breast Cancer. *Ann. Surg.* **1994**, *220*, 391–401. [CrossRef]

38. Tsioulias, G.J.; Wood, T.F.; Morton, N.L.; Bilchik, A.J. Lymphatic mapping and focused analysis of sentinel lymph nodes upstage gastrointestinal neoplasms. *Arch. Surg.* **2000**, *135*, 926–932. [CrossRef]

39. Gasparri, M.L.; Caserta, D.; Panici, P.B.; Papadia, A.; Mueller, M.D. Surgical staging in endometrial cancer. *J. Cancer Res. Clin. Oncol.* **2019**, *145*, 213–221. [CrossRef]

40. Tajima, Y.; Murakami, M.; Yamazaki, K.; Masuda, Y.; Kato, M.; Sato, A.; Goto, S.; Otsuka, K.; Kato, T.; Kusano, M. Sentinel Node Mapping Guided by Indocyanine Green Fluorescence Imaging During Laparoscopic Surgery in Gastric Cancer. *Ann. Surg. Oncol.* **2010**, *17*, 1787–1793. [CrossRef]

41. McMasters, K.M.; Tuttle, T.; Carlson, D.J.; Brown, C.M.; Noyes, R.D.; Glaser, R.L.; Vennekotter, D.J.; Turk, P.S.; Tate, P.S.; Sardi, A.; et al. Sentinel Lymph Node Biopsy for Breast Cancer: A Suitable Alternative to Routine Axillary Dissection in Multi-Institutional Practice When Optimal Technique Is Used. *J. Clin. Oncol.* **2000**, *18*, 2560–2566. [CrossRef]

42. Kato, N.; Takahashi, M.; Tsuji, T.; Ihara, S.; Brautigam, M.; Miyazawa, T. Dose-Dependency and Rate of Decay of Efficacy of Resovist on MR Images in a Rat Cirrhotic Liver Model. *Investig. Radiol.* **1999**, *34*, 551–557. [CrossRef]

43. Douek, M.; On behalf of the SentiMAG Trialists Group; Klaase, J.; Monypenny, I.; Kothari, A.; Zechmeister, K.; Brown, D.; Wyld, L.; Drew, P.; Garmo, H.; et al. Sentinel Node Biopsy Using a Magnetic Tracer Versus Standard Technique: The SentiMAG Multicentre Trial. *Ann. Surg. Oncol.* **2014**, *21*, 1237–1245. [CrossRef]

44. Ghilli, M.; Carretta, E.; Di Filippo, F.; Battaglia, C.; Fustaino, L.; Galanou, I.; Rucci, P.; Fantini, M.; Roncella, M. The superparamagnetic iron oxide tracer: A valid alternative in sentinel node biopsy for breast cancer treatment. *Eur. J. Cancer Care* **2015**, *26*, e12385. [CrossRef]

45. Houpeau, J.-L.; Chauvet, M.-P.; Guillemin, F.; Bendavid-Athias, C.; Charitansky, H.; Kramar, A.; Giard, S. Sentinel lymph node identification using superparamagnetic iron oxide particles versus radioisotope: The French Sentimag feasibility trial. *J. Surg. Oncol.* **2016**, *113*, 501–507. [CrossRef]

46. Karakatsanis, A.; Christiansen, P.M.; Fischer, L.; Hedin, C.; Pistioli, L.; Sund, M.; Rasmussen, N.R.; Jørnsgård, H.; Tegnelius, D.; Eriksson, S.; et al. The Nordic SentiMag trial: A comparison of super paramagnetic iron oxide (SPIO) nanoparticles versus Tc(99) and patent blue in the detection of sentinel node (SN) in patients with breast cancer and a meta-analysis of earlier studies. *Breast Cancer Res. Treat.* **2016**, *157*, 281–294. [CrossRef]

47. Piñero-Madrona, A.; Torró-Richart, J.; de León-Carrillo, J.; de Castro-Parga, G.; Navarro-Cecilia, J.; Domínguez-Cunchillos, F.; Román-Santamaría, J.; Fuster-Diana, C.; Pardo-García, R. Superparamagnetic iron oxide as a tracer for sentinel node biopsy in breast cancer: A comparative non-inferiority study. *Eur. J. Surg. Oncol.* **2015**, *41*, 991–997. [CrossRef]

48. Rubio, I.; Diaz-Botero, S.; Esgueva, A.; Rodriguez, R.; Cortadellas, T.; Cordoba, O.; Espinosa-Bravo, M. The superparamagnetic iron oxide is equivalent to the Tc99 radiotracer method for identifying the sentinel lymph node in breast cancer. *Eur. J. Surg. Oncol.* **2015**, *41*, 46–51. [CrossRef]

49. Thill, M.; Kurylcio, A.; Welter, R.; van Haasteren, V.; Grosse, B.; Berclaz, G.; Polkowski, W.; Hauser, N. The Central-European SentiMag study: Sentinel lymph node biopsy with superparamagnetic iron oxide (SPIO) vs. radioisotope. *Breast* **2014**, *23*, 175–179. [CrossRef]

50. Elrefai, A.L.; Sasada, I. Magnetic particle detection in unshielded environment using orthogonal fluxgate gradiometer. *J. Appl. Phys.* **2015**, *117*, 17C114. [CrossRef]

51. Karo, H.; Sasada, I. Superparamagnetic nanoparticle detection system by using a fundamental mode orthogonal fluxgate (FM-OFG) gradiometer. *AIP Adv.* **2017**, *7*, 56716. [CrossRef]

52. Paperno, E. Suppression of magnetic noise in the fundamental-mode orthogonal fluxgate. *Sens. Actuators A Phys.* **2004**, *116*, 405–409. [CrossRef]

53. Sasada, I. Orthogonal fluxgate mechanism operated with dc biased excitation. *J. Appl. Phys.* **2002**, *91*, 7789. [CrossRef]

54. Cousins, A.; Balalis, G.L.; Thompson, S.K.; Morales, D.F.; Mohtar, A.; Wedding, A.B.; Thierry, B. Novel Handheld Magnetometer Probe Based on Magnetic Tunnelling Junction Sensors for Intraoperative Sentinel Lymph Node Identification. *Sci. Rep.* **2015**, *5*, 10842. [CrossRef]

55. Visscher, M.; Waanders, S.; Krooshoop, H.; Haken, B.T. Selective detection of magnetic nanoparticles in biomedical applications using differential magnetometry. *J. Magn. Magn. Mater.* **2014**, *365*, 31–39. [CrossRef]

56. Waanders, S.; Visscher, M.; Wildeboer, R.R.; Oderkerk, T.O.B.; Krooshoop, H.J.G.; Haken, B.T. A handheld SPIO-based sentinel lymph node mapping device using differential magnetometry. *Phys. Med. Biol.* **2016**, *61*, 8120–8134. [CrossRef]

57. Sekino, M.; Kuwahata, A.; Ookubo, T.; Shiozawa, M.; Ohashi, K.; Kaneko, M.; Saito, I.; Inoue, Y.; Ohsaki, H.; Takei, H.; et al. Handheld magnetic probe with permanent magnet and Hall sensor for identifying sentinel lymph nodes in breast cancer patients. *Sci. Rep.* **2018**, *8*, 1–9. [CrossRef]

58. Mellor, K.L.; Powell, A.; Lewis, W.G. Laparoscopic Surgery's 100 Most Influential Manuscripts: A Bibliometric Analysis. *Surg. Laparosc. Endosc. Percutaneous Tech.* **2018**, *28*, 13–19. [CrossRef]

59. Mihara, K.; Matsuda, S.; Nakamura, Y.; Aiura, K.; Kuwahata, A.; Chikaki, S.; Sekino, M.; Kusakabe, M.; Suzuki, S.; Fuchimoto, D.; et al. Intraoperative laparoscopic detection of sentinel lymph nodes with indocyanine green and superparamagnetic iron oxide in a swine gallbladder cancer model. *PLoS ONE* **2021**, *16*, e0248531. [CrossRef]

60. Johnson, L.; Pinder, S.E.; Douek, M. Deposition of superparamagnetic iron-oxide nanoparticles in axillary sentinel lymph nodes following subcutaneous injection. *Histopathology* **2012**, *62*, 481–486. [CrossRef]

61. Takeuchi, H.; Oyama, T.; Kamiya, S.; Nakamura, R.; Takahashi, T.; Wada, N.; Saikawa, Y.; Kitagawa, Y. Laparoscopy-assisted Proximal Gastrectomy with Sentinel Node Mapping for Early Gastric Cancer. *World J. Surg.* **2011**, *35*, 2463–2471. [CrossRef]

62. Kitagawa, Y.; Takeuchi, H.; Takagi, Y.; Natsugoe, S.; Terashima, M.; Murakami, N.; Fujimura, T.; Tsujimoto, H.; Hayashi, H.; Yoshimizu, N.; et al. Sentinel Node Mapping for Gastric Cancer: A Prospective Multicenter Trial in Japan. *J. Clin. Oncol.* **2013**, *31*, 3704–3710. [CrossRef]

63. Kuwahata, A.; Tanaka, R.; Matsuda, S.; Amada, E.; Irino, T.; Mayanagi, S.; Chikaki, S.; Saito, I.; Tanabe, N.; Kawakubo, H.; et al. Development of Magnetic Probe for Sentinel Lymph Node Detection in Laparoscopic Navigation for Gastric Cancer Patients. *Sci. Rep.* **2020**, *10*, 1798. [CrossRef] [PubMed]

64. Imai, K.; Kawaharada, Y.; Ogawa, J.-I.; Saito, H.; Kudo, S.; Takashima, S.; Saito, Y.; Atari, M.; Ito, A.; Terata, K.; et al. Development of a New Magnetometer for Sentinel Lymph Node Mapping Designed for Video-Assisted Thoracic Surgery in Non–Small Cell Lung Cancer. *Surg. Innov.* **2015**, *22*, 401–405. [CrossRef] [PubMed]

65. Zada, A.; Peek, M.C.L.; Ahmed, M.; Anninga, B.; Baker, R.; Kusakabe, M.; Sekino, M.; Klaase, J.M.; Haken, B.T.; Douek, M. Meta-analysis of sentinel lymph node biopsy in breast cancer using the magnetic technique. *Br. J. Surg.* **2016**, *103*, 1409–1419. [CrossRef] [PubMed]

66. Taruno, K.; Kurita, T.; Kuwahata, A.; Yanagihara, K.; Enokido, K.; Katayose, Y.; Nakamura, S.; Takei, H.; Sekino, M.; Kusakabe, M. Multicenter clinical trial on sentinel lymph node biopsy using superparamagnetic iron oxide nanoparticles and a novel handheld magnetic probe. *J. Surg. Oncol.* **2019**, *120*, 1391–1396. [CrossRef]

67. He, M.; Jiang, Z.; Wang, C.; Hao, Z.; An, J.; Shen, J. Diagnostic value of near-infrared or fluorescent indocyanine green guided sentinel lymph node mapping in gastric cancer: A systematic review and meta-analysis. *J. Surg. Oncol.* **2018**, *118*, 1243–1256. [CrossRef]

68. Liberale, G.; Bohlok, A.; Bormans, A.; Bouazza, F.; Galdon, M.G.; El Nakadi, I.; Bourgeois, P.; Donckier, V. Indocyanine green fluorescence imaging for sentinel lymph node detection in colorectal cancer: A systematic review. *Eur. J. Surg. Oncol.* **2018**, *44*, 1301–1306. [CrossRef]

69. Currie, A.; Brigic, A.; Thomas-Gibson, S.; Suzuki, N.; Moorghen, M.; Jenkins, J.; Faiz, O.; Kennedy, R. A pilot study to assess near infrared laparoscopy with indocyanine green (ICG) for intraoperative sentinel lymph node mapping in early colon cancer. *Eur. J. Surg. Oncol.* **2017**, *43*, 2044–2051. [CrossRef]

70. Black, C.; Marotti, J.; Zarovnaya, E.; Paydarfar, J. Critical evaluation of frozen section margins in head and neck cancer resections. *Cancer* **2006**, *107*, 2792–2800. [CrossRef]

71. Esbona, K.; Li, Z.; Wilke, L.G. Intraoperative Imprint Cytology and Frozen Section Pathology for Margin Assessment in Breast Conservation Surgery: A Systematic Review. *Ann. Surg. Oncol.* **2012**, *19*, 3236–3245. [CrossRef]

72. Hatta, H.; Tsuneyama, K.; Kondo, T.; Takano, Y. Development of an Ultrasound-emitting Device for Performing Rapid Immunostaining Procedures. *J. Histochem. Cytochem.* **2010**, *58*, 421–428. [CrossRef]

73. Leong, A.S.-Y.; Daymon, M.E.; Milios, J. Microwave irradiation as a form of fixation for light and electron microscopy. *J. Pathol.* **1985**, *146*, 313–321. [CrossRef]

74. Onishi, T.; Matsuda, S.; Nakamura, Y.; Kuramoto, J.; Tsuruma, A.; Sakamoto, S.; Suzuki, S.; Fuchimoto, D.; Onishi, A.; Chikaki, S.; et al. Magnetically Promoted Rapid Immunofluorescence Staining for Frozen Tissue Sections. *J. Histochem. Cytochem.* **2019**, *67*, 575–587. [CrossRef]

75. Sekino, M.; Kuwahata, A.; Fujita, S.; Matsuda, S.; Kaneko, M.; Chikaki, S.; Sakamoto, S.; Saito, I.; Handa, H.; Kusakabe, M. Development of an optimized dome-shaped magnet for rapid magnetic immunostaining. *AIP Adv.* **2020**, *10*, 025317. [CrossRef]

76. McGuigan, A.; Kelly, P.; Turkington, R.C.; Jones, C.; Coleman, H.G.; McCain, R.S. Pancreatic cancer: A review of clinical diagnosis, epidemiology, treatment and outcomes. *World J. Gastroenterol.* **2018**, *24*, 4846–4861. [CrossRef]

77. Novis, D.A.; Zarbo, R.J. Interinstitutional comparison of frozen section turnaround time. A College of American Pathologists Q-Probes study of 32868 frozen sections in 700 hospitals. *Arch. Pathol. Lab. Med.* **1997**, *121*, 559–567.

78. Dadfar, S.M.; Camozzi, D.; Darguzyte, M.; Roemhild, K.; Varvarà, P.; Metselaar, J.; Banala, S.; Straub, M.; Güvener, N.; Engelmann, U.; et al. Size-isolation of superparamagnetic iron oxide nanoparticles improves MRI, MPI and hyperthermia performance. *J. Nanobiotechnol.* **2020**, *18*, 1–13. [CrossRef]

Article

Dual Targeting with Cell Surface Electrical Charge and Folic Acid via Superparamagnetic Fe$_3$O$_4$@Cu$_{2-x}$S for Photothermal Cancer Cell Killing

Zicheng Deng [1,2,3], Jou Lin [1], Sergey L. Bud'ko [4], Brent Webster [1], Tanya V. Kalin [3,5], Vladimir V. Kalinichenko [2,3,5,*] and Donglu Shi [1,*]

[1] The Materials Science and Engineering Program, College of Engineering and Applied Science, University of Cincinnati, Cincinnati, OH 45221, USA; dengzh@mail.uc.edu (Z.D.); lin2jo@mail.uc.edu (J.L.); websteba@mail.uc.edu (B.W.)
[2] Center for Lung Regenerative Medicine, Cincinnati Children's Hospital Medical Center, 3333 Burnet Avenue, Cincinnati, OH 45229, USA
[3] Division of Pulmonary Biology, Cincinnati Children's Hospital Medical Center, 3333 Burnet Avenue, Cincinnati, OH 45229, USA; tatiana.kalin@cchmc.org
[4] Division of Materials Science and Engineering, Ames Laboratory, Ames, IA 50011, USA; budko@ameslab.gov
[5] Department of Pediatrics, College of Medicine, University of Cincinnati and Cincinnati Children's Hospital Medical Center, 3333 Burnet Avenue, Cincinnati, OH 45229, USA
* Correspondence: Vladimir.Kalinichenko@cchmc.org (V.V.K.); donglu.shi@uc.edu (D.S.)

Citation: Deng, Z.; Lin, J.; Bud'ko, S.L.; Webster, B.; Kalin, T.V.; Kalinichenko, V.V.; Shi, D. Dual Targeting with Cell Surface Electrical Charge and Folic Acid via Superparamagnetic Fe$_3$O$_4$@Cu$_{2-x}$S for Photothermal Cancer Cell Killing. *Cancers* **2021**, *13*, 5275. https://doi.org/10.3390/cancers13215275

Academic Editors: Moriaki Kusakabe and Akihiro Kuwahata

Received: 11 September 2021
Accepted: 19 October 2021
Published: 21 October 2021

Publisher's Note: MDPI stays neutral with regard to jurisdictional claims in published maps and institutional affiliations.

Simple Summary: There are two critical issues in cancer hyperthermia: (1) photothermal effect and (2) cancer cell targeting efficiency. While the former can be addressed by rendering the nano carriers with significant IR absorptions, the latter is dealt with using a novel dual-targeting strategy. In this study, the Fe$_3$O$_4$ nanoparticle was coated with a shell of Cu$_{2-x}$S; the resulting Fe$_3$O$_4$@Cu$_{2-x}$S exhibited strong IR absorption for enhanced photothermal cancer cell killing. The Fe$_3$O$_4$@Cu$_{2-x}$S nanoparticles are surface functionalized with amphiphilic polyethylenimine (LA-PEI) and Folic acid-TPGS (FA-TPGS) for two purposes: (1) the PEI surface coating renders the particles positively charged, enabling them to effectively bind with negatively-charged cancer cells for more intimate nano/bio contact resulting in much stronger cancer cell ablation; (2) the folic acid modification further increases the targeting efficiency via the folic receptors on the cancer cell surface. Dual-targeting with the surface electrical charge and the tumor-specific folic acid synergistically facilitates both passive and active targeting for significantly improved photothermal killing.

Abstract: A major challenge in cancer therapy is to achieve high cell targeting specificity for the highest therapeutic efficacy. Two major approaches have been shown to be quite effective, namely, (1) bio-marker mediated cell targeting, and (2) electrical charge driven cell binding. The former utilizes the tumor-specific moieties on nano carrier surfaces for active targeting, while the latter relies on nanoparticles binding onto the cancer cell surfaces due to differences in electrical charge. Cancer cells are known for their hallmark metabolic pattern: high rates of glycolysis that lead to negatively charged cell surfaces. In this study, the nanoparticles of Fe$_3$O$_4$@Cu$_{2-x}$S were rendered positively charged by conjugating their surfaces with different functional groups for strong electrostatic binding onto the negatively-charged cancer cells. In addition to the positively charged surfaces, the Fe$_3$O$_4$@Cu$_{2-x}$S nanoparticles were also modified with folic acid (FA) for biomarker-based cell targeting. The dual-targeting approach synergistically utilizes the effectiveness of both charge- and biomarker-based cell binding for enhanced cell targeting. Further, these superparamagnetic Fe$_3$O$_4$@Cu$_{2-x}$S nanoparticles exhibit much stronger IR absorptions compared to Fe$_3$O$_4$, therefore much more effective in photothermal therapy.

Keywords: superparamagnetic nanoparticles; cancer cell photothermal therapy; surface charge targeting; folic acid targeting; vitamin E TPGS modification

---

1. Introduction

Hyperthermia therapy has been shown to be an effective and efficient cancer treatment when applied locally to kill the cancer cells in a tumor-isolated fashion without adverse effects on healthy cells and tissues [1]. The key in a successful photothermal therapy is the targeted delivery of therapeutic agents, such as the photothermal nanoparticles, to tumors in a precision manner. Upon application of near-infrared (NIR) laser, the nanoparticles, typically gold or iron oxide, that are taken up by the tumor, can raise temperature to hyperthermic levels (~45 °C) for ablation of the targeted cancer cells [2]. In recent years, a variety of nanoparticles have been developed with multifunctionalities for medical diagnosis and therapeutics, among which the iron-oxide nanoparticles exhibit pronounced photothermal effects and, therefore, are most widely applied for photothermal therapy (PTT) [3–6]. Specifically, the superparamagnetic Fe_3O_4 nanoparticles have been extensively studied for biomedical applications, such as gene or drug delivery, magnetic resonance imaging (MRI), and magnetically-guided targeting [7–12]. With these unique nanoparticles, various cancer therapeutic strategies have been developed utilizing some of their fascinating properties, such as chemical stability, bio-compatibility, and strong photothermal effects. Recently, enhanced NIR absorption has been observed in modified iron-oxide nanoparticles for much stronger photothermal effects [13–15]. As is well-known, the Fe_3O_4 nanoparticles have a strong UV absorption, but it gradually decreases in the visible region without any noticeable NIR peaks [16].

Our previous works have shown that, by decorating the Fe_3O_4 nanoparticles with the $Cu_{2-x}S$ shells, the modified $Fe_3O_4@Cu_{2-x}S$ nanoparticles exhibit pronounced IR absorptions with broad absorption peaks near 1160 nm. This is due to the fact that the $Cu_{2-x}S$ nanocrystal has a tunable localized surface plasmon resonance (LSPR) in the IR region [16,17]. The enhanced IR absorption in $Fe_3O_4@Cu_{2-x}S$ can provide much greater photothermal effect in cancer hyperthermia therapy.

To achieve maximum therapeutic effects, cell targeting is critically required for high uptake of the nanoparticles into tumor cells via systemic intravenous administration. Selective delivery of therapeutic agents into tumor lesions has been a key challenge for the successful management of cancers. To address this critical issue in cell targeting, an electrical charge-based targeting method has been developed [18]. This unique targeting method is based on the so-called Warburg effect that characterizes the cancer cells with high rate of glycolysis. Normal cells typically depend on the mitochondrial oxidative phosphorylation process to generate Adenosine triphosphate (ATP). However, all cancer cells exhibit negative surface charges that are associated with their metabolic behavior: they constantly secrete lactic acid, resulting in the cross-membrane movement of lactate, an end product of the glycolysis pathway in hypoxia. Therefore, the increase of glycolysis levels in cancer cells causes increased glucose uptake and lactate secretion levels, exceeding that of normal cells [19]. The cross-membrane movement of lactate in cancer cells also causes the loss of labile inorganic cations that form lactate salts and acids [19]. Consequently, the cancer cell surfaces are left with a net of negative electrical charges [18,20,21]. If the nanoparticles can be rendered positively charged, they are able to electrostatically bind onto the cancer cells for the detection, targeting, and effective cell killing via PTT. In our previous work, we succeeded in photothermal therapy with only charge-based cell targeting. The positively-charged Fe_3O_4 nanoparticles were found to bind onto cancer cells more efficiently compared to the negatively-charged counterparts. This is due to the Coulomb force attraction between the nanoparticles and tumor cells with the opposite charges [18,19,22]. Surface charge-mediated cancer cell targeting has also been utilized to engineer a nanoprobe for the detection of circulating tumor cells (CTC) in clinical blood [23,24].

Furthermore, active targeting has been achieved by using the tumor-specific ligands to target the cell surface molecules or receptors [25]. Folic acid (FA) is considered one of the most suitable targeting ligands for cancer therapy due to the folate receptor being overexpressed on many cancer cell types [26–28]. Recent research on the magnetic nanoparticles has shown that FA modification is an effective strategy to enhance cancer cell targeting efficiency [29–33].

To enhance PTT efficiency in this study, we carried out photothermal experiments with two new strategies: (1) the Fe_3O_4 nanoparticles were surface-modified with CuS to develop $Fe_3O_4@Cu_{2-x}S$ with enhanced NIR absorptions for stronger photothermal cell killing; (2) in addition to bio-targeting with FA, the nanoparticles were rendered positively charged to achieve the so-called dual cell targeting for increased cell-specific binding. It is well-known that Fe_3O_4 has strong UV absorption, but shows no peaks in the IR region. For more efficient photothermal heating, we have developed a core–shell structure with $Cu_{2-x}S$ forming a shell on the core of Fe_3O_4. In this hybrid structure, while Fe_3O_4 provides the superparamagnetic property for the photothermal effect, the $Cu_{2-x}S$ shell on Fe_3O_4 renders the system with pronounced IR absorptions for further enhancement of the PTT efficacy. These supermagnetic nanoparticles have been widely used in medical diagnostics and therapeutics, such as the magnetic resonance imaging (MRI), photodynamic therapy, and magnetic targeting [34–37].

The novel concept is schematically depicted in Figure 1. As shown in this figure, a cationic amphiphilic polymer: lauric acid–polyethylenimine (LA-PEI) is coated on the $Fe_3O_4@Cu_{2-x}S$ nanoparticles and stabilized by the D-α-tocopheryl polyethylene glycol succinate (Vitamin E TPGS or TPGS) to form the positively-charged nanoparticles: PEI-$Fe_3O_4@Cu_{2-x}S$. TPGS is a widely used adjuvant in drug delivery which has been approved by the FDA [38]. This biocompatible amphiphilic molecule can be used as a surface stabilizer for enhanced drug stability [38–40]. Furthermore, the surfaces of the $Fe_3O_4@Cu_{2-x}S$ nanoparticles are conjugated with the folic acid (FA-PEI-$Fe_3O_4@Cu_{2-x}S$) to increase the targeting efficiency on the folate receptor-expressing cell lines.

Figure 1a shows the schematic pathway for preparation of the positively charged $Fe_3O_4@Cu_{2-x}S$ nanoparticles. The Fe_3O_4 nanoparticles are synthesized through a thermal decomposition process. The as-synthesized Fe_3O_4 nanoparticles are then coated with a $Cu_{2-x}S$ layer. The hydrophobic Fe_3O_4 or $Fe_3O_4@Cu_{2-x}S$ are stabilized with the amphiphilic polymers and TPGS in order to transfer the nanoparticles from organic solvent to an aqueous phase. Figure 1b illustrates the Warburg effect in cancer cells. As shown in this figure, the negative surface charges are created due to secretion of lactic acid by cancer cells. Figure 1c depicts the concept of the dual-targeting via both surface charges and biomarkers. PEI, as a cationic polymer, provides the positive charges on the nanoparticle surfaces enabling their binding onto the negatively-charged cancer cells. The folic acid modified on TPGS further increases the interaction between the nanoparticles and cancer cells with the folate receptor overexpression.

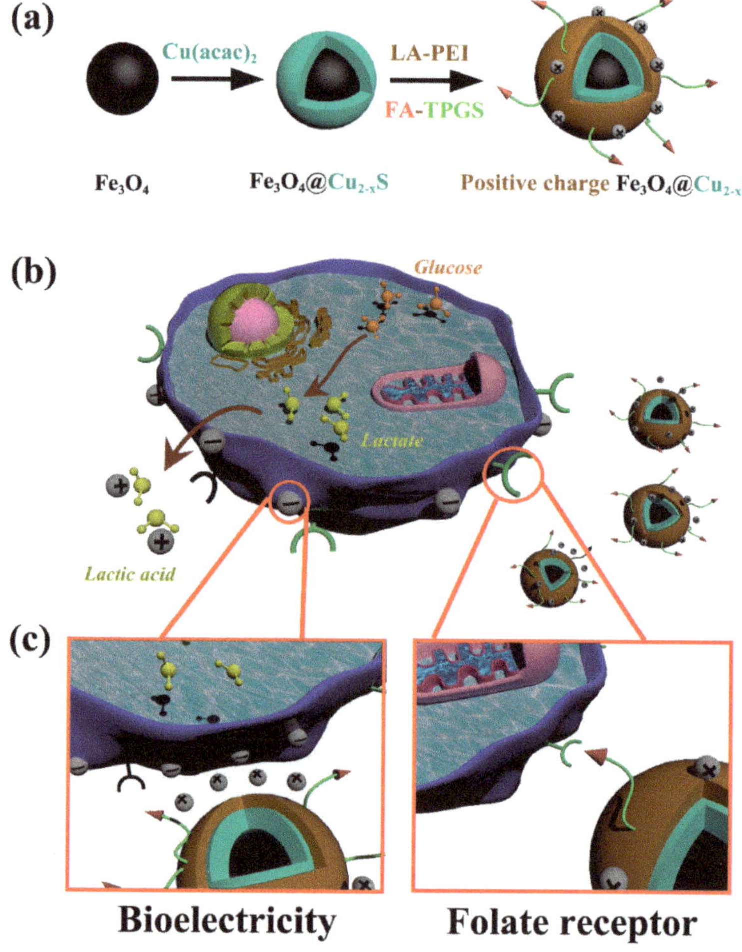

Figure 1. (**a**) Pathway for preparation of the nanoparticles; (**b**) schematic illustration of the Warburg effect, and (**c**) the strategies for electrical charge- and biomarker-mediated cancer targeting (passive and active targeting) via nanoparticles.

2. Materials and Methods

2.1. General

All chemicals for nanoparticle synthesis, including iron (III) acetylacetonate (Fe (acac)$_3$, $\geq$99.9%), copper (II) acetylacetonate (Cu (acac)$_2$, $\geq$99.9%), oleylamine (70%), sulfur (99.998%), N-methyl-2-pyrrolidone (99.5%), polyethylenimine (PEI, branched, Mw 600), D-α-tocopheryl polyethylene glycol succinate (Vitamin E TPGS or TPGS), polymer(isobutylene-alt- maleic anhydride) (Mw = 6000), Hexadecylamine (98%), folic acid (FA, $\geq$97%), and 4-dimethylaminopyridine (DMAP), were purchased from Sigma-Aldrich Inc. (St. Louis, MO, USA). 1-Ethyl-3-(3-dimethylaminopropyl) carbodiimide hydrochloride (EDC), N-Hydroxysuccinimide (NHS) lauric acid (LA), and the organic solvents including chloroform, cyclohexane, and tetrahydrofuran, were purchased from Fisher Scientific Inc. (Hampton, NH, USA).

The cell culture materials, including RPMI-1640 medium, Dulbecco's Modified Eagle's medium (DMEM), fetal bovine serum (FBS), penicillin-streptomycin and 0.25% trypsin-EDTA, were purchased from Thermo Fisher Scientific, Waltham, MA, USA. Dulbecco's Phosphate Buffered Saline (DPBS), and phosphate-buffered saline (PBS) were purchased from Corning Corp, Corning, NY, USA. Cell Counting Kit-8 (CCK-8), Cyanine5 NHS ester, Cyanine5 amine (non-sulfonated), Calcein-AM, and Propidium Iodide (PI) were purchased from Apexbio Technology LLC (Houston, TX, USA).

2.2. Synthesis of Fe_3O_4 and $Fe_3O_4@Cu_{2-x}S$ Nanoparticles

Fe_3O_4 and $Fe_3O_4@Cu_{2-x}S$ nanoparticles were synthesized as described in previous studies with modification [16,17]. Briefly, a certain amount of Fe (acac)$_3$ in the NMP/oleylamine mixture (4:3, v/v) was injected into a preheated oleylamine at 300 °C under a nitrogen protection. Keeping the system at 300 °C for 10 min with stirring, it was cooled down to 60 °C for collecting the Fe_3O_4 nanoparticle which was washed by methanol. The dried Fe_3O_4 nanoparticle was dispersed in the chloroform until use.

For $Cu_{2-x}S$ coating, a certain amount of sulfur solution in an oleylamine/cyclohexane mixture (6:5, v/v) was injected into the as-synthesized Fe_3O_4 nanoparticle at 70 °C. Subsequently, Cu (acac)$_2$ was dissolved in an oleylamine/chloroform mixture (1:4, v/v). This very mixture was then injected into the reaction system and kept at 70 °C for 0.5 h with stirring to obtain the $Fe_3O_4@Cu_{2-x}S$ nanoparticles. The collected $Fe_3O_4@Cu_{2-x}S$ nanoparticles were washed with methanol and the dried $Fe_3O_4@Cu_{2-x}S$ nanoparticles were dispersed in chloroform until use.

2.3. LA-PEI and Folate Modified TPGS Synthesis

The amphiphilic PEI was developed from modified hydrophilic fatty acid molecules via EDC/NHS coupling, as described previously [41–44]. Briefly, the molar ratios of EDC to LA and NHS to EDC were respectively set at 1.25:1 and 1.25:1. They were mixed in the ethanol with 10% MES buffer (100 mM, pH = 6). After 15 min of reaction at 40 °C, solvated PEI (0.25 eqv. molar to lauric acid) was quickly added into the solution and allowed to react for 24 h at 40 °C. The product of LA-PEI was purified by dialysis for three days.

The synthesis of the folate-modified TPGS followed a modified procedure from a previously reported method [45]. FA, CDI, and DMAP (with molar ratio of 1:1.2:0.5) were dissolved in DMSO (with a FA concentration of 20 mg/mL) and stirred at room temperature for 24 h. TPGS (1 eqv. molar to FA) was then added to the reaction system for another 24 h. The product of FA-TPGS was purified by dialysis (MWCO 1 kDa) for three days.

2.4. Polymer Coating of the Fe_3O_4 and $Fe_3O_4@Cu_{2-x}S$ Nanoparticles

The hydrophobic Fe_3O_4 and $Fe_3O_4@Cu_{2-x}S$ nanoparticles were transferred from organic to aqueous solution by coating the amphiphilic polymers onto the particle surfaces [46]. For the positively-charge nanoparticles, the Fe_3O_4 or $Fe_3O_4@Cu_{2-x}S$ nanoparticles were dissolved in chloroform (1 mL) and added to deionized water (10 mL) containing LA-PEI and FA-TPGS. After sonication for 30 min, the chloroform in the oil-in-water emulsion was evaporated. The extra polymers were removed by dialysis for 48 h. For comparison, the amphiphilic polymer coating on the negatively charged nanoparticles was developed according to the previous reports [16,47,48]. Subsequently, 272 mg of polymer (isobutylene-alt-maleic anhydride) and 320 mg of hexadecylamine were dissolved in THF and heated to 60 °C. It was kept at 60 °C until a cloudy mixture became transparent and all THF had been evaporated. The resulting polymer was dissolved again in anhydrous chloroform. For the negatively charged polymer coating, the Fe_3O_4 or $Fe_3O_4@Cu_{2-x}S$ nanoparticles were mixed with TPGS in chloroform at a mass ratio of 5 to 2 under sonication for a homogeneous mixture. The mixture was then added into a polymer solution (100 mg/mL in chloroform) with negative charges and ultrasonicated for another 5 min. Upon rotary evaporation of the organic solvent, the nanoparticles were dissolved in an aqueous sodium borate buffer (SBB, pH 12) and ultrasonicated for 15 min. The extra polymers were removed by dialysis and the final products were kept at 4 °C until use. For the Cy5 fluorescence dye labeled nanoparticles, the Cy5-NHS ester or Cy5-amine were added to the PEI nanoparticles or to the EDC-NHS negatively charge polymer coated nanoparticles at a mass ratio of 1:100 (dye to nanoparticle). The extra unreacted fluorescence dye was removed over three days of dialysis (MWCO 20 kDa).

2.5. Nanoparticle Characterizations

The hydrodynamic diameter and surface potential were determined by dynamic light scattering (DLS) using a Zetasizer Nano-ZS (Malvern, Malvern, UK). For the photothermal experiments, samples were irradiated by using an 808 nm laser (Q-BAIHE, Shenzhen, China) with power of 2 W/cm^2. The temperature was measured and recorded by using an infrared camera (FLIR E6). The power density of the solar simulator was calibrated by an optical power meter (1919-R, Newport Corporation, Irvine, CA, USA). Nanoparticle size was determined by transmission electron microscopy (CM-20 TEM). The absorption and transmittance spectra were obtained by using a UV–VIS NIR spectrometer Lambda 900 (PerkinElmer Inc., Waltham, MA, USA). The X-ray diffraction analysis was acquired by X-ray Diffractometer (X'Pert MPD).

2.6. Cell Lines and Culture Conditions

Three cancer cell lines were used in these studies: RD 769 Mouse Rhabdomyosarcoma, A549 Human Lung Adenocarcinoma, and MDA-MB-231 Human Breast Carcinoma. The non-malignant CCD-19Lu Human Lung Fibroblast Cell Line was used for comparison with cancer cells. The A549 and CCD-19Lu cell lines were purchased from American Type Culture Collection (ATCC). The RD 769 rhabdomyosarcoma cell line was a kind gift from Dr. Timothy Cripe (Nationwide Children's Hospital, OH). The MDA-MB-231 breast cancer cell line was a kind gift from Dr. Jun-Lin Guan (University of Cincinnati, Cincinnati, OH, USA). The RD 769 and MDA-MB-231 cell lines were cultured in Dulbecco's modified Eagle's medium (DMEM). The A549 cell line was cultured in Kaighn's Modification of Ham's F-12 Medium (F12K), and the CCD-19Lu cell line was cultured in Eagle's Minimum Essential Medium (EMEM). All mediums were supplemented with 10% fetal bovine serum (FBS) and 1% antibiotic–antimycotic. Cells were maintained at 37 °C in 5% CO_2 humidified atmosphere.

2.7. Calcein-AM/PI Assay

After photothermal treatment, a mixture of Calcein-AM (4 µM) and Propidium Iodide (8 µM) in DPBS was added to the cells and incubated for 10 min for co-staining. The live cells were labeled in green color by Calcein-AM and the dead cells were labeled in red color by PI. The EVOS M7000 fluorescence microscope was used to examine the live/dead cells.

2.8. Confocal Microscopy Imaging

The cells were seeded in 8 chamber-slides 24 h before the experiments. Nanoparticles labeled with Cy5 were incubated with the cells at 37 °C and the excess of NPs was removed by washing with PBS. The cells were fixed using 4% PFA (Paraformaldehyde, Electron Microscopy Sciences, Hatfield, PA, USA), and the cell nuclei were counterstained with DAPI (4′,6-diamidino-2-phenylindole, blue). The cells and nanoparticles were prepared, imaged and analyzed using a Nikon A1R GaAsP inverted confocal microscope as described [49,50].

2.9. Flow Cytometry Analysis

The cells were seeded in 24-well plates and incubated for 24 h before the flow cytometry analysis. The nanoparticles were added to the cells and incubated at 37 °C for 5 min. The cells were subsequently rinsed with PBS, trypsinized, and transferred to tubes. Cell-associated fluorescence was determined using a BD LSR II flow cytometer, and the data were analyzed using the FlowJo software, as previously described [51,52].

2.10. Real-Time Quantitative RT-PCR (qRT-PCR)

Total RNA was extracted from cultured cells using the RN easy micro-Kit (Qiagen, Germantown, MD, USA) as described [53,54]. The cDNA was generated using iScript cDNA Synthesis Kits (BIO-RAD, Hercules, CA, USA). Quantitative real-time RT-PCR (qRT-PCR) was performed according to the TaqMan Gene Expression Assay protocols (Invitrogen, Waltham, CA, USA) [55,56].

2.11. Photothermal Conversion Efficiency

The photothermal conversion efficiencies (η) of Fe_3O_4 and $Fe_3O_4@Cu_{2-x}S$ nanoparticles were calculated using the equations developed by Roper et al. [19,57]. The photothermal conversion efficiency can be expressed by the following:

$$\eta = \frac{hS(T_{Max}-T_{sur})-Q_s}{I(1-10^{-A_{808}})} \tag{1}$$

where h is the heat transfer coefficient (W $\times$ m$^{-2}\times°$C^{-1}); S is the surface area of the container (m^2); T_{Max} is the maximum temperature of the solution ($°$C); T_{Sur} is the surrounding temperature; Q_s is the energy input by the sample cuvette and the solution (W), I is the incident laser power (W), and A_{808} is the absorbance of the Fe_3O_4 and $Fe_3O_4@Cu_{2-x}S$ nanoparticles in the standard rectangular glass cell with lid at the wavelength of 808 nm.

The value of hS is obtained by the following equation [16,19,57]:

$$hS = \frac{m_{H_2O}C_{P,\,H_2O}}{\tau_S} \tag{2}$$

where m_{H_2O} and $C_{P,\,H_2O}$ are respectively the mass (g) and heat capacity (J/g$\times°$C) of the sample. τ_S is the sample system time (s) which is given by [57]:

$$\tau_S = -\frac{t}{\ln\theta} \tag{3}$$

where θ is defined as the ratio of $(T - T_{Sur})$ to $(T_{Max} - T_{Sur})$, and T is the solution temperature ($°$C). In this research, the heat capacity of water is 4.18 J/g, the mass of the solution is 0.1 g, the incident laser power is 0.5 W, A_{808} was determined to be 0.53071, and Q_s is 0.005 W.

2.12. In Vitro Photothermal Cancer Killing Efficiency

The in vitro cancer cell killing efficiency was assessed by using the Cell Counting Kit (CCK-8, Apexbio Technology LLC). The cells were seeded on 96-well plates 24 h prior to the photothermal experiments. The nanoparticles were diluted to different concentrations in the DPBS and incubated with the cells for 5 min at 37 °C. Excess NPs were removed and replaced by PBS. The cells were irradiated with an 808 nm laser (2 W cm^{-2}) for 5 min. After that, 10 μL CCK-8 was added to the plate and incubated at 37 °C for 3 h. The assay absorbance was measured at a wavelength of 450 nm in a Microplate Reader. The viability of the cell was calculated by the equations below:

$$\text{Cell viability (\%)} = [(As - Ab)/(Ac - Ab)] \times 100$$

As = Absorbance of treated cell
Ac = Absorbance of untreated cell
Ab = Absorbance of blank background

3. Results and Discussion

Figure 2a shows the TEM images of the Fe_3O_4 and $Fe_3O_4@Cu_{2-x}S$ nanoparticles without surface modification. The average sizes of the Fe_3O_4 and $Fe_3O_4@Cu_{2-x}S$ nanoparticles were ~10 nm and ~15 nm, respectively. The X-ray powder diffraction patterns of both Fe_3O_4

and $Fe_3O_4@Cu_{2-x}S$ are shown in Figure 2b. As can be seen in this figure, all diffraction peaks can be assigned to Fe_3O_4 with the crystal planes identified. In addition, the peaks of the (103) and (110) planes are identified for CuS respectively at $2\theta = 31.8°$ and $48.1°$.

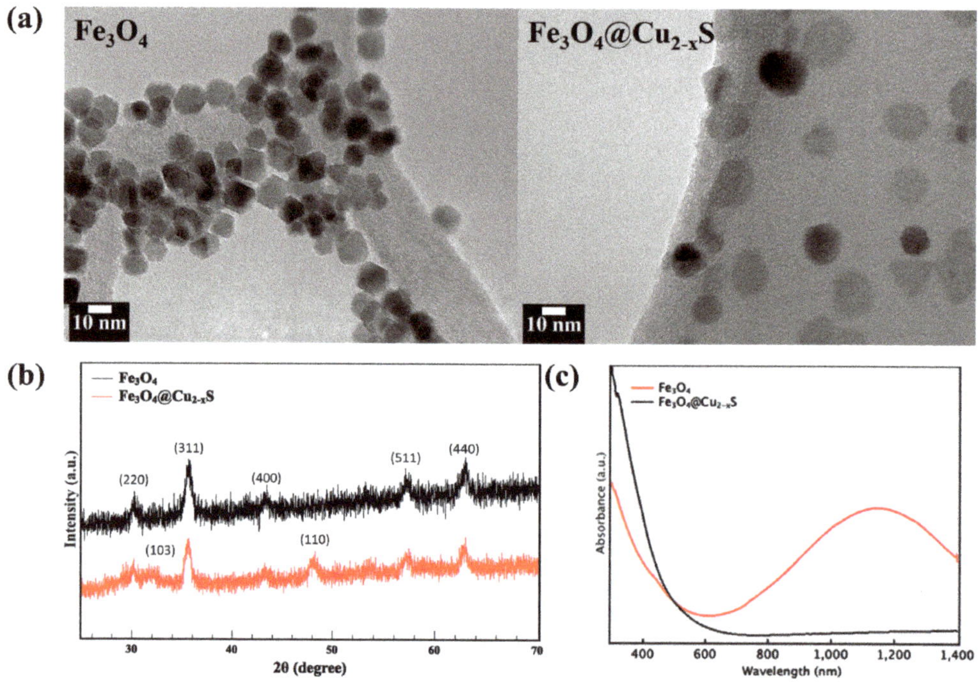

Figure 2. (**a**) TEM images of the Fe_3O_4 and $Fe_3O_4@Cu_{2-x}S$ nanoparticles; (**b**) powder X-ray diffraction patterns of the Fe_3O_4 and $Fe_3O_4@Cu_{2-x}S$ nanoparticles, and (**c**) UV−vis NIR absorption spectra for solutions of the Fe_3O_4 and $Fe_3O_4@Cu_{2-x}S$ nanoparticles dispersed in toluene.

The UV−vis NIR absorption spectra of Fe_3O_4 and $Fe_3O_4@Cu_{2-x}S$ solutions are shown in Figure 2c. As shown in this figure, Fe_3O_4 is characterized by a strong UV absorption, but no peak is observed in the IR region. With a $Cu_{2-x}S$ shell on Fe_3O_4, however, there is a pronounced IR absorption at 1160 nm in $Fe_3O_4@Cu_{2-x}S$. The enhanced IR absorption can be utilized for creating strong photothermal effects in PTT.

The amphiphilic polymer coating was designed to stabilize the nanoparticles in aqueous solutions and control the surface functionalization, including surface charge and targeting ligands. Vitamin E TPGS-Folic acid (FA-TPGS) (Figure S1a) and lauric acid-Polyethylenimine (LA-PEI) (Figure S1b) were synthesized for nanoparticle surface modifications. The cationic polymer PEI provides the positive surface charges for the nanoparticles, combined with folic acid modification to enhance the nanoparticles' interactions with cancer cells efficiently. The structures of the amphiphilic polymers were characterized with ^{1}H-NMR (Figure S2). The ^{1}H-NMR of FA shows a peak in the region around 12 ppm corresponding to the proton signal of the carboxyl groups of FA (Figure S2a) which is not seen in the ^{1}H-NMR FA-TPGS (Figure S2c). Similarly, the proton signal of the carboxyl groups of LA (Figure S2d) is also not shown in the reaction products (Figure S2f). These results indicate that the carboxyl group on FA and LA was successfully coupled to TPGS and PEI, respectively. Other characteristic peaks in the reactants are present in the final products, indicating successful synthesis of FA-TPGS (Figure S2c) and LA-PEI (Figure S2f).

Thermogravimetric analysis (TGA) was used to quantify the polymer coating on the $Fe_3O_4@Cu_{2-x}S$ nanoparticles. As shown in Figure 3a, the total weight loss of FA-PEI-$Fe_3O_4@Cu_{2-x}S$ is 7.55%. With surface modification, the weight loss in temperatures ranging from 200 °C to 450 °C increases to 42.76%. The increase in weight loss in this region is due to the decomposition of the amphiphilic polymers coating. Further weight loss from 450 °C to 600 °C is from the char residue, which is 11.1%. The TGA data demonstrate the mass ratio of the polymer coating to $Fe_3O_4@Cu_{2-x}S$ to be approximately 1:1.

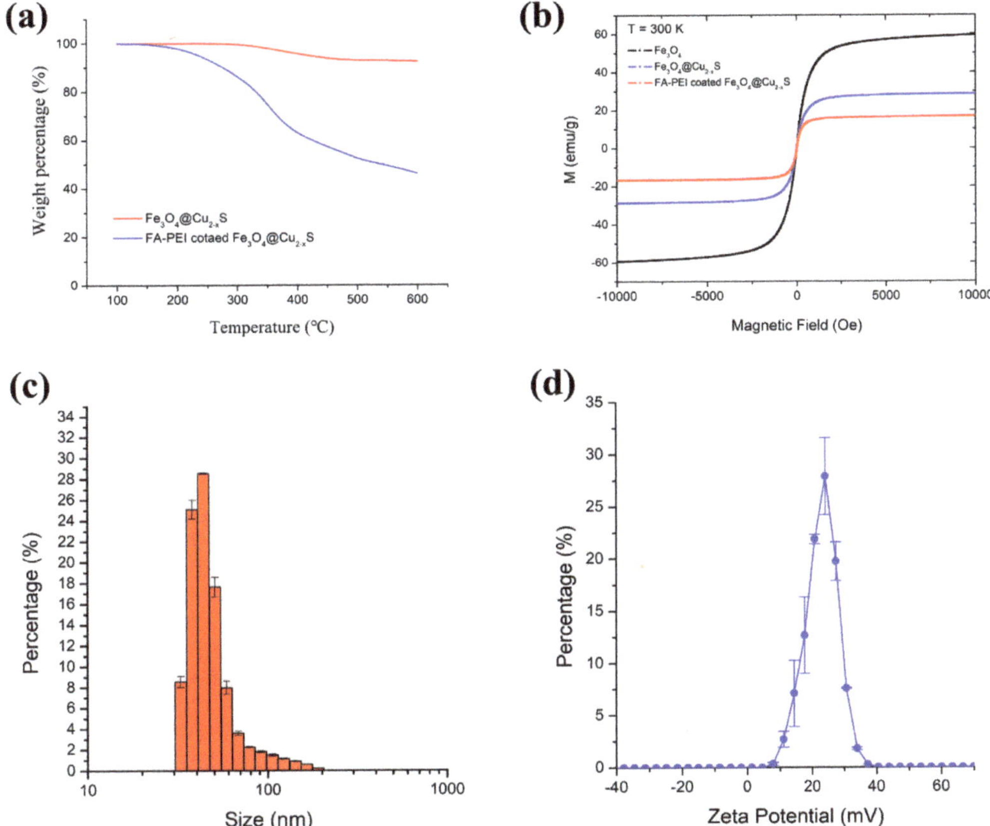

Figure 3. (**a**) Thermogravimetric analysis (TGA) curves of uncoated $Fe_3O_4@Cu_{2-x}S$ and FA-PEI- $Fe_3O_4@Cu_{2-x}S$; (**b**) magnetic hysteresis loops of Fe_3O_4, $Fe_3O_4@Cu_{2-x}S$, and FA-PEI- $Fe_3O_4@Cu_{2-x}S$; (**c**) size distribution of FA-PEI- $Fe_3O_4@Cu_{2-x}S$ nanoparticle, and (**d**) surface zeta potential of FA-PEI- $Fe_3O_4@Cu_{2-x}S$.

The magnetic properties of the Fe_3O_4, $Fe_3O_4@Cu_{2-x}S$, and coated $Fe_3O_4@Cu_{2-x}S$ were characterized by vibrating-sample magnetometer (VSM). The magnetic hysteresis curves are shown in Figure 3b. As shown in this figure, the saturation magnetization of Fe_3O_4 is 59. 4 emu/g, considerably larger than that of $Fe_3O_4@Cu_{2-x}S$ (28.6 emu/g). The reduction in magnetization in $Fe_3O_4@Cu_{2-x}S$ is attributed to the non-magnetic $Cu_{2-x}S$ component on the surfaces of the Fe_3O_4 nanoparticles. The saturation magnetization is reduced to 16.67 emu/g after coating $Fe_3O_4@Cu_{2-x}S$ nanoparticles with non-magnetic polymers. However, all nanoparticles show superparamagnetic behavior reflected by the highly reversible hysteresis curves regardless of the surface modifications by $Cu_{2-x}S$ or polymer coating.

The hydrodynamic size and the surface charge were determined using Dynamic light scattering (DLS). The size distribution of the FA-PEI- $Fe_3O_4@Cu_{2-x}S$ nanoparticles is majorly in the range of 37–50 nm (Figure 3c) with a surface charge of 27.33 $\pm$ 0.69 mV (Figure 3d). Upon surface modification, the average hydrodynamic diameter extends to 192.37 $\pm$ 2.15 nm, which is in an appropriate average range (180–220 nm) for accumulating readily in tumor vasculature as compared to those in the previously reported studies on medical diagnosis and therapeutics [10,40,58–61]. For comparison, the negatively-charged nanoparticles were used as control in the photothermal cancer cell killing experiments, following the procedures reported previously [16,47,48]. The size and zeta potential distributions of the negatively-charged nanoparticles are shown in Figure S3a,b. The average hydrodynamic diameter and the surface charge of the negatively charged nanoparticles are 142.27 $\pm$ 2.71 nm and -31.5 ± 1.23 mV, respectively. The polymer weight percentage of the negatively-charged nanoparticles (determined by TGA) is 55.33% (Figure S3c), which is 12.57% higher than that of the FA-PEI-$Fe_3O_4@Cu_{2-x}S$ nanoparticles.

Figure 4a,b show the photothermal properties of the FA-PEI functionalized Fe_3O_4 and $Fe_3O_4@Cu_{2-x}S$ nanoparticles with varied concentrations under 808-nm laser irradiation (2 W/cm^2). The nanoparticles were dispersed in water and placed in a 96-well plate (100 uL aqueous solution). After the solution was irradiated for five minutes, the light source was turned off, and the temperatures were measured by infrared thermal camera. Figure 4a,b show, respectively, the temperature increases as function of time for the surface-functionalized Fe_3O_4 (Figure 4a) and $Fe_3O_4@Cu_{2-x}S$ (Figure 4b) nanoparticles of various concentrations irradiated by 808 nm laser (2 W/cm^2). As shown in these figures, the temperature increases for both particle systems at the beginning are rather rapid due to the photothermal effects of the nanoparticles, but leveling off after 1 min as a result of heat loss through the environment. The light is turned off at 5 min and temperatures are thereafter decreasing rapidly for all concentrations. As can also be seen in these figures, the heating curves are consistent with the particle concentrations that the highest temperature reaches 56 °C for FA-PEI- Fe_3O_4 at the concentration of 0.6 g/mL and 68 °C for FA-PEI-$Fe_3O_4@Cu_{2-x}S$ at the same concentration. This significant increase in temperature in the latter is due to the pronounced IR absorbance in the $Fe_3O_4@Cu_{2-x}S$ solution. Therefore, the $Fe_3O_4@Cu_{2-x}S$ nanoparticles are expected to exert a much stronger photothermal effect than that of the Fe_3O_4 counterpart.

The photostability of the functionalized Fe_3O_4 and $Fe_3O_4@Cu_{2-x}S$ nanoparticles were characterized by three on/off cycles of laser irradiation (Figure 4c,d). By turning light off and on every 10 min, the heating curves show consistent increases and decreases after several cycles indicating good photostability of the nanoparticles. The negatively-charged $Fe_3O_4@Cu_{2-x}S$ nanoparticles also show similar photothermal effects and photostabilities as shown in Figure S3d,e. However, the negatively-charged nanoparticles show weaker photothermal effect due to its higher polymer to particle ratio in comparison to the positively-charged FA-PEI- $Fe_3O_4@Cu_{2-x}S$ at the same concentration. The photothermal conversion efficiency at 0.15 mg/mL is calculated by Equations (1)–(3) and the results are shown in Table S1.

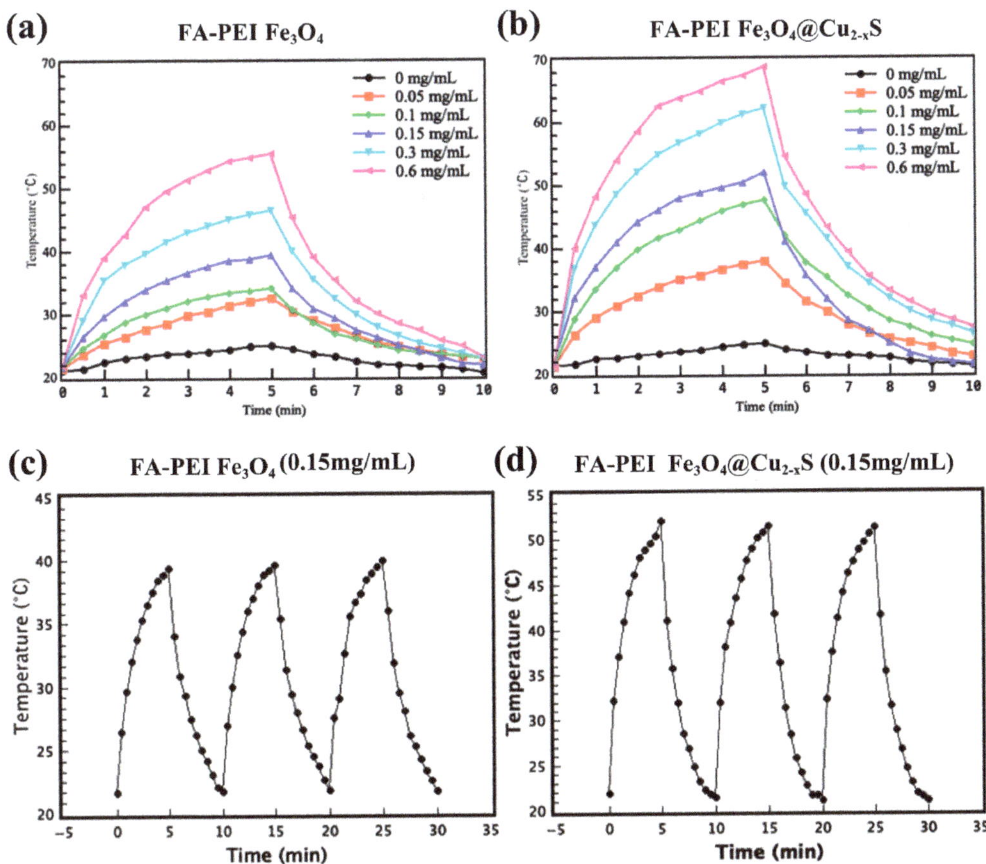

Figure 4. Temperature vs. time curves of (**a**) FA-PEI- Fe_3O_4 and (**b**) FA-PEI- $Fe_3O_4@Cu_{2-x}S$ at different concentrations. The temperature vs. time curves of (**c**) FA-PEI- Fe_3O_4 and (**d**) FA-PEI- $Fe_3O_4@Cu_{2-x}S$ for three on/off cycles.

The cancer cell binding efficiencies of different nanoparticles were assessed by incubating the cells with the Cy5-fluorescent-dye-labeled nanoparticles for 5 min, and extra nanoparticles were removed by washing with PBS. The $Fe_3O_4@Cu_{2-x}S$ nanoparticles with the negative-charged polymer coating were used as comparison. Compared with the negative $Fe_3O_4@Cu_{2-x}S$ nanoparticle treated cancer cells (which hardly displays any Cy5 signal), the positive PEI-$Fe_3O_4@Cu_{2-x}S$ treated cancer cells show visible Cy5 signals around the cancer cells (Figure 5a–c). With FA functionalization, the FA-PEI $Fe_3O_4@Cu_{2-x}S$ nanoparticles show a significant fluorescence increase on the cancer cells (Figure 5a–c). The normal cell line (CCD-19Lu) interacted with neither the negative, nor the positive or folate modified positive nanoparticles due to their neutral surfaces. Therefore, the Cy5 signal from all three nanoparticles is not observed on CCD-19Lu cells (Figure 5d). These experimental results indicate strong electrostatic interactions between the charged nanoparticles and the cancer cells.

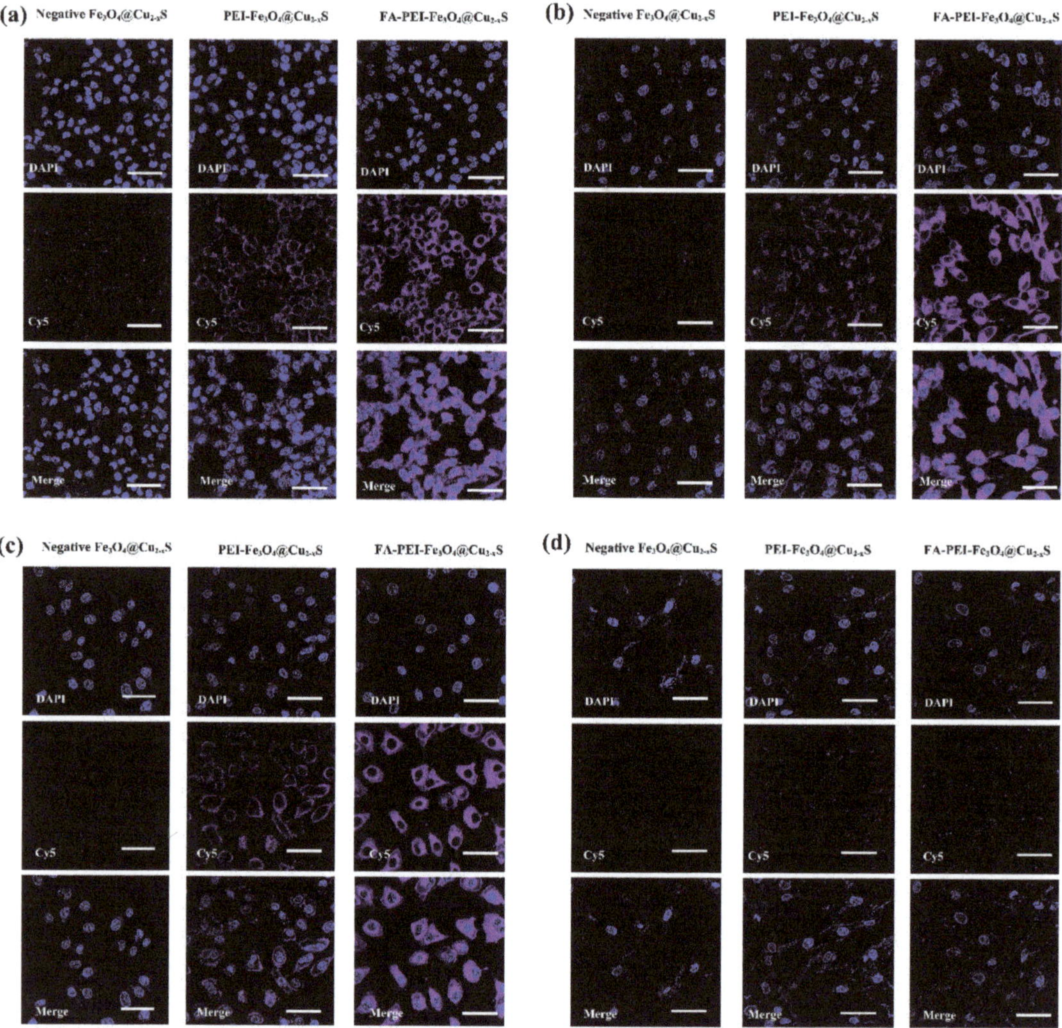

Figure 5. Cy5 labeled nanoparticle binding to the cancer cells. (**a**) RD 769, (**b**) MDA-MB-231, (**c**) A549, and (**d**) CCD-19Lu. DAPI was used to stain the cell nucleus and Cy5 was labeled on the nanoparticles. Scale bar, 50 μm.

The quantification of the nanoparticle binding to the cancer cells was determined by flow cytometry. As shown in Figure 6a–c, the positively-charged nanoparticles exhibit higher Cy5 fluorescence compared to the negatively-charged $Fe_3O_4@Cu_{2-x}S$ nanoparticles in all three cancer cell lines. Figure 6d shows the quantification of the median fluorescence intensity (MFI). Compared to the negatively-charged $Fe_3O_4@Cu_{2-x}S$ nanoparticles, the intensities of Cy5 fluorescence from the positively-charged $Fe_3O_4@Cu_{2-x}S$ nanoparticles bound onto RD 769, MDA-MB-231 and A459 cells are 7.08-fold, 4.57-fold, and 13.18-fold, respectively (Figure 5e). After FA modification, the Cy5 signals from all three cancer cell lines are further increased to 2.26-fold, 2.97-fold, and 1.72-fold respectively, compared to treatment with the positively-charged nanoparticles without folic acid modification (Figure 5e). The binding efficiency of FA-modified nanoparticles is dependent on folate receptor levels in cancer cell lines. The folate receptor (FOLR1) expression levels in the human cell lines were examined by RT-qPCR and shown in Figure S4. The FOLR1 expression levels in MDA-MB-231 and A549 are, respectively, 11.31-fold and 2.26-fold higher than CCD-19Lu. The highest FOLR1 level of MDA-MB-231 significantly improved the binding efficiency with FA-PEI-Fe$_3$O$_4$@Cu$_{2-x}$S.

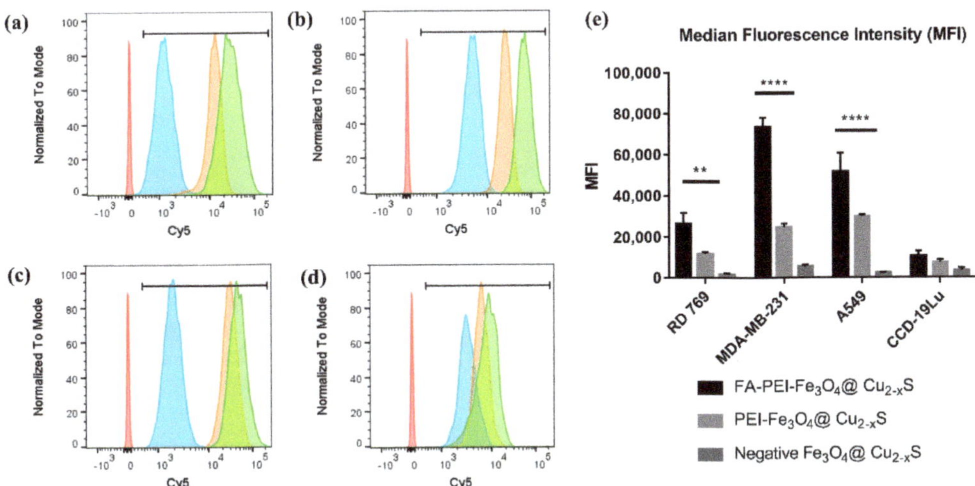

Figure 6. Flow cytometry histogram profiles of (**a**) RD 769 cell line; (**b**) MDA-MB-231 cell line; (**c**) A549 cell line; (**d**) CCD-19Lu cell line with different nanoparticles: negatively charged Fe$_3$O$_4$@Cu$_{2-x}$S (blue), PEI- Fe$_3$O$_4$@Cu$_{2-x}$S (orange) and FA-PEI- Fe$_3$O$_4$@Cu$_{2-x}$S (green), and (**e**) median fluorescence intensities (MFI) of the RD 769, MDA-MB-231, and A549 cell lines. (** $p < 0.01$, **** $p < 0.0001$).

In normal CCD-19Lu cells, the nanoparticles with the positively charged surfaces and folic acid modification only increased the cell binding slightly. Cell binding efficiencies of both PEI-Fe$_3$O$_4$@Cu$_{2-x}$S and FA-PEI-Fe$_3$O$_4$@Cu$_{2-x}$S were found to be insignificant when compared with the negatively-charged Fe$_3$O$_4$@Cu$_{2-x}$S (Figure 5e). The flow cytometry results indicate that the mouse cancer cells have less nanoparticle binding than human cell lines (Figure 5e). However, the positively-charged surfaces and folic acid modification still increased the targeting efficiency. In human cancer cell lines, the positively-charged nanoparticle binding efficiencies on cancer cell lines are 3.37-fold for MDA-MB-231 and 4.08-fold for A549, both are higher than that on the normal cell line (CCD-19Lu) (Figure 5e). Upon folic acid modification, the nanoparticle binding efficiencies on cancer cell lines are further increased to 7.15-fold (MDA-MB-231) and 5.02-fold (A549) compared with the normal cell line (CCD-19Lu) (Figure 5e). Fluorescent microscopy and flow cytometry data demonstrate that the FA-PEI-Fe$_3$O$_4$@Cu$_{2-x}$S has high targeting efficiency to the cancer cell lines, as shown in Figure 5e.

The Calcein-AM/PI live-dead staining is shown in Figure 7. Under the same conditions (5 min incubation, 2 W cm^{-2} 808 nm laser irradiation for 5 min), the negatively-charged Fe$_3$O$_4$@Cu$_{2-x}$S with laser irradiation did not cause significant cell death in all three cancer cell lines. After treatment with positively-charged PEI-Fe$_3$O$_4$@Cu$_{2-x}$S nanoparticles, however, the cancer cells showed higher red fluorescence intensity, indicating effective cancer cell killing by the photothermal effect. The folic acid modification further increased the cancer cell killing efficiency due to higher nanoparticle-cancer cell binding efficiency. Figure 7a shows the toxicities of different nanoparticles (0.32 mg/mL) incubated with RD 769 (Figure S5a), MDA-MB-231 (Figure S5b), A549 tumor cells (Figure S5c), and normal CCD-19Lu cells (Figure S5d) without laser irradiation. As shown in Figure 7, with folic acid modification, the nanoparticle toxicities to the cancer cells are higher than those without FA modification (Figure 6a–c), due to the higher binding efficiency of FA-PEI-Fe$_3$O$_4$@Cu$_{2-x}$S. However, the cell death caused by the FA-PEI-Fe$_3$O$_4$@Cu$_{2-x}$S nanoparticles without laser treatment is significantly lower than that with laser, indicating strong photothermal effect of FA-PEI-Fe$_3$O$_4$@Cu$_{2-x}$S. For normal CCD-19Lu cells, none of the nanoparticles exhibited high toxicity, with or without laser (Figures 6d and S5d) due to insignificant particle binding on normal cells since they are practically neutral compared to the negatively charged cancer cells. The 808-nm laser irradiation treatment without nanoparticles did not show significant cytotoxicity for either cancer or normal cell lines (Figure S5e).

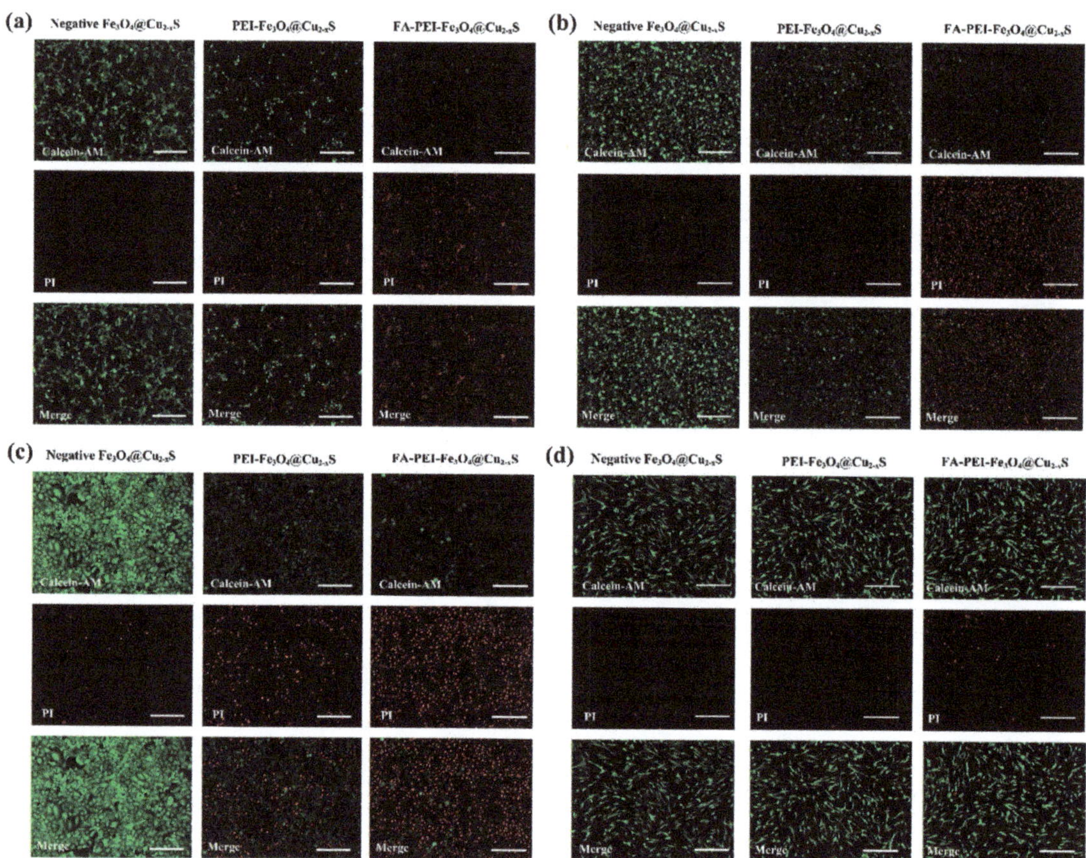

Figure 7. In vitro photothermal therapy effect of the negatively-charged Fe$_3$O$_4$@Cu$_{2-x}$S, positively-charged PEI-Fe$_3$O$_4$@Cu$_{2-x}$S, and FA-PEI- Fe$_3$O$_4$@Cu$_{2-x}$S at 0.32 mg/mL on (**a**) RD 769 cell line; (**b**) MDA-MB-231 cell line; (**c**) A549 cell line, and (**d**) CCD-19Lu cell line with 5 min 2 W cm^{-2} 808-nm laser irradiation. Calcein-AM/PI live-dead staining was used to stain live (green) and dead (red) cells. Scale bar, 275 μm.

The cell viabilities at different nanoparticle concentrations are shown in Figure 8. As shown in Figure 8a–c, with laser treatment, the FA-PEI-Fe$_3$O$_4$@Cu$_{2-x}$S nanoparticles inflict the strongest photothermal cancer cell killing in all three cancer cell lines due to the highest cell binding efficiency, followed by the PEI-Fe$_3$O$_4$@Cu$_{2-x}$S nanoparticles. In contrast, the cancer cell killing was negligible after photothermal treatment with the negatively-charged Fe$_3$O$_4$@Cu$_{2-x}$S nanoparticles, likely, due to weak nanoparticle binding on cancer cell surfaces. The viabilities of cancer cells without laser treatment were significantly higher than the laser-treated counterpart groups with the same nanoparticle concentrations (Figure S6a–d). Compared with cancer cells, the normal cells show much higher cell survival rate for all groups. These quantitative data show that the FA-PEI-Fe$_3$O$_4$@Cu$_{2-x}$S nanoparticles have much greater cancer cell photothermal killing efficiency with negligible influence on normal cells.

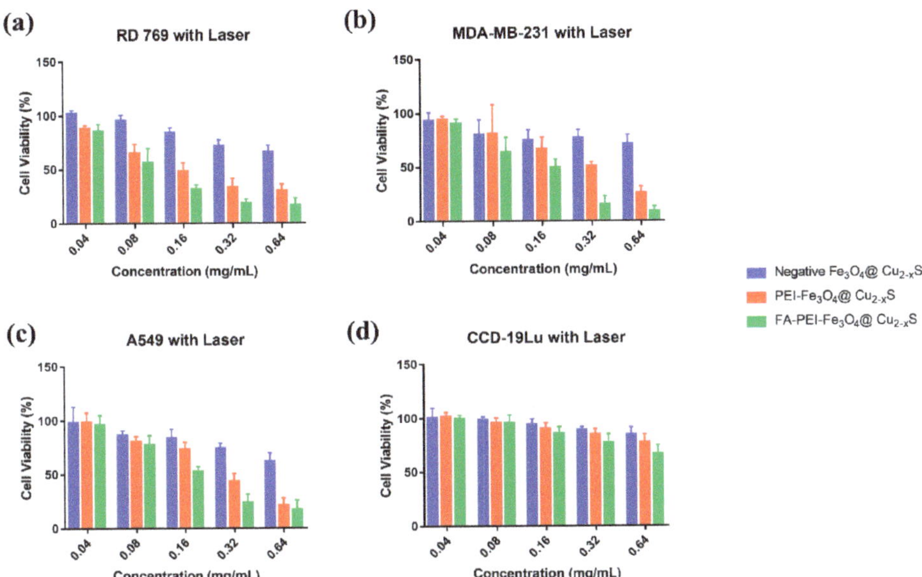

Figure 8. Cell viability vs. concentration of different nanoparticles for cancer cell lines (**a**) RD 769; (**b**) MDA-MB-231; (**c**) A549 and a normal cell line, and (**d**) CCD-19Lu under photothermal treatments (2 W cm^{-2} 808-nm laser irradiation for 5 min).

Based on the in vitro data in this study, we have shown that the nanoparticle surfaces conjugated with both positive electrical charge and folic acid have significantly enhanced cancer cell binding, leading to improved photothermal killing efficiency. The outcomes of this study can be applied to other nano-carrier systems for more effective photothermal therapy. The dual-targeting concept will require future in vivo experiments to demonstrate its validity in preclinical settings.

4. Conclusions

We have synthesized both Fe_3O_4 and $Fe_3O_4@Cu_{2-x}S$ nanoparticles and compared their characteristics in optical absorption and photothermal effect for enhanced photothermal cancer therapy. By modifying the particle surfaces of Fe_3O_4 with CuS, we have developed the $Fe_3O_4@Cu_{2-x}S$ nanoparticles that exhibit pronounced IR abortions that contribute to much stronger photothermal effect in cancer cell killing compared to the Fe_3O_4 counterparts. As a result, the photothermal conversion efficiency of $Fe_3O_4@Cu_{2-x}S$ has increased by 29.18%, while that of Fe_3O_4 is only 22.99%. Both Fe_3O_4 and $Fe_3O_4@Cu_{2-x}S$ are surface-modified with polymer coatings for dual targeting with cell surface electrical charge and folic acid. As all cancer cell surfaces are negatively charged due to high glycolysis rates, rendering the positively-charged nanoparticles enables efficient binding onto cancer cells for enhanced photothermal cancer cell killing. The cationic polymer coating on the nanoparticles has been found to facilitate the nanoparticle binding to cancer cells rapidly due to charge difference between the nanoparticles (positive) and the cancer cells (negative). The folic acid modification on the charged nanoparticle surfaces has further enhanced the nanoparticle targeting efficiency via folate receptor, which is overexpressed in cancer cells (active targeting). With the unique dual targeting strategy, the FA-PEI-$Fe_3O_4@Cu_{2-x}S$ nanoparticles show much higher cancer cell binding and subsequent photothermal cancer cell killing without noticeable toxicity to normal cells under the same conditions. In contrast, the negatively-charged $Fe_3O_4@Cu_{2-x}S$ nanoparticles show insignificant cell binding and photothermal toxicity due to repulsive force between the nanoparticles and cancer cells, since both have the same electrical charge. The experimental results from this study show a promise in photothermal cancer therapy by dual targeting of cancer cells via conjugating both the positive surface charge and the tumor-specific biomarkers on the nanoparticle surfaces.

Supplementary Materials: The following are available online at https://www.mdpi.com/article/10.3390/cancers13215275/s1, Figure S1: The chemical structures and synthesis procedures of (**a**) FA-TPGS and (**b**) LA-PEI. Figure S2: 1H-NMR characterization of the coating polymers (**a**) Folic acid, (**b**) TPGS, (**c**) FA-TPGS in DMSO-d6, (**d**) Lauric acid, and (**e**) PEI, and (**f**) LA-PEI in Chloroform-d. Figure S3: (**a**) Size distribution of the negative $Fe_3O_4@Cu_{2-x}S$ nanoparticle, (**b**) surface zeta potential of negative $Fe_3O_4@Cu_{2-x}S$, (**c**) thermogravimetric analysis (TGA) curves of uncoated $Fe_3O_4@Cu_{2-x}S$ and negative $Fe_3O_4@Cu_{2-x}S$, (**d**) temperature vs. time curves for negative $Fe_3O_4@Cu_{2-x}S$, and (**e**) temperature vs. time curves of negative $Fe_3O_4@Cu_{2-x}S$ for three on/off cycles. Figure S4: FOLR1 expression levels in different cell lines. Human FOLR1 mRNA levels in tumor cells were measured by qRT-PCR and normalized to Beta-Actin mRNA. Figure S5: In vitro toxicity of negatively-charged $Fe_3O_4@Cu_{2-x}S$, positively-charged PEI- $Fe_3O_4@Cu_{2-x}S$, and FA-PEI- $Fe_3O_4@Cu_{2-x}S$ at 0.32 mg/mL on (**a**) RD 769 cell line, (**b**) MDA-MB-231 cell line, (**c**) A549 cell line, (**d**) CCD-19Lu cell line without laser, and (**e**) 5 min 2 W cm^{-2} 808 nm laser irradiation treatment for different cells without nanoparticle. Scale bar: 275 μm. Figure S6: Cell viability vs. concentration of different nanoparticles for cancer cells (**a**) RD 769, (**b**) MDA-MB-231, (**c**) A549 and (**d**) non-malignant CCD-19Lu cells without laser treatment. Table S1: Photothermal conversion efficiencies of negative $Fe_3O_4@Cu_{2-x}S$, PEI-$Fe_3O_4@Cu_{2-x}S$, and FA-PEI-$Fe_3O_4@Cu_{2-x}S$ at concentration of 0.15 mg/mL.

Author Contributions: Conceptualization, D.S.; Data curation, Z.D., J.L., S.L.B. and B.W.; Investigation, Z.D.; Project administration, D.S.; Resources, T.V.K.; Supervision, V.V.K. and D.S. All authors have read and agreed to the published version of the manuscript.

Funding: This work was supported by NIH Grants HL141174 (to V.V.K.), HL149631 (to V.V.K.) and HL152973 (to V.V.K. and T.V.K.).

Institutional Review Board Statement: Not applicable.

Informed Consent Statement: Not applicable.

Data Availability Statement: Data available in a publicly accessible repository.

Conflicts of Interest: The authors declare no conflict of interest.

References

1. Crezee, J.; Franken, N.; Oei, A. Hyperthermia-based anti-cancer treatments. *Cancers* **2021**, *13*, 1240. [CrossRef]
2. Dunn, A.W.; Zhang, Y.; Mast, D.; Pauletti, G.M.; Xu, H.; Zhang, J.; Ewing, R.C.; Shi, D. In-vitro depth-dependent hyperthermia of human mammary gland adenocarcinoma. *Mater. Sci. Eng. C* **2016**, *69*, 12–16. [CrossRef]
3. Wang, R.; Yang, H.; Fu, R.; Su, Y.; Lin, X.; Jin, X.; Du, W.; Shan, X.; Huang, G. Biomimetic upconversion nanoparticles and gold nanoparticles for novel simultaneous dual-modal imaging-guided photothermal therapy of cancer. *Cancers* **2020**, *12*, 3136. [CrossRef] [PubMed]
4. Qin, J.; Lian, J.; Wu, S.; Wang, Y.; Shi, D. Recent advances in nanotechnology for breast cancer therapy. *Nano LIFE* **2019**, *9*, 1940003. [CrossRef]
5. Nicosia, A.; Cavallaro, G.; Costa, S.; Utzeri, M.A.; Cuttitta, A.; Giammona, G.; Mauro, N. Carbon nanodots for on demand chemophotothermal therapy combination to elicit necroptosis: Overcoming apoptosis resistance in breast cancer cell lines. *Cancers* **2020**, *12*, 3114. [CrossRef] [PubMed]
6. Nocito, G.; Calabrese, G.; Forte, S.; Petralia, S.; Puglisi, C.; Campolo, M.; Esposito, E.; Conoci, S. Carbon dots as promising tools for cancer diagnosis and therapy. *Cancers* **2021**, *13*, 1991. [CrossRef]
7. Marcelo, G.; Lodeiro, C.; Capelo, J.L.; Lorenzo, J.; Oliveira, E. Magnetic, fluorescent and hybrid nanoparticles: From synthesis to application in biosystems. *Mater. Sci. Eng. C* **2020**, *106*, 110104. [CrossRef]
8. Li, A.; Zuo, L. Construction of anti-EpCAM drug-loaded immunomagnetic balls and its application in diagnosis of breast cancer. *Nano LIFE* **2019**, *9*, 1940006. [CrossRef]
9. Wang, Z.; Qiao, R.; Tang, N.; Lu, Z.; Wang, H.; Zhang, Z.; Xue, X.; Huang, Z.; Zhang, S.; Zhang, G.; et al. Active targeting theranostic iron oxide nanoparticles for MRI and magnetic resonance-guided focused ultrasound ablation of lung cancer. *Biomaterials* **2017**, *127*, 25–35. [CrossRef]

10. Sun, S.; Wang, Y.; Zhou, R.; Deng, Z.; Han, Y.; Han, X.; Tao, W.; Yang, Z.; Shi, C.; Hong, D.; et al. Targeting and regulating of an oncogene via nanovector delivery of microRNA using patient-derived xenografts. *Theranostics* **2017**, *7*, 677–693. [CrossRef]

11. Wang, Y.; Sun, S.; Zhang, Z.; Shi, D. Nanomaterials for cancer precision medicine. *Adv. Mater.* **2018**, *30*, e1705660. [CrossRef] [PubMed]

12. Wang, Z.; Chang, Z.; Lu, M.; Shao, D.; Yue, J.; Yang, D.; Zheng, X.; Li, M.; He, K.; Zhang, M.; et al. Shape-controlled magnetic mesoporous silica nanoparticles for magnetically-mediated suicide gene therapy of hepatocellular carcinoma. *Biomaterials* **2018**, *154*, 147–157. [CrossRef] [PubMed]

13. Giustini, A.J.; Petryk, A.A.; Cassim, S.M.; Tate, J.A.; Baker, I.; Hoopes, P.J. Magnetic nanoparticle hyperthermia in cancer treatment. *Nano LIFE* **2010**, *1*, 17–32. [CrossRef]

14. Dunn, A.; Ehsan, S.M.; Mast, D.; Pauletti, G.M.; Xu, H.; Zhang, J.; Ewing, R.C.; Shi, D. Photothermal effects and toxicity of Fe_3O_4 nanoparticles via near infrared laser irradiation for cancer therapy. *Mater. Sci. Eng. C* **2015**, *46*, 97–102. [CrossRef]

15. Kolosnjaj-Tabi, J.; Kralj, S.; Griseti, E.; Nemec, S.; Wilhelm, C.; Sangnier, A.P.; Bellard, E.; Fourquaux, I.; Golzio, M.; Rols, M.-P. Magnetic silica-coated iron oxide nanochains as photothermal agents, disrupting the extracellular matrix, and eradicating cancer cells. *Cancers* **2019**, *11*, 2040. [CrossRef]

16. Tian, Q.; Hu, J.; Zhu, Y.; Zou, R.; Chen, Z.; Yang, S.; Li, R.; Su, Q.; Han, Y.; Liu, X. Sub-10 nm $Fe_3O_4@Cu_{2-x}S$ core–shell nanoparticles for dual-modal imaging and photothermal therapy. *J. Am. Chem. Soc.* **2013**, *135*, 8571–8577. [CrossRef]

17. Lin, J.; Zhao, Y.; Shi, D. Optical thermal insulation via the photothermal effects of Fe_3O_4 and $Fe_3O_4@Cu_{2-x}S$ thin films for energy-efficient single-pane windows. *MRS Commun.* **2020**, *10*, 155–163. [CrossRef]

18. Chen, B.; Le, W.; Wang, Y.; Li, Z.; Wang, D.; Lin, L.; Cui, S.; Hu, J.J.; Hu, Y.; Yang, P.; et al. Targeting negative surface charges of cancer cells by multifunctional nanoprobes. *Theranostics* **2016**, *6*, 1887–1898. [CrossRef]

19. Han, X.; Deng, Z.; Yang, Z.; Wang, Y.; Zhu, H.; Chen, B.; Cui, Z.; Ewing, R.C.; Shi, D. Biomarkerless targeting and photothermal cancer cell killing by surface-electrically-charged superparamagnetic Fe_3O_4 composite nanoparticles. *Nanoscale* **2017**, *9*, 1457–1465. [CrossRef]

20. Shi, D. Cancer cell surface negative charges: A bio-physical manifestation of the warburg effect. *Nano LIFE* **2017**, *7*, 1771001. [CrossRef]

21. Wang, Y.; Han, X.; Cui, Z.; Shi, D. Bioelectricity, its fundamentals, characterization methodology, and applications in nano-bioprobing and cancer diagnosis. *Adv. Biosyst.* **2019**, *3*, e1900101. [CrossRef] [PubMed]

22. Zhao, J.; Wu, S.; Qin, J.; Shi, D.; Wang, Y. Electrical-charge-mediated cancer cell targeting via protein corona-decorated superparamagnetic nanoparticles in a simulated physiological environment. *ACS Appl. Mater. Interfaces* **2018**, *10*, 41986–41998. [CrossRef] [PubMed]

23. Wu, S.; Wang, Y.; Shi, D. Positively charged magnetic nanoparticles for capture of circulating tumor cells from clinical blood samples. *Nano LIFE* **2020**, *10*, 1971001. [CrossRef]

24. Wu, S.; Gu, L.; Qin, J.; Zhang, L.; Sun, F.; Liu, Z.; Wang, Y.; Shi, D. Rapid label-free isolation of circulating tumor cells from patients' peripheral blood using electrically charged Fe_3O_4 nanoparticles. *ACS Appl. Mater. Interfaces* **2020**, *12*, 4193–4203. [CrossRef] [PubMed]

25. Deng, Z.; Kalin, G.T.; Shi, D.; Kalinichenko, V.V. Nanoparticle delivery systems with cell-specific targeting for pulmonary diseases. *Am. J. Respir. Cell Mol. Biol.* **2021**, *64*, 292–307. [CrossRef] [PubMed]

26. Yoo, J.; Park, C.; Yi, G.; Lee, D.; Koo, H. Active targeting strategies using biological ligands for nanoparticle drug delivery systems. *Cancers* **2019**, *11*, 640. [CrossRef]

27. Frigerio, B.; Bizzoni, C.; Jansen, G.; Leamon, C.P.; Peters, G.J.; Low, P.S.; Matherly, L.H.; Figini, M. Folate receptors and transporters: Biological role and diagnostic/therapeutic targets in cancer and other diseases. *J. Exp. Clin. Cancer Res.* **2019**, *38*, 1–12. [CrossRef]

28. Yu, J.; He, X.; Wang, Z.; Wang, Y.; Liu, S.; Li, X.; Huang, Y. Combining PD-L1 inhibitors with immunogenic cell death triggered by chemo-photothermal therapy via a thermosensitive liposome system to stimulate tumor-specific immunological response. *Nanoscale* **2021**, *13*, 12966–12978. [CrossRef]

29. Nam, K.C.; Han, Y.S.; Lee, J.-M.; Kim, S.C.; Cho, G.; Park, B.J. Photo-functionalized magnetic nanoparticles as a nanocarrier of photodynamic anticancer agent for biomedical theragnostics. *Cancers* **2020**, *12*, 571. [CrossRef]

30. Nejadshafiee, V.; Naeimi, H.; Goliaei, B.; Bigdeli, B.; Sadighi, A.; Dehghani, S.; Lotfabadi, A.; Hosseini, M.; Nezamtaheri, M.S.; Amanlou, M.; et al. Magnetic bio-metal–organic framework nanocomposites decorated with folic acid conjugated chitosan as a promising biocompatible targeted theranostic system for cancer treatment. *Mater. Sci. Eng. C* **2019**, *99*, 805–815. [CrossRef]

31. Moradi, R.; Mohammadzadeh, R.; Akbari, A. Kappa-carrageenan crosslinked magnetic folic acid-conjugated chitosan nanocomposites for arginase encapsulation, delivery and cancer therapy. *Nano LIFE* **2021**, *11*, 2140005. [CrossRef]

32. Wang, F.; Pauletti, G.M.; Wang, J.; Zhang, J.; Ewing, R.C.; Wang, Y.; Shi, D. Dual surface-functionalized janus nanocomposites of polystyrene/$Fe_3O_4@SiO_2$ for simultaneous tumor cell targeting and stimulus-induced drug release. *Adv. Mater.* **2013**, *25*, 3485–3489. [CrossRef] [PubMed]

33. Jin, Z.; Nguyen, K.T.; Go, G.; Kang, B.; Min, H.-K.; Kim, S.-J.; Kim, Y.; Li, H.; Kim, C.-S.; Lee, S.; et al. Multifunctional nanorobot system for active therapeutic delivery and synergistic chemo-photothermal therapy. *Nano Lett.* **2019**, *19*, 8550–8564. [CrossRef] [PubMed]

34. Zhou, M.; Li, J.; Liang, S.; Sood, A.K.; Liang, D.; Li, C. CuS nanodots with ultrahigh efficient renal clearance for positron emission tomography imaging and image-guided photothermal therapy. *ACS Nano* **2015**, *9*, 7085–7096. [CrossRef]

35. Cheng, G.; Zong, W.; Guo, H.; Li, F.; Zhang, X.; Yu, P.; Ren, F.; Zhang, X.; Shi, X.; Gao, F.; et al. Programmed size-changeable nanotheranostic agents for enhanced imaging-guided chemo/photodynamic combination therapy and fast elimination. *Adv. Mater.* **2021**, *33*, 2100398. [CrossRef]

36. Wang, S.; Lin, J.; Wang, Z.; Zhou, Z.; Bai, R.; Lu, N.; Liu, Y.; Fu, X.; Jacobson, O.; Fan, W.; et al. Core-satellite polydopamine-gadolinium-metallofullerene nanotheranostics for multimodal imaging guided combination cancer therapy. *Adv. Mater.* **2017**, *29*, 1701013. [CrossRef]

37. Liu, Y.; Bhattarai, P.; Dai, Z.; Chen, X. Photothermal therapy and photoacoustic imaging via nanotheranostics in fighting cancer. *Chem. Soc. Rev.* **2019**, *48*, 2053–2108. [CrossRef]

38. Yang, C.; Wu, T.; Qi, Y.; Zhang, Z. Recent advances in the application of vitamin E TPGS for drug delivery. *Theranostics* **2018**, *8*, 464–485. [CrossRef]

39. Cheng, X.; Zeng, X.; Li, D.; Wang, X.; Sun, M.; He, L.; Tang, R. TPGS-grafted and acid-responsive soy protein nanogels for efficient intracellular drug release, accumulation, penetration in 3D tumor spheroids of drug-resistant cancer cells. *Mater. Sci. Eng. C* **2019**, *102*, 863–875. [CrossRef]

40. Cheng, W.; Liang, C.; Xu, L.; Liu, G.; Gao, N.; Tao, W.; Luo, L.; Zuo, Y.; Wang, X.; Zhang, X.; et al. TPGS-functionalized polydopamine-modified mesoporous silica as drug nanocarriers for enhanced lung cancer chemotherapy against multidrug resistance. *Small* **2017**, *13*, 1–12. [CrossRef]

41. Bolte, C.; Ustiyan, V.; Ren, X.; Dunn, A.W.; Pradhan, A.; Wang, G.; Kolesnichenko, O.A.; Deng, Z.; Zhang, Y.; Shi, D.; et al. Nanoparticle delivery of proangiogenic transcription factors into the neonatal circulation inhibits alveolar simplification caused by hyperoxia. *Am. J. Respir. Crit. Care Med.* **2020**, *202*, 100–111. [CrossRef] [PubMed]

42. Dunn, A.W.; Kalinichenko, V.V.; Shi, D. Highly efficient in vivo targeting of the pulmonary endothelium using novel modifications of polyethylenimine: An importance of charge. *Adv. Health Mater.* **2018**, *7*, e1800876. [CrossRef]

43. Sun, F.; Wang, G.; Pradhan, A.; Xu, K.; Gomez-Arroyo, J.; Zhang, Y.; Kalin, G.T.; Deng, Z.; Vagnozzi, R.J.; He, H.; et al. Nanoparticle delivery of STAT3 alleviates pulmonary hypertension in a mouse model of alveolar capillary dysplasia. *Circulation* **2021**, *144*, 539–555. [CrossRef] [PubMed]

44. Pradhan, A.; Dunn, A.; Ustiyan, V.; Bolte, C.; Wang, G.; Whitsett, J.A.; Zhang, Y.; Porollo, A.; Hu, Y.-C.; Xiao, R.; et al. The S52F FOXF1 mutation inhibits STAT3 signaling and causes alveolar capillary dysplasia. *Am. J. Respir. Crit. Care Med.* **2019**, *200*, 1045–1056. [CrossRef] [PubMed]

45. Luiz, M.T.; Viegas, J.S.R.; Abriata, J.P.; Tofani, L.B.; Vaidergorn, M.D.M.; Emery, F.D.S.; Chorilli, M.; Marchetti, J.M. Docetaxel-loaded folate-modified TPGS-transfersomes for glioblastoma multiforme treatment. *Mater. Sci. Eng. C* **2021**, *124*, 112033. [CrossRef] [PubMed]

46. Xiang, H.; Xue, F.; Yi, T.; Tham, H.P.; Liu, J.-G.; Zhao, Y. $Cu_{2-x}S$ nanocrystals cross-linked with chlorin e6-functionalized polyethylenimine for synergistic photodynamic and photothermal therapy of cancer. *ACS Appl. Mater. Interfaces* **2018**, *10*, 16344–16351. [CrossRef]

47. Xu, Y.; Qin, Y.; Palchoudhury, S.; Bao, Y. Water-soluble iron oxide nanoparticles with high stability and selective surface functionality. *Langmuir* **2011**, *27*, 8990–8997. [CrossRef]

48. Lin, C.-A.J.; Sperling, R.A.; Li, J.K.; Yang, T.-Y.; Li, P.-Y.; Zanella, M.; Chang, W.H.; Parak, W.J. Design of an amphiphilic polymer for nanoparticle coating and functionalization. *Small* **2008**, *4*, 334–341. [CrossRef]

49. Wang, X.; Bhattacharyya, D.; Dennewitz, M.B.; Kalinichenko, V.V.; Zhou, Y.; Lepe, R.; Costa, R.H. Rapid hepatocyte nuclear translocation of the Forkhead Box M1B (FoxM1B) transcription factor caused a transient increase in size of regenerating transgenic hepatocytes. *Gene Expr.* **2003**, *11*, 149–162. [CrossRef]

50. Kalinichenko, V.V.; Gusarova, G.A.; Shin, B.; Costa, R.H. The Forkhead Box F1 transcription factor is expressed in brain and head mesenchyme during mouse embryonic development. *Gene Expr. Patterns* **2003**, *3*, 153–158. [CrossRef]

51. Ren, X.; Zhang, Y.; Snyder, J.; Cross, E.R.; Shah, T.A.; Kalin, T.V.; Kalinichenko, V.V. Forkhead Box M1 transcription factor is required for macrophage recruitment during liver repair. *Mol. Cell. Biol.* **2010**, *30*, 5381–5393. [CrossRef]

52. Cai, Y.; Bolte, C.; Le, T.; Goda, C.; Xu, Y.; Kalin, T.V.; Kalinichenko, V.V. FOXF1 maintains endothelial barrier function and prevents edema after lung injury. *Sci. Signal.* **2016**, *9*, ra40. [CrossRef] [PubMed]

53. Kim, I.-M.; Zhou, Y.; Ramakrishna, S.; Hughes, D.E.; Solway, J.; Costa, R.H.; Kalinichenko, V.V. Functional characterization of evolutionarily conserved DNA regions in Forkhead Box F1 gene locus. *J. Biol. Chem.* **2005**, *280*, 37908–37916. [CrossRef] [PubMed]

54. Kalin, T.V.; Meliton, L.; Meliton, A.Y.; Zhu, X.; Whitsett, J.A.; Kalinichenko, V.V. Pulmonary mastocytosis and enhanced lung inflammation in mice heterozygous null for the Foxf1 gene. *Am. J. Respir. Cell Mol. Biol.* **2008**, *39*, 390–399. [CrossRef] [PubMed]

55. Bolte, C.; Zhang, Y.; Wang, I.-C.; Kalin, T.V.; Molkentin, J.D.; Kalinichenko, V.V. Expression of Foxm1 transcription factor in cardiomyocytes is required for myocardial development. *PLoS ONE* **2011**, *6*, e22217. [CrossRef]

56. Ustiyan, V.; Wert, S.E.; Ikegami, M.; Wang, I.-C.; Kalin, T.V.; Whitsett, J.A.; Kalinichenko, V.V. Foxm1 transcription factor is critical for proliferation and differentiation of Clara cells during development of conducting airways. *Dev. Biol.* **2012**, *370*, 198–212. [CrossRef]

57. Roper, D.K.; Ahn, W.; Hoepfner, M. Microscale heat transfer transduced by surface plasmon resonant gold nanoparticles. *J. Phys. Chem. C* **2007**, *111*, 3636–3641. [CrossRef]

58. Chen, H.-A.; Ma, Y.-H.; Hsu, T.-Y.; Chen, J.-P. Preparation of peptide and recombinant tissue plasminogen activator conjugated poly (lactic-Co-glycolic acid) (PLGA) magnetic nanoparticles for dual targeted thrombolytic therapy. *Int. J. Mol. Sci.* **2020**, *21*, 2690. [CrossRef]
59. Uthaman, S.; Pillarisetti, S.; Hwang, H.S.; Mathew, A.P.; Huh, K.M.; Rhee, J.H.; Park, I.-K. Tumor microenvironment-regulating immunosenescence-independent nanostimulant synergizing with near-infrared light irradiation for antitumor immunity. *ACS Appl. Mater. Interfaces* **2021**, *13*, 4844–4852. [CrossRef] [PubMed]
60. Gibori, H.; Eliyahu, S.; Krivitsky, A.; Ben-Shushan, D.; Epshtein, Y.; Tiram, G.; Blau, R.; Ofek, P.; Lee, J.S.; Ruppin, E.; et al. Amphiphilic nanocarrier-induced modulation of PLK1 and miR-34a leads to improved therapeutic response in pancreatic cancer. *Nat. Commun.* **2018**, *9*, 1–18. [CrossRef]
61. Zheng, X.; Pan, D.; Chen, M.; Dai, X.; Cai, H.; Zhang, H.; Gong, Q.; Gu, Z.; Luo, K. Tunable hydrophile–lipophile balance for manipulating structural stability and tumor retention of amphiphilic nanoparticles. *Adv. Mater.* **2019**, *31*, e1901586. [CrossRef] [PubMed]

Article

Magnetically Guided Localization Using a Guiding-Marker System® and a Handheld Magnetic Probe for Nonpalpable Breast Lesions: A Multicenter Feasibility Study in Japan

Tomoko Kurita [1,*], Kanae Taruno [2], Seigo Nakamura [2,3], Hiroyuki Takei [1], Katsutoshi Enokido [4], Takashi Kuwayama [2], Yoko Kanada [2], Sadako Akashi-Tanaka [2], Misaki Matsuyanagi [4], Meishi Hankyo [1], Keiko Yanagihara [1], Takashi Sakatani [5], Kentaro Sakamaki [6], Akihiro Kuwahata [7,8], Masaki Sekino [7] and Moriaki Kusakabe [9,10]

[1] Department of Breast Surgery and Oncology, Nippon Medical School Hospital, Tokyo 113-8603, Japan; takei-hiroyuki@nms.ac.jp (H.T.); m-fanchiang@nms.ac.jp (M.H.); keikof@nms.ac.jp (K.Y.)

[2] Division of Breast Surgical Oncology, Department of Surgery, Showa University School of Medicine, Tokyo 142-8666, Japan; ktaruno@med.showa-u.ac.jp (K.T.); seigonak@med.showa-u.ac.jp or seigonak@gmail.com (S.N.); kuwayama@med.showa-u.ac.jp (T.K.); yoko02410198@yahoo.co.jp (Y.K.); sakashi@med.showa-u.ac.jp (S.A.-T.)

[3] Tianjin's Clinical Research Center for Cancer, Tianjin Medical University Cancer Institute and Hospital, Tianjin 300060, China

[4] Division of Breast Surgical Oncology, Department of Surgery, Showa University Fujigaoka Hospital, Kanagawa 227-8501, Japan; enotoshi@med.showa-u.ac.jp (K.E.); misakimatsuyanagi@yahoo.co.jp (M.M.)

[5] Department of Diagnostic Pathology, Nippon Medical School Hospital, Tokyo 113-8603, Japan; takashi-sakatani@nms.ac.jp

[6] Center for Data Science, Yokohama City University, Yokohama 236-0027, Japan; sakamaki@yokohama-cu.ac.jp

[7] Graduate School of Engineering, University of Tokyo, Tokyo 113-8656, Japan; akihiro.kuwahata.b1@tohoku.ac.jp or kuwahata@bee.t.u-tokyo.ac.jp (A.K.); sekino@g.ecc.u-tokyo.ac.jp (M.S.)

[8] School of Engineering, Tohoku University, Miyagi 980-8579, Japan

[9] Graduate School of Agricultural and Life Sciences, University of Tokyo, Tokyo 113-8657, Japan; amkusa@g.ecc.u-tokyo.ac.jp or kusa3matrix@gmail.com

[10] Department of Research and Development, Matrix Cell Research Institute Inc., Tokyo 101-0025, Japan

[*] Correspondence: tomoko28@nms.ac.jp

Citation: Kurita, T.; Taruno, K.; Nakamura, S.; Takei, H.; Enokido, K.; Kuwayama, T.; Kanada, Y.; Akashi-Tanaka, S.; Matsuyanagi, M.; Hankyo, M.; et al. Magnetically Guided Localization Using a Guiding-Marker System® and a Handheld Magnetic Probe for Nonpalpable Breast Lesions: A Multicenter Feasibility Study in Japan. *Cancers* **2021**, *13*, 2923. https://doi.org/10.3390/cancers13122923

Academic Editor: David Hardisson

Received: 4 May 2021
Accepted: 8 June 2021
Published: 11 June 2021

Publisher's Note: MDPI stays neutral with regard to jurisdictional claims in published maps and institutional affiliations.

Simple Summary: In this multicenter feasibility study, non-palpable breast lesions in 89 patients were localized using a handheld cordless magnetic probe (TAKUMI) and a magnetic marker (Guiding-Marker System®). Additionally, a dye was injected subcutaneously under ultrasound guidance to indicate the extent of the tumor. Consequently, a magnetic marker was detected in all resected specimens, and the initial surgical margin was positive only in five (6.1%) of 82 patients. Thus, the magnetic guiding localization system with ultrasound guidance is useful for the detection and excision of non-palpable breast lesions.

Abstract: Accurate pre-operative localization of nonpalpable lesions plays a pivotal role in guiding breast-conserving surgery (BCS). In this multicenter feasibility study, nonpalpable breast lesions were localized using a handheld magnetic probe (TAKUMI) and a magnetic marker (Guiding-Marker System®). The magnetic marker was preoperatively placed within the target lesion under ultrasound or stereo-guidance. Additionally, a dye was injected subcutaneously to indicate the extent of the tumor excision. Surgeons checked for the marker within the lesion using a magnetic probe. The magnetic probe could detect the guiding marker and accurately localize the target lesion intraoperatively. All patients with breast cancer underwent wide excision with a safety margin of ≥5 mm. The presence of the guiding-marker within the resected specimen was the primary outcome and the pathological margin status and re-excision rate were the secondary outcomes. Eighty-seven patients with nonpalpable lesions who underwent BCS, from January to March of 2019 and from January to July of 2020, were recruited. The magnetic marker was detected in all resected specimens. The surgical margin was positive only in 5/82 (6.1%) patients; these patients underwent re-excision. This feasibility study demonstrated that the magnetic guiding localization system is useful for the detection and excision of nonpalpable breast lesions.

Keywords: nonpalpable breast lesion; breast cancer; magnetic maker; magnetic probe; surgery

1. Introduction

The introduction of mammographic screening programs has led to the identification of an increased number of nonpalpable breast lesions. Currently, in developed countries, approximately 20% to 30% of detected breast cancer cases are nonpalpable [1,2]. Moreover, due to the development of neoadjuvant chemotherapy, there has been an increase in the number of patients in whom a complete response was obtained [3]. These lesions are often difficult to identify preoperatively and intraoperatively. Several techniques have been developed for the localization of nonpalpable breast cancers [1,4–7]. The two most established techniques for pre-operative localization of nonpalpable breast lesions are wire-guided localization (WGL) and radioactive seed localization (RSL). WGL involves the percutaneous implantation of a hooked wire under image guidance to mark the center or outer edges of target lesions. Although WGL is the most commonly used method [8], it has several disadvantages, including mechanical stimulation of wire plucking, kinking, and patient discomfort. On the other hand, RSL involves implanting a small radioactive seed to identify the lesion and/or its borders. RSL overcomes many of the disadvantages of WGL, but it requires a strict nuclear regimen, which is its main limitation for applicability at hospitals.

Recently, a non-wire, non-radioactive localization technique, known as the magnetically guided localization (MGL) method, has been developed as an alternative to WGL and RSL. The MGL method uses a handheld magnetic probe with a cord (Sentimag, Endomagnetics Ltd., Cambridge, England and Wales) and a magnetic marker (Magseed, Endomagnetics Ltd., Cambridge, England and Wales) for localization of nonpalpable lesions [5,9–13]. Currently, the MGL method appears to be a feasible and safe method of breast lesion localization. However, they are commercially not available in Japan. On the other hand, in Japan, a handheld cordless magnetic probe (TAKUMI, Matrix cell Research Institute Inc., Tokyo, Japan) has been developed to detect sentinel lymph nodes (SLNs) in breast cancer patients [14,15]. In the identification rates of SLNs, the MGL method is not inferior to the gamma probe and dye-guided method [15–17]. In this study, we evaluated the feasibility of an occult lesion localization technique using the handheld cordless magnetic probe (TAKUMI) and the magnetic marker (Guiding-Marker System®, Hakko Co., Ltd., Nagano, Japan). The primary outcome was the successful identification of the guiding marker in the excised specimen. Surgical margin status and re-excision rates were evaluated as the secondary outcomes.

2. Materials and Methods

2.1. Patients

Patients with nonpalpable breast cancer, who were histologically diagnosed using core needle biopsy (CNB) or vacuum assisted biopsy (VAB), at three hospitals (Nippon Medical School Hospital, Showa University Hospital, and Showa University Fujigaoka Hospital) were enrolled. In this study, two types of magnetic probes (TAKUMI) were used; the first type of magnetic probe was used from January 2019 to March 2019 (Figure 1a: generation 1 (Gen.1)), and the second type of magnetic probe from January 2020 to July 2020 (Figure 1b: generation 2 (Gen.2)).

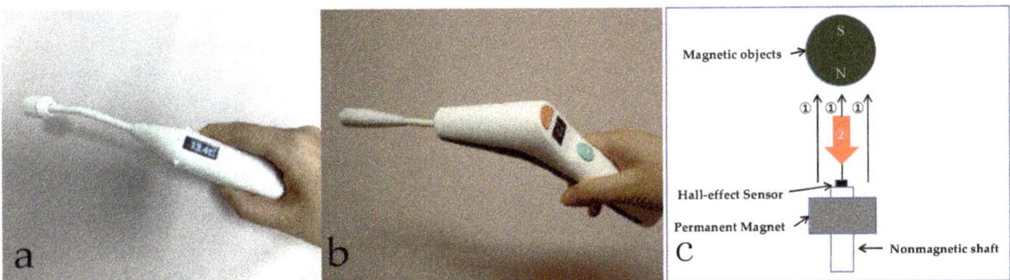

Figure 1. The magnetic probe TAKUMI (Matrix cell Research Institute Inc., Tokyo, Japan): (**a**) First type: Gen.1; (**b**) Second type: Gen.2; (**c**) The detection mechanism of TAKUMI. ① The objects are magnetized by the magnetic fields from the permanent magnet. ② The newly generated magnetic field from the magnetized objects is detected by the Hall effect sensor.

The female patients, aged $\geq$20 years, who met the following criteria were included: underwent breast-conserving surgery (BCS) for breast cancer; received neoadjuvant chemotherapy; and underwent tumor excision for indeterminate or suspicious on CNB or VAB. The patients who were pregnant, had inflammatory breast cancer, underwent breast implant insertion, and/or had a metal allergy were excluded. The breast lesion, which could not be palpated during pre-operative examination by the surgeons, was defined as a nonpalpable breast lesion. Written informed consent was obtained from all participants.

2.2. Materials

The TAKUMI is a handheld cordless magnetic probe equipped with a permanent magnet and a Hall effect sensor for detecting magnetic objects. The objects are magnetized by the magnetic fields from the permanent magnet, and the newly generated magnetic field from the magnetized objects is detected by the sensor. The value of the detected signal is visible on a small display, and sounds are produced according to the detected values (Figure 1a–c). It was developed at the University of Tokyo under a grant from the Japan Agency for Medical Research and Development. The second type of TAKUMI (Gen.2) is commercially available for sentinel lymph node biopsy and mammary occult lesion localization, which has regulatory approval in Europe for medical device safety (CE marking of conformity, NB:0344, EC certificate No.: 4201663CE01). The second type of TAKUMI (Gen.2) has been improved with the addition of a push-button and easy battery replacement.

The Guiding-Marker System® was used as a magnetic marker. It consists of a stainless-steel hook connected to a 30-cm long 5-0 nylon monofilament suture (Figure 2a). The tip of the marker is bent, and the size of the marker is φ 0.28 mm $\times$ 10 mm in length (Figure 2b). The Guiding-Marker System® has been used for thoracoscopic resection of pulmonary nodules [18,19] and for MRI-guided breast lesion mapping [20]. The guiding marker was inserted into the center of the target lesions under ultrasound guidance, using a 21-gauge 10 cm long steel needle. It is similar as the insertion of a conventional breast maker. The flexible nylon suture is not associated with mechanical stimulation of wire plucking, kinking, and patient discomfort seen with WGL.

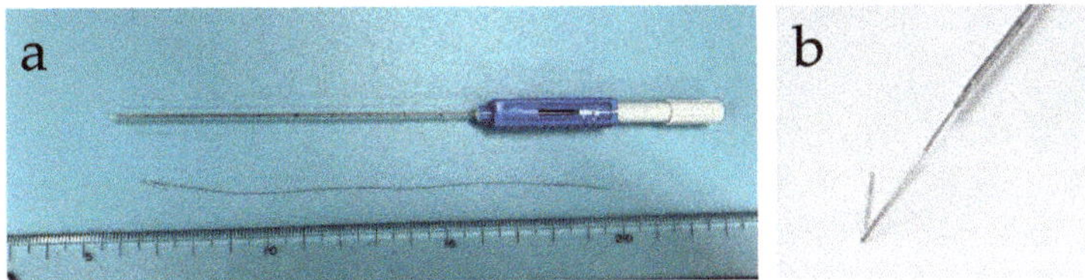

Figure 2. The Guiding-Marker System® (Hakko, Nagano, JAPAN). (**a**) It consists of a stainless-steel hook connected to a nylon thread and a 21-gauge 10 cm long steel needle. (**b**) The tip of the marker is bent, and the size of the marker is φ 0.28 mm $\times$ 10 mm.

2.3. Surgical Procedures

A 21-gauge 10-cm long steel needle was used to insert the Guiding-Marker System® into the center of the target lesions under ultrasound guidance before surgery (Figure 3a). Only in cases with microcalcification, stereotaxic mammography was used for guidance. The markers were placed the day before surgery, or before surgery under anesthesia. An ultrasound and TAKUMI were used to confirm whether the marker was located within the lesion, and to mark on the skin by ink before incision. The visibility of the puncture needle and the marker was ensured under ultrasound during the procedure (Figure 3a–c). The procedure of the Guiding-Marker System® insertion was performed by experienced breast surgeons.

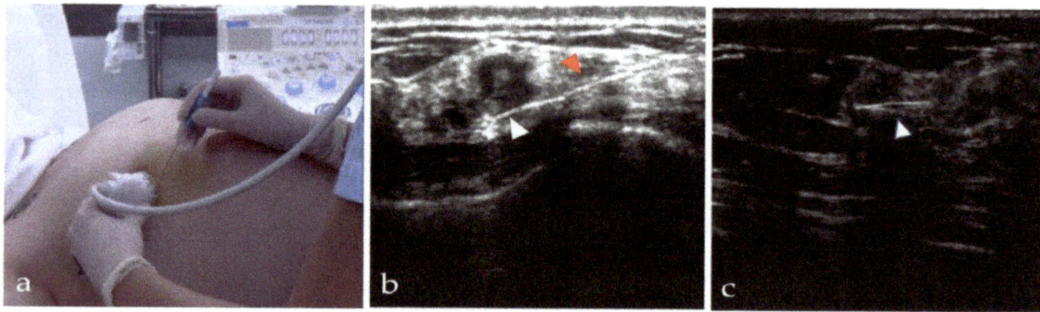

Figure 3. (a–c) The procedure of the Guiding-Marker System® insertion: (a) A needle was inserted using the ultrasound guidance; (b,c) A needle and a guiding marker have clear visibility under the ultrasound. The white arrowheads in **b** and **c** indicate guiding markers. The red arrowhead in **b** indicates a puncture needle.

After sentinel lymph node (SLN) biopsy, a small amount of sterile gentian violet or indigocarmine with gel was injected subcutaneously at several points, at least 5 mm from the edge of the tumor under sonographic guidance, to indicate the ductal spread of the tumor to be excised (Figure 4b).

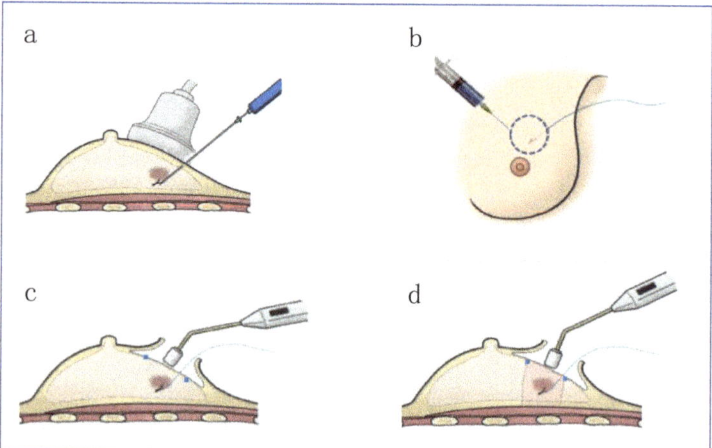

Figure 4. The surgical procedures: (a) A marker was inserted into the center of the target lesions using ultrasound guidance; (b) A small amount of sterile gentian violet with gel was injected subcutaneously, at least 5 mm from the edge of the tumor under sonographic guidance; (c) TAKUMI was used to detect the magnetic marker in the area to be excised; (d) The tumor was resected cylindrically.

Then, a skin incision was made over/outside the lesion or areolar line, and TAKUMI in a sterile bag was used to confirm the magnetic marker within the dye-marked area (Figure 4c). Since a magnetic probe reacts to materials containing iron, titanium muscle retractors and surgical equipment were used while checking the position of the lesion intraoperatively. The tumor was resected cylindrically (Figure 4d). Once the specimen was excised, the magnetic probe was used to confirm the presence of a magnetic marker in the resected breast tissue. A radiograph of the specimen was taken to confirm the presence of the lesion and the magnetic marker within the resected specimen (Figure 5). All surgical procedures were performed by experienced breast surgeons.

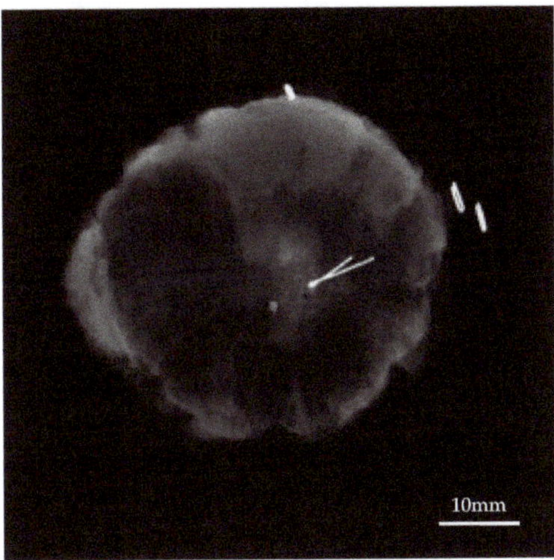

Figure 5. The specimen mammogram showing the calcified lesion and a magnetic marker within the resected specimen.

2.4. Pathological Examination

The surgical margin of the resected breast tissue, depending on the physicians' discretion, was histologically examined on frozen section during the surgery. If the surgical margin was positive for cancer cells on the frozen section, additional breast tissue, corresponding to the positive surgical margin, was resected. On the other hand, a second surgery was performed when the surgical margin was positive for invasive carcinoma on the ink margin on the formalin-fixed paraffin-embedded section. Therefore, the margin status, in most cases, was determined on initially resected breast tissue. Margins were classified based on consensus guidelines [21,22]. For invasive cancer, the margin was considered positive if the tumor was found on the ink margin. For ductal carcinoma in situ (DCIS), the margin was positive if DCIS was present on the ink margin and it was close if DCIS was found within <2 mm.

3. Results

Eighty-seven patients were recruited in the study, including 39 patients from January 2019 to March 2019 and 48 patients from January 2020 to July 2020. Eighty-two patients (94.3%) underwent partial mastectomy for breast cancer treatment, five (5.7%) underwent tumor excision for diagnostic purposes. Patient demographics and tumor characteristics are shown in Table 1. The mean age was 54.4 years (range, 33–88 years). Localization markers were placed for masses (71 patients) under sonographic guidance and for calcification (16 patients) under mammographic guidance. Histologically, 64 patients had invasive carcinoma, and 23 had DCIS.

Table 1. Patients and tumor characteristics.

	Gen.1	Gen.2	Total
No. of patients	n = 39	n = 48	n = 87
Ages, y/o (mean)	40–76 (54.4)	33–88 (54.4)	33–88 (54.4)
Menopausal status (Pre/Post/unknown)	14/25/0	22/23/3	36/48/3
Tumor status			
Tumor/low echoic lesion	34 (87.2%)	37 (77.1%)	71 (81.6%)
Microcalcification	5 (12.8%)	11(22.9%)	16 (18.4%)
The size of the lesion			
Clinical tumor size, exclude microcalcification, mm (mean)	0–23 (10.2)	5–30 (11.8)	0–33 (11.1)
Pathological size of lesion, mm (mean)	0–60 (18.4)	0–40 (14.6)	0–60 (16.8)
Histological type of lesion			
Carcinoma in situ	6/39 (15.4%)	17/48 (35.4%)	23/87 (26.4%)
IDC	31/39 (79.5%)	27/48	58/87 (66.6%)
ILC	0/39 (0%)	3/48 (6.3%)	3/87 (3.5%)
Other invasive carcinoma	2/39 (5.1%)	1/48 (2.1%)	3/87 (3.5%)
Histological subtype of breast cancer			
Luminal A	29/39 (74.4%)	41/48 (85.4%)	70/87 (80.5%)
Luminal B	3/39 (7.7%)	4/48 (8.3%)	7/87 (8.0%)
HER2 enriched	3/39 (7.7%)	2/48(4.2%)	5/87 (5.7%)
Triple negative	4/39 (10.2%)	1/48 (2.1%)	5/87(5.7%)
Neoadjuvant chemotherapy	6/39 (15.4%)	3/48 (6.3%)	9/87 (10.3%)
Pathological complete response	3/39 (7.7%)	0/48 (0%)	3/87 (3.5%)

IDC: invasive ductal carcinoma; ILC: invasive lobular carcinoma.

The study outcomes are reported in Table 2. Technical success was defined as the presence of the magnetic marker in the excised specimen on the postoperative specimen radiograph. In 85 out of 87 patients (97.7%), the magnetic markers were detectable on magnetic probing, both during and after the surgical resection of the target lesion. However, in two cases, the magnetic marker, which could not be detected before surgery, was detected intraoperatively. In all cases, guiding markers were removed during the initial surgical operation. The mean size of the lesion was 11.1 mm along the longest dimension, ranging from 0 mm to 33 mm, including non-invasive carcinoma. The size of lesions with pathological complete response after chemotherapy were measured as 0 mm. The mean weight of all excised specimens was 39.9 g, ranging from 2 g to 184 g, depending on the extent of the lesion. Five cases of tumor excision for diagnostic purposes are excluded from the analysis on margins status.

Table 2. The results of final pathology and the clinical outcome.

		Gen.1	Gen.2	Total
Surgery	Partial mastectomy	37/39 (94.9%)	43/45 (93.8%)	82/87 (94.3%)
	Tumor excision	2/39 (5.1%)	3/45 (6.2%)	5/87 (5.7%)
Specimen weight, g (mean)		2–131 (38.7)	5–184 (41.9)	2–184 (39.9)
Detectable rate of the Guiding-Marker System®	Transcutaneous	38/39 (97.4%)	47/48 (97.9%)	85/87 (97.7%)
	Intramammary	39/39 (100%)	48/48 (100%)	87/87 (100%)
Removal rate of the marker in the specimen		39/39 (100%)	48/48 (100%)	87/87 (100%)
Histological examination of surgical margin	Intraoperative frozen section	7 /37 (18.9%)	9/45 (20.0%)	16 /82 (19.5%)
	Postoperative permanent section	30/37 (82.1%)	36/45 (80.0%)	66/82 (80.4%)
Positive margin status on initial resection	Intraoperative re-excision	3/37 (8.1%)	2 /45(4.4%)	5/82 (6.1%)
	Re-excision on second operation	1/37 (2.7%)	0 /45(0%)	1/82 (1.2%)
Radiation	Whole breast radiation	39/39 (100%)	48/48 (100%)	87/87 (100%)
	Boost radiation	1/37 (2.7%)	5/45 (11.1%)	6/82 (7.5%)

Five (6.1%) of 82 patients, diagnosed with breast cancer on histopathology, were histologically diagnosed with positive surgical margins; four (4.9%) underwent re-excision due to positive surgical margin on the frozen section and one (1.2%) underwent a second surgery due to positive margin on the permanent section. Postoperatively, all patients received breast radiation. Six patients (7.5%) underwent boost radiation therapy because of close margin. There is no difference between the first type (Gen.1) and second type (Gen.2) of TAKUMI in the results of the clinical outcomes. No complications, including allergies and pathological responses to the marker, were observed in marker placement, both preoperatively, intraoperatively, and postoperatively. There are no complaints from patients about discomfort with mechanical stimulation of wire plucking, kinking.

4. Discussion

The localization of nonpalpable breast lesions has increasingly become an important component of BCS. The two most established techniques for pre-operative localization of nonpalpable breast lesions are WGL and RSL. However, WGL and RSL, requiring a wire and radioisotope, respectively, have several disadvantages. Hence, the ideal method would be a non-wire, non-radioactive localization method that does not require an energy source. MGL appeared as an effective alternative to WGL and RSL that overcomes the disadvantages of these techniques.

A feasibility study of the MGL technique was performed in the UK [9]. They used Sentimag as a magnetic probe and Magseed as a magnetic tracer. Consequently, all 20 patients with nonpalpable breast cancer underwent successful surgical excision guided by a magnetic probe; however, a surgical re-excision was required in two (10%) patients with positive surgical margins. In a review and pooled analysis of 1559 surgical excisions, Gera et al. [23] reported a successful localization and retrieval rate of 99.9% and a relatively low re-excision rate of 11.3%. Micha et al. [24] reported that the Magseed group obtained the better satisfaction for clinicians in terms of the technical aspects, and also decreased the anxiety of patients between localization and surgery, with comparison to the WGL group. Thus, several studies have demonstrated that MGL using Magseed is an easy, sensitive, and effective localization method [5,10,23,25]. The use of Magseed technology for lesion localization gained US Food and Drug Administration approval in 2016. MGL has been a beneficial addition to the rapidly developing breast localization technologies.

In Japan, neither the magnetic probe (Sentimag) nor the magnetic marker (Magseed) is commercially available. However, a novel handheld cordless magnetic probe (TAKUMI) has been developed to detect the SLNs in breast cancer patients [14,15]. Compared to Sentimag, there is an important difference in operation principle to be realized in the compact shape of TAKUMI. Sentimag utilizes an alternating current (AC) magnetic field generated by an AC power supply. In contrast, TAKUMI utilizes direct current (DC) magnetic fields generated by a permanent magnet without any AC power supply. TAKUMI could be operated with a small battery, therefore, that would make it possible to realize the compact shape and the cordless device.

In BCS, there is a conflict between obtaining an adequate excision margin around the tumor and not removing too much tissue, which may result in breast deformity [26]. Although the MGL technique could detect the guiding marker and give us an accurate localization of the target lesions during the surgery, it could not indicate the extent of the lesion. In several studies evaluating MSL using Magseed, the rate of positive surgical margins ranged from 10% to 16% [11–13,25]. These results are not so different from those with WGL (20–70%) [27] and RSL (7–27%) [1,28]. On the other hand, ultrasound guided BCS can overcome the problem of positive surgical margins, although it may miss the location of a small tumor during surgery [29]. Therefore, we verified the MGL method for nonpalpable breast lesions by using a handheld cordless magnetic probe (TAKUMI) and magnetic marker (Guiding-Marker System®).

In this study, a dye was injected subcutaneously under ultrasound guidance to indicate the extent of the tumor excision. Consequently, the surgical margin was positive only in

5 (6.1%) of 82 patients, and 6 (7.5%) of 82 patients underwent boost radiation therapy because of close margin. No complication, including allergies and pathological response to the marker, were observed in marker placement preoperatively, intraoperatively, or postoperatively. Thus, the MGL with ultrasound guidance is useful for the detection and excision of nonpalpable breast lesions. However, the present study is a single-arm feasibility study. To further evaluate the usefulness of the magnetic marker and probe method, we would like to conduct a double-blind study, comparing it with conventional methods with a greater sample size and more surgeons.

There are several limitations associated with the use of MGL, including limited use of metal instruments during surgery and limited depth of detection. Because of the magnetic property of the probe, metal instruments containing iron should be kept away from the tip of the detection probe [9]. In this study, titanium muscle retractors and surgical equipment were used while checking the position of the lesion intraoperatively. Recently, a new non-radioactive wireless localization system, called the Magnetic Occult Lesion Localization Instrument (MOLLI), has been developed in the Sunnybrook Odette Cancer Center. The effect of surgical instruments on MOLLI function is minimal, and does not impact its accuracy or reliability [30]. On the other hand, Magseed, which is made of stainless-steel (approximately φ 1 mm $\times$ 5 mm in length), can only detect up to a depth of 4 cm, whereas the MOLLI system, which has a custom-made MOLLI marker (φ 1.6 mm $\times$ 3.8 mm in length), can detect up to a depth of 53 mm [5]. The Guiding-Marker System® in this study consists of a stainless-steel hook (φ 0.28 mm $\times$ 10 mm in length) connected to a 30-cm long 5-0 nylon monofilament suture. In preclinical studies (not published) the Guiding-Marker System® was detectable up to depths of 31 mm when the marker was oriented parallel to the longitudinal axis of the probe. The detectable depth was 21 mm when the marker was oriented perpendicularly. In this study, the magnetic marker could not be detected transcutaneously in 2 cases due to the detectable depth limit, however, these were detected intramammary intraoperatively. The nylon suture helps to avoid losing the marker, so that all the markers were removed during the initial surgical operation. The nylon suture resolved the patient's discomfort compare to the wire on WGL method. The Guiding-Marker System® was originally developed for the preoperative localization of pulmonary tumors [18], but it may be too shallow to detect a deeply located breast lesion. Technological development of the magnetic marker could resolve this limitation.

5. Conclusions

We evaluated the feasibility of an occult lesion localization technique using a handheld cordless magnetic probe (TAKUMI) and a magnetic marker (Guiding-Marker System®). The high rate of successful localization and low rate of re-excision support the effectiveness of MGL. The MGL method is a reliable, accurate, and convenient system for localizing nonpalpable breast lesions. This technique does not have disadvantages that are commonly associated with WGL and RGL. However, MGL also has several limitations, such as a limited ability to use surgical metal instruments and limited depth of detection during surgery. Therefore, we need to develop more effective technologies that can, besides localization, determine the extent of the tumor excision.

Author Contributions: T.K. (Tomoko Kurita) and K.T. served as joint first authors, each with an equal contribution to the manuscript. Protocol advice was provided by S.N. and H.T., and patient recruitment was provided by K.E., T.K. (Takashi Kuwayama), Y.K., S.A.-T., M.M., M.H. and K.Y. Pathological advice was provided by T.S., and statistical advice was provided by K.S. The technical support and development of magnetic probe was provided by A.K., M.S. and M.K. All authors have read and agreed to the published version of the manuscript.

Funding: This project was funded by the Japan Agency for Medical Research and Development, Medi-cal-engineering cooperation project (Development of medical devices and systems for advanced medical services, grant number 18he0902007h0004, Development of medical devices through collaboration between medicine and industry, grant number 18he1302026j0001 and 19he1302026j0002).

Institutional Review Board Statement: The study was conducted according to the guidelines of the Declaration of Helsinki, and approved by the Institutional Review Board of Nippon Medical school Foundation (CRB3180001, CRB3180001 protocol code nms2018-0801-01 approved 15 December 2018, nms2019-1001approved 6 January 2020). The study was registered at https://jrct.niph.go.jp/re (protocol code jRCTs032180422 (accessed on 30 December 2018), protocol code jRCTs032190189 (accessed on 17 January 2020).

Informed Consent Statement: Informed consent was obtained from all subjects involved in the study.

Data Availability Statement: The data presented in this study are available in this article.

Acknowledgments: The authors thank Takayuki Nakagawa, Ph.D., Kohei Saeki, Ph.D., Norio Tanabe and Itsuro Sai-to, Ph.D. for technical support and development of magnetic probe, and Ryoko Kobayashi and Kayoko Maeda for data management.

Conflicts of Interest: The authors declare no conflict of interest.

References

1. Sharek, D.; Zuley, M.L.; Zhang, J.Y.; Soran, A.; Ahrendt, G.M.; Ganott, M.A. Radioactive Seed Localization Versus Wire Localization for Lumpectomies: A Comparison of Outcomes. *Am. J. Roentgenol.* **2015**, *204*, 872–877. [CrossRef]
2. Lynge, E.; Ponti, A.; James, T.; Májek, O.; von Euler-Chelpin, M.; Anttila, A.; Fitzpatrick, P.; Frigerio, A.; Kawai, M.; Scharpantgen, A.; et al. Variation in detection of ductal carcinoma in situ during screening mammography: A survey within the International Cancer Screening Network. *Eur. J. Cancer* **2014**, *50*, 185–192. [CrossRef] [PubMed]
3. Morrow, M.; Khan, A.J. Locoregional Management After Neoadjuvant Chemotherapy. *J. Clin. Oncol.* **2020**, *38*, 2281–2289. [CrossRef] [PubMed]
4. Benson, J.R.; Jatoi, I.; Keisch, M.; Esteva, F.J.; Makris, A.; Jordan, V.C. Early breast cancer. *Lancet* **2009**, *373*, 1463–1479. [CrossRef]
5. Hong, N.L.; Wright, F.C.; Semple, M.; Nicolae, A.M.; Ravi, A. Results of a phase I, non-randomized study evaluating a Magnetic Occult Lesion Localization Instrument (MOLLI) for excision of non-palpable breast lesions. *Breast Cancer Res. Treat.* **2019**, *179*, 671–676. [CrossRef]
6. Cox, C.E.; Garcia-Henriquez, N.; Glancy, M.J.; Whitworth, P.; Cox, J.M.; Themar-Geck, M.; Prati, R.; Jung, M.; Russell, S.; Appleton, K.; et al. Pilot Study of a New Nonradioactive Surgical Guidance Technology for Locating Nonpalpable Breast Lesions. *Ann. Surg. Oncol.* **2016**, *23*, 1824–1830. [CrossRef] [PubMed]
7. Aydogan, T.; Sezgin, E.; Ilvan, S.; Yilmaz, O.C.; Aslan, N.; Golshan, M.; Aydogan, F. Comparison of Radio-guided Occult Lesion Localization (ROLL) and Magnetic Occult Lesion Localization (MOLL) for Non-palpable Lesions: A Phantom Model Study. *Clin. Breast Cancer* **2020**, *20*, e9–e13. [CrossRef]
8. Hall, F.M.; Frank, H.A. Preoperative localization of nonpalpable breast lesions. *AJR Am. J. Roentgenol.* **1979**, *132*, 101–105. [CrossRef] [PubMed]
9. Ahmed, M.; Anninga, B.; Goyal, S.; Young, P.; Pankhurst, Q.A.; Douek, M.; on behalf of the MagSNOLL Trialists Group. Magnetic sentinel node and occult lesion localization in breast cancer (MagSNOLL Trial). *BJS* **2015**, *102*, 646–652. [CrossRef]
10. Harvey, J.R.; Lim, Y.; Murphy, J.; Howe, M.; Morris, J.; Goyal, A.; Maxwell, A.J. Safety and feasibility of breast lesion localization using magnetic seeds (Magseed): A multi-centre, open-label cohort study. *Breast Cancer Res. Treat.* **2018**, *169*, 531–536. [CrossRef]
11. Price, E.R.; Khoury, A.L.; Esserman, L.J.; Joe, B.N.; Alvarado, M.D. Initial Clinical Experience with an Inducible Magnetic Seed System for Preoperative Breast Lesion Localization. *Am. J. Roentgenol.* **2018**, *210*, 913–917. [CrossRef] [PubMed]
12. Zacharioudakis, K.; Down, S.; Bholah, Z.; Lee, S.; Khan, T.; Maxwell, A.J.; Howe, M.; Harvey, J. Is the future magnetic? Magseed localisation for non palpable breast cancer. A multi-centre non randomised control study. *Eur. J. Surg. Oncol.* **2019**, *45*, 2016–2021. [CrossRef] [PubMed]
13. Pieszko, K.; Wichtowski, M.; Cieciorowski, M.; Jamont, R.; Murawa, D. Evaluation of the nonradioactive inducible magnetic seed system Magseed for preoperative localization of nonpalpable breast lesions—Initial clinical experience. *Contem. Oncol.* **2020**, *24*, 51–54. [CrossRef] [PubMed]
14. Sekino, M.; Kuwahata, A.; Ookubo, T.; Shiozawa, M.; Ohashi, K.; Kaneko, M.; Saito, I.; Inoue, Y.; Ohsaki, H.; Takei, H.; et al. Handheld magnetic probe with permanent magnet and Hall sensor for identifying sentinel lymph nodes in breast cancer patients. *Sci. Rep.* **2018**, *8*, 11–95. [CrossRef] [PubMed]
15. Taruno, K.; Kurita, T.; Kuwahata, A.; Yanagihara, K.; Enokido, K.; Katayose, Y.; Nakamura, S.; Takei, H.; Sekino, M.; Kusakabe, M. Multicenter clinical trial on sentinel lymph node biopsy using superparamagnetic iron oxide nanoparticles and a novel handheld magnetic probe. *J. Surg. Oncol.* **2019**, *120*, 1391–1396. [CrossRef]
16. Zada, A.; Peek, M.C.L.; Ahmed, M.; Anninga, B.; Baker, R.; Kusakabe, M.; Sekino, M.; Klaase, J.M.; Haken, B.T.; Douek, M. Meta-analysis of sentinel lymph node biopsy in breast cancer using the magnetic technique. *BJS* **2016**, *103*, 1409–1419. [CrossRef]
17. Douek, M.; Klaase, J.; Monypenny, I.; Kothari, A.; Zechmeister, K.; Brown, D.; Wyld, L.; Drew, P.; Garmo, H.; Agbaje, O.; et al. Sentinel node biopsy using a magnetic tracer versus standard technique: The SentiMAG multicentre trial. *Ann. Surg. Oncol.* **2013**, *21*, 1237–1245. [CrossRef]

18. Kanazawa, S.; Ando, A.; Yasui, K.; Mitani, M.; Hiraki, Y.; Shimizu, N.; Hamanaka, D.; Kojima, K. Localization of pulmonary nodules for thoracoscopic resection: Experience with a system using a short hookwire and suture. *Am. J. Roentgenol.* **1998**, *170*, 332–334. [CrossRef]

19. Miyoshi, K.; Toyooka, S.; Gobara, H.; Oto, T.; Mimura, H.; Sano, Y.; Kanazawa, S.; Date, H. Clinical outcomes of short hook wire and suture marking system in thoracoscopic resection for pulmonary nodules. *Eur. J. Cardio Thorac. Surg.* **2009**, *36*, 378–382. [CrossRef]

20. Nakamura, S. Image-guided breast conserving surgery based on 3D-MRI. *Nihon Geka Gakkai Zasshi* **2002**, *103*, 794–798.

21. Moran, M.S.; Schnitt, S.J.; Giuliano, A.E.; Harris, J.R.; Khan, S.A.; Horton, J.; Klimberg, S.; Chavez-MacGregor, M.; Freedman, G.; Houssami, N.; et al. Society of Surgical Oncology–American Society for Radiation Oncology Consensus Guideline on Margins for Breast-Conserving Surgery With Whole-Breast Irradiation in Stages I and II Invasive Breast Cancer. *Ann. Surg. Oncol.* **2014**, *21*, 704–716. [CrossRef]

22. Morrow, M.; Van Zee, K.J.; Solin, L.J.; Houssami, N.; Chavez-MacGregor, M.; Harris, J.R.; Horton, J.; Hwang, S.; Johnson, P.L.; Marinovich, M.L.; et al. Society of Surgical Oncology–American Society for Radiation Oncology–American Society of Clinical Oncology Consensus Guideline on Margins for Breast-Conserving Surgery With Whole-Breast Irradiation in Ductal Carcinoma In Situ. *J. Clin. Oncol.* **2016**, *34*, 4040–4046. [CrossRef]

23. Gera, R.; Tayeh, S.; Al-Reefy, S.; Mokbel, K. Evolving Role of Magseed in Wireless Localization of Breast Lesions: Systematic Review and Pooled Analysis of 1,559 Procedures. *Anticancer. Res.* **2020**, *40*, 1809–1815. [CrossRef] [PubMed]

24. Micha, A.E.; Sinnett, V.; Downey, K.; Allen, S.; Bishop, B.; Hector, L.R.; Patrick, E.P.; Edmonds, R.; Barry, P.A.; Krupa, K.D.C.; et al. Patient and clinician satisfaction and clinical outcomes of Magseed compared with wire-guided localisation for impalpable breast lesions. *Breast Cancer* **2021**, *28*, 196–205. [CrossRef] [PubMed]

25. Schermers, B.; van der Hage, J.; Loo, C.; Peeters, M.V.; Winter-Warnars, H.; van Duijnhoven, F.; Haken, B.T.; Muller, S.; Ruers, T.J.M. Feasibility of magnetic marker localisation for non-palpable breast cancer. *Breast* **2017**, *33*, 50–56. [CrossRef] [PubMed]

26. Noguchi, M.; Yokoi-Noguchi, M.; Ohno, Y.; Morioka, E.; Nakano, Y.; Kosaka, T.; Kurita, T. Oncoplastic breast conserving surgery: Volume replacement vs. volume displacement. *Eur. J. Surg. Oncol.* **2016**, *42*, 926–934. [CrossRef] [PubMed]

27. Besic, N.; Zgajnar, J.; Hocevar, M.; Rener, M.; Frkovic-Grazio, S.; Snoj, N.; Lindtner, J. Breast biopsy with wire localization: Factors influencing complete excision of nonpalpable carcinoma. *Eur. Radiol.* **2002**, *12*, 2684–2689. [CrossRef]

28. van der Noordaa, M.; Pengel, K.; Groen, E.; van Werkhoven, E.; Rutgers, E.; Loo, C.; Vogel, W.; Peeters, M.V. The use of radioactive iodine-125 seed localization in patients with non-palpable breast cancer: A comparison with the radioguided occult lesion localization with 99m technetium. *Eur. J. Surg. Oncol.* **2015**, *41*, 553–558. [CrossRef]

29. Gerrard, A.D.; Shrotri, A. Surgeon-led Intraoperative Ultrasound Localization for Nonpalpable Breast Cancers: Results of 5 Years of Practice. *Clin. Breast Cancer* **2019**, *19*, e748–e752. [CrossRef]

30. Nicolae, A.; Dillon, J.; Semple, M.; Hong, N.L.; Ravi, A. Evaluation of a Ferromagnetic Marker Technology for Intraoperative Localization of Nonpalpable Breast Lesions. *Am. J. Roentgenol.* **2019**, *212*, 727–733. [CrossRef]

cancers

MDPI

Article

Optimization of SPIO Injection for Sentinel Lymph Node Dissection in a Rat Model

Mirjam C. L. Peek [1,†], Kohei Saeki [2,*,†], Kaichi Ohashi [3], Shinichi Chikaki [3], Rose Baker [4], Takayuki Nakagawa [2], Moriaki Kusakabe [2], Michael Douek [1] and Masaki Sekino [3]

[1] Division of Cancer Studies, King's College London, London SE1 9RT, UK; mirjam_peek@hotmail.com (M.C.L.P.); michael.douek@kcl.ac.uk (M.D.)
[2] Graduate School of Agricultural and Life Sciences, University of Tokyo, Tokyo 113-8657, Japan; anakaga@g.ecc.u-tokyo.ac.jp (T.N.); amkusa@mail.ecc.u-tokyo.ac.jp (M.K.)
[3] Department of Electrical Engineering and Information Systems, Graduate School of Engineering, University of Tokyo, Tokyo 113-8654, Japan; ka1.ohashi.702@gmail.com (K.O.); chikaki@g.ecc.u-tokyo.ac.jp (S.C.); sekino@g.ecc.u-tokyo.ac.jp (M.S.)
[4] Department of Statistics, University of Salford, Salford M5 4WU, UK; r.d.baker@salford.ac.uk
* Correspondence: ko-saeki@vet.ous.ac.jp
† These authors contributed equally to the work.

Simple Summary: In this study, the following injection characteristics were evaluated to optimize magnetic tracer uptake in the sentinel lymph nodes (SLN) in a rat hindleg model: (a) iron dose, (b) effect of dilution, (c) effect of injecting at different time courses and (d) effect of massaging the injection site. In conclusion, injection dose and time were primary factors for the SLN iron uptake. The result from this study will provide a background for magnetic procedures.

Citation: Peek, M.C.L.; Saeki, K.; Ohashi, K.; Chikaki, S.; Baker, R.; Nakagawa, T.; Kusakabe, M.; Douek, M.; Sekino, M. Optimization of SPIO Injection for Sentinel Lymph Node Dissection in a Rat Model. *Cancers* **2021**, *13*, 5031. https://doi.org/10.3390/cancers13195031

Academic Editors: Elisabetta Abruzzese and Carlos S. Moreno

Received: 29 July 2021
Accepted: 5 October 2021
Published: 8 October 2021

Publisher's Note: MDPI stays neutral with regard to jurisdictional claims in published maps and institutional affiliations.

Abstract: The magnetic technique, consisting of a magnetic tracer and a handheld magnetometer, is a promising alternative technique for sentinel lymph node dissection (SLND) and was shown to be non-inferior to the standard technique in terms of identification rates. In this study, injection characteristics (iron dose, dilution, time course and massaging) were evaluated to optimize magnetic tracer uptake in the sentinel lymph nodes (SLN) in a rat hindleg model. 202 successful SLNDs were performed. Iron uptake in the SLN is proportional (10% utilization rate) to the injection dose between 20 and 200 µg, showing a plateau uptake of 80 µg in the SLN around 1000 µg injection. Linear regression showed that time had a higher impact than dilution, on the SLN iron uptake. Massaging showed no significant change in iron uptake. The amount of residual iron at the injection site was also proportional to the injection dose without any plateau. Time was a significant factor for wash-out of residual iron. From these results, preoperative injection may be advantageous for SLN detection as well as reduction in residual iron at the injection site by potential decrease in required injection dose.

Keywords: sentinel lymph node dissection; sentinel lymph node; super paramagnetic iron-oxide particles; rat model; magnetic technique; magnetic tracer

1. Introduction

Sentinel lymph node dissection (SLND), also referred to as sentinel lymph node biopsy, is the standard of care for clinically and radiologically node-negative breast cancer patients to stage the axilla and determine if cancer has spread to regional lymph nodes [1,2]. The current standard technique consists of a radioisotope and blue dye which are injected subcutaneously in the breast on the day of surgery [3]. However, this technique has some drawbacks including strict regulations regarding the use of radioisotopes and complications related to the use of blue dye [4]. The magnetic technique consisting of magnetic nanoparticles and a handheld magnetometer is a promising alternative technique for SLND. This magnetic technique was shown to be non-inferior to the standard technique in several trials and meta-analyses [4–17]. There are different magnetic nanoparticles on the market but the most commonly used are Sienna+ (Endomag Ltd., Cambridge, UK), Magtrace (former SiennaXP, Endomag Ltd., Cambridge, UK) and Resovist (Schering AG, Berlin, Germany) with iron concentrations of 28.0, 28.0 and 27.9 mg/mL, respectively [18,19].

Recently, it was discovered that residual magnetic nanoparticles at the injection site can lead to susceptibility artifacts on breast MRI [20]. This could be an issue in patients who need post-operative MRI such as BRCA carriers, patients who have SLND prior to primary treatment or patients whose tumors are not visible on mammography. Therefore, it is important to reduce the volume of magnetic nanoparticles injected to optimize this technique.

Most clinical trials used a magnetic nanoparticle volume of 2.0 mL diluted with saline to 5.0 mL, injected on the morning of or just prior to surgery followed by a five-minute massage of the injection site [4–10]. Hersi et al. [16] performed a study comparing injections at lower iron doses and different injecting timings, and Rubio et al. [19] compared Magtrace injections of 1.0, 1.5 and 2.0 mL. Recently, studies have successfully employed injection days or even weeks prior to surgery, revealing that preoperative injection is also feasible [21–23]. Knowledge from the previous studies indicates that iron dose, time course of injection, dilution, and massaging could affect iron uptake by the sentinel lymph nodes (SLN). Optimization of these factors is important in order to reduce the residual magnetic nanoparticles left at the injection site.

In this study, injection characteristics were evaluated to optimize magnetic tracer uptake in the SLN in a rat model. We performed multiple experiments to determine (a) the optimum iron dose, (b) the effect of dilution, (c) the effect of injecting at different time courses and (d) the effect of massaging the injection site.

2. Materials and Methods

Animal studies were approved by the local Ethics Board (Accession Number P15–124) and the experiments were performed between December 2015 and March 2016 at the University of Tokyo according to the guidelines of the institution and ensured humane care of animals. Authors adhered to the ARRIVE guidelines [24] (Data S1). No sample size calculation was performed and the number of animals in each group (n = 4–5) was determined according to the previous publication [25]. Female Sprague Dawley rats of ten weeks old (Nihon SLC, Shizuoka, Japan) (approximately 200 g) were randomly allocated to experiments. A flow diagram of the entire study can be found in Figure S1.

2.1. Dose Increase Experiment

A set volume of 100 μL was injected bilaterally in the subcutis between the second and third digits of the hind legs (Figure 1a). Resovist was injected manually on the right side and saline (control) on the left side. The magnetic tracer was diluted with saline to set the iron dose at 2, 10, 20, 40, 100, 200, 1000, 2000, 2790 and 4000 μg per 100 μL. In order to obtain the 4000 μg of iron per 100 μL of solution, two 0.8 mL vials of Resovist were centrifuged at 20 degrees Celsius for twelve hours with a relative centrifugal field of 20,000 G. After centrifugation, 0.4 mL of aqueous supernatant was carefully and aseptically discarded, and residual liquid and sediment were mixed thoroughly using a vortex mixer. A small sample (2 μL) was evaluated using a superconducting quantum interference device (SQUID) and an iron concentration of 1.4 times the standard Resovist concentration, 40.0 mg/mL, was found. All samples were evaluated for aggregation of nanoparticles using the fiber-optics particle analyzer with autosampler (FPAR-1000AS, Otsuka Electronics Co. Ltd., Osaka, Japan).

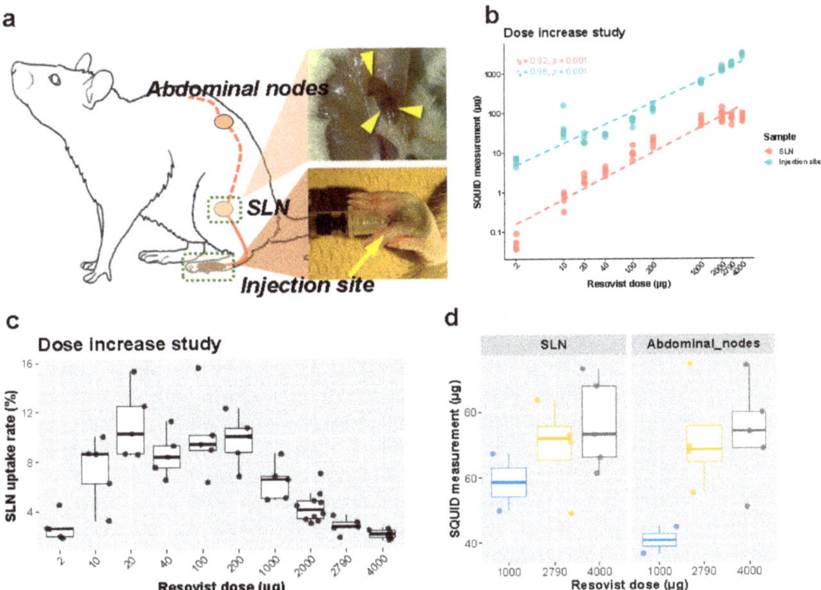

Figure 1. Schematic illustration of an animal study and iron dose increase experiment. (**a**) Schematic illustration of an animal study. (**b**) Iron accumulation measured by a superconducting quantum interference device (SQUID) in the SLNs (red) and at the injection sites (blue) 24 h after Resovist injection. Both axes are logarithmic scales. $n = 5$ each group, except for $n = 10$ of the 2000 μg group. (**c**) The SLN uptake rate in the dose increase experiment. Each measurement was overlaid on a box plot. (**d**) Iron accumulation in the secondary nodes in the 1000 μg ($n = 2$), 2790 μg ($n = 4$), and 4000 μg ($n = 5$) groups. Corresponding results from the SLN measurements are redisplayed on the left for comparison. SLN: sentinel lymph node.

Prior to injection, all rats were marked, shaved and anesthetized using a mixture of 2–3% Isoflurane (Wako Pure Chemical Industries, Osaka, Japan) and 300 mL/min air via an automatic delivery system (Isoflurane Vaporizer SN-487, Shinano Seisakusho, Tokyo Japan), first using an induction chamber and, after the rats were anesthetized properly, through a mouthpiece during the procedure. Each iron dose was injected in five rats except for 2000 μg, which was injected in two groups of five rats by two researchers (M. Peek and K. Saeki). This was done to confirm reproducibility of the results between researchers, making the total number of rats used for this experiment 55. After injection, the rats were placed back in their cages for recovery.

After 24 h, the rats were anesthetized and euthanized by cervical dislocation. In rats, the popliteal lymph nodes are the primary and dominant SLNs draining the distal hindleg including the injection site. Following euthanasia, SLND of the popliteal nodes was performed bilaterally in a prone position. The caudal skin of the stifle joint was incised, and the popliteal lymph nodes were anatomically located within the thigh muscles and dissected (Figure 1a). Collected lymph nodes were placed in formalin.

In selected animals from the 1000 ($n = 2$), 2000 ($n = 4$) and 4000 ($n = 5$) μg iron injection groups, abdominal lymph nodes were also harvested as secondary lymph nodes: Briefly, the animals were placed in dorsal recumbency, and celiotomy was performed. At the caudal furcation of the descending aorta, iliac lymph nodes were identified and resected.

For all animals, the hindlegs were amputated bilaterally at the tarsal joints, weighed, and placed in a drying oven for 48 h at 80 degrees Celsius. The dry weight of the distal legs was determined prior to powdering them using TissueLyser (20 Hz, 2 min; Qiagen, Hilden, Germany), for further analysis. The excised sentinel lymph nodes (SLN) and the powdered rat legs were analyzed by SQUID to determine the amount of iron within the samples. The SLN uptake rate was subsequently calculated as (Iron accumulation in SLN)/(Injected iron dose) × 100.

2.2. Dilution and Time-Course Experiment

Resovist equivalent to 200 μg iron (7.17 μL) was diluted with saline two-, five-, and ten-fold, which resulted in final volumes of 14.34 μL, 35.84 μL, and 71.68 μL, respectively. All samples were evaluated for aggregation of nanoparticles using the FPAR-1000AS. A set iron dose of 200 μg was chosen, as any higher iron dose would already reach the plateau level uptake, thereby enabling the evaluation of dilution and time on the iron uptake.

All rats were marked, shaved and anesthetized using the same method as described above. Indicated Resovist solutions were injected bilaterally in the subcutis between the second and third digits of the hind legs, using an automated injection pump (MCIP-Jr, Minato Concept, Tokyo, Japan). The injection duration was set at 15 s independent of differences in injection volumes. During injection, the minimum and maximum pressures were recorded. SLND was performed after 10 and 30 min and 1, 6 and 24 h. Each sampling was performed bilaterally on two rats, giving four datasets per harvesting time point per dilution, a total of 80 datasets in 40 rats.

After injection, rats were placed back in their cages for recovery and SLND was performed after the indicated time frames. All rats were anesthetized and euthanized by cervical dislocation and bilateral SLND of the popliteal nodes was performed, as described for the dose increasing experiments.

As for the animals euthanized at 24 h after injection, abdominal nodes were excised in addition to the popliteal SLNs. The excised lymph nodes were placed in formalin and analyzed with SQUID. The distal hindlegs of the rats were processed as described above and analyzed with SQUID.

2.3. Massage Experiment

The rats were anesthetized as described above. Resovist was diluted 10 times with saline, and 71.7 μL of the solution (equivalent to 200 μg iron) was manually injected bilaterally in five rats; on the right side, this was followed by a five-minute massage of the injection site. The massage was manually performed with a one-second hold and one-second release cycle on the subcutaneous dome initiated by the injection. Rats were placed back in their cages for recovery. After 30 min, the rats were anesthetized and euthanized by cervical dislocation and SLND of the popliteal nodes was performed, as described for the dose increasing experiments. Distal hindlegs were processed and both injection sites and SLNs were analyzed with SQUID, as described above.

2.4. MRI Experiments

Imaging was performed using a 7.0 T BioSpec high-field small animal MRI system (Bruker Biospin, Germany). T1-weighted (T1W) MRI images with FLASH sequence were acquired in axial orientation without fat suppression and with the following parameters: TR/TE = 892.3/5.4 ms; FOV = 60 × 60 mm; matrix = 256 × 256; slice thickness = 1.0 mm; inter-slice distance = 1.0 mm; FA = 40 degrees; isotropic in-plane resolution = 0.14 mm. The maximum diameter of the artifacts at the SLNs caused by magnetic nanoparticles was recorded.

MRI was performed in rats who were injected with 2, 20, 40, 100, 200 and 2000 μg of iron (five rats per group) during the iron increasing experiments, and two age-matched untreated rats (control). MRI was performed to evaluate the size of the artifacts at the SLNs caused by magnetic nanoparticles. The animals were euthanized 24 h after injection, immediately followed by MRI scanning and harvesting of the SLNs.

For a single rat, continuous MRI scans were performed to visualize the uptake of magnetic nanoparticles within the SLNs. The rat was anesthetized using an intravenous injection of alpha-chloralose (approximately 50 mg/kg/h, to effect), placed in a prone position and breathing was monitored whilst scanning. A bilateral injection was performed with neat (7.17 μL; left) and 10 times diluted (71.7 μL; right) Resovist, both equivalent to 200 μg iron. MRI was performed every three minutes for the first hour and every ten minutes for the subsequent four hours. After five hours, the rat was euthanized, and SLND was performed on the popliteal, groin and abdominal nodes. The nodes were placed in formalin, and the distal hindlegs were processed with SQUID, as described above.

2.5. SQUID Measurements

The magnetic moment of the magnetic tracer contained in the extracted SLNs and powdered distal hindlegs were measured using SQUID (MPMS-5S, Quantum Design Inc., San Diego, CA, USA). Each node was placed in a capsule in the middle of a plastic cylinder and placed in the machine. The SQUID consisted of a uniform measurement magnetic field with a range between −300 to +300 mT (−3000 to +3000 Oe) and a magnetic field detection coil to measure the change of interlinked magnetic flux.

The measurement region showed linear curves for living tissue due to the diamagnetic effect, and non-linear curves for living tissue containing the magnetic tracer due to the mixed diamagnetic and super-paramagnetic effects. To determine the magnetic moment of the magnetic tracer contained in the SLNs, the mixed signal was separated using the least squares method (Mathematica, Wolfram Research Inc., Champaign, IL, USA) and the non-linear curve was extracted.

2.6. Statistical Analysis

Nonparametric (two-sided) tests were used wherever possible. The correlation between iron accumulation in the SLNs and the injection sites, and the injected dose was calculated using the Spearman (rank-based) correlation. Statistical differences in iron uptake between massage and no-massage groups were assessed using the Mann–Whitney U-test. The Spearman correlation was also used to examine associations between MRI artifact size and iron uptake, dilution or time of injection. A parametric analysis had to be used to simultaneously regress iron accumulation in the SLNs on dilution, time, and the amount of iron at the injection site. This multiple regression is described later.

All statistical analysis and visualization were performed using IBM SPSS Statistics Version 23 (SPSS Inc., Chicago, IL, USA) and results were reported according to the SAMPL guideline [26]. No criteria for exclusion was set and all the collected data were included.

3. Results

SLND was performed in 202 procedures (101 rats) except for two control animals in MRI experiments, and at least one popliteal lymph node (hereafter, SLN) was obtained with each procedure (Figure 1a). The higher dose injection of Resovist resulted in apparent brown discoloration of the SLNs (Figure 1a). Abdominal nodes (hereafter, the secondary nodes) were harvested in eleven rats from the dose increase experiment and eight rats from the 24 h harvesting time point group of the dilution and time-course experiments. Furthermore, a total of 202 hindlegs were amputated at the tarsal joint, dried and powdered for SQUID analysis.

In the dose increase experiments, the fiber-optics particle analyzer showed no aggregation for samples of 1000 μg or higher. Aggregation was found for samples with an iron dose of 200 μg and 100 μg (98.9 ± 1.4 and 127.0 ± 15.9, respectively, a particle size higher than neat Resovist 2790 μg: 73.1 ± 4.09 nm). In the dilution and time course experiments, no aggregation was found for the two-times diluted sample. Aggregation was detected in the five- and 10-times diluted samples (80.0 ± 5.7 and 86.8 ± 0.71, respectively, a particle size higher than neat Resovist: 73.1 ± 4.09 nm).

3.1. Dose Increase Experiment

3.1.1. Iron Uptake in the SLNs

SQUID analysis of the SLNs demonstrated a plateau uptake of iron in the SLNs of approximately 80 µg (Figure 1b, Table S1). This plateau was reached by injecting an iron volume of 1000 µg or higher. Prior to this plateau, the amount of iron taken up by the SLNs increased with increasing amounts of iron injected, which resulted in an overall significant strong correlation between the injected iron dose and the amount of iron in the SLNs (Spearman's r_s (56) = 0.92, $p < 0.001$) (Figure 1b, Table S1). By calculation, it was evident that the utilization rate (i.e., iron in node/iron injected) between injected iron volumes of 10–200 µg was approximately 10%, with the highest utilization rates at an injected iron volume of 20 µg (11.0%), compared to 100 µg (10.0%) and 200 µg (9.7%) (Figure 1c). The utilization rate dropped significantly to 6% upon injection of 1000 µg iron, compared to 20-µg injection (Figure 1c). For the 4000 µg injection, this ratio dropped to 2%. The secondary nodes contained similar amounts of iron to in the SLNs under the investigated conditions (1000-, 2790- and 4000-µg injections) (Figure 1d).

3.1.2. Iron at the Injection Site

The amount of residual iron after 24 h at the injection site increased as the injected iron dose increased, with a statistically strong correlation (Spearman's r_s (56) = 0.96, $p < 0.001$, Figure 1b, Table S1). Iron at the injection site did not show any plateau accumulation as observed with SLNs, which indicates that an excessive injection of iron only results in an increased residue of iron at the injection site (Figure 1b,c).

3.1.3. Difference between Researchers Performing Experiments

Experiments with the 2000 µg iron group were repeated in two settings by different researchers (M. Peek and K. Saeki) to determine the inter-researcher variability. The median iron amounts in the SLNs were 87.5 µg (range, 57.9–106.0 µg; K.S.) and 73.6 µg (range, 61.7–139.0 µg; M.P.). The median iron amounts at the injection site were 1192.5 µg (range, 1101.5–1243.1 µg; K.S.) and 1316.4 µg (range 1170.8–1594.4 µg; M.P.). No statistical difference was found between the researchers about iron amounts in the SLNs (Mann–Whitney U = 12.0, $p = 1.00$) and iron amounts at the injection site (Mann–Whitney U = 6.0, $p = 0.22$), supporting experimental consistency. In the other experiments, the two researchers performed procedures interchangeably and were not distinguished.

3.2. Dilution and Time-Course Experiment

3.2.1. Iron Uptake in the SLNs

SQUID analysis showed an increase in iron uptake in the SLNs over time (Figure 2a–c, Table S2). An increase in iron uptake was seen six hours after injection (Figure 2c). The first hour showed no large difference in iron uptake. Although the effect of dilution was not apparent at most time points, the five- and ten-fold dilution group demonstrated an increased iron uptake after six hours, with little difference in iron uptake at 24 h (Figure 2b). Neat Resovist showed barely any uptake within the first hour after injection whilst all diluted volumes were actively taken up by the SLNs. The highest iron uptake was seen with the ten-fold diluted tracer after 10 min, 30 min, and 6 h and the lowest overall uptake was seen with neat Resovist. The uptake at 24 h was, however, very similar with all dilution volumes.

3.2.2. Iron at the Injection Site

At the injection site, the amount of iron per dilution group showed no difference, however over time, the iron amount slowly tended to decrease in all dilution groups (Figure 2d–f, Table S2).

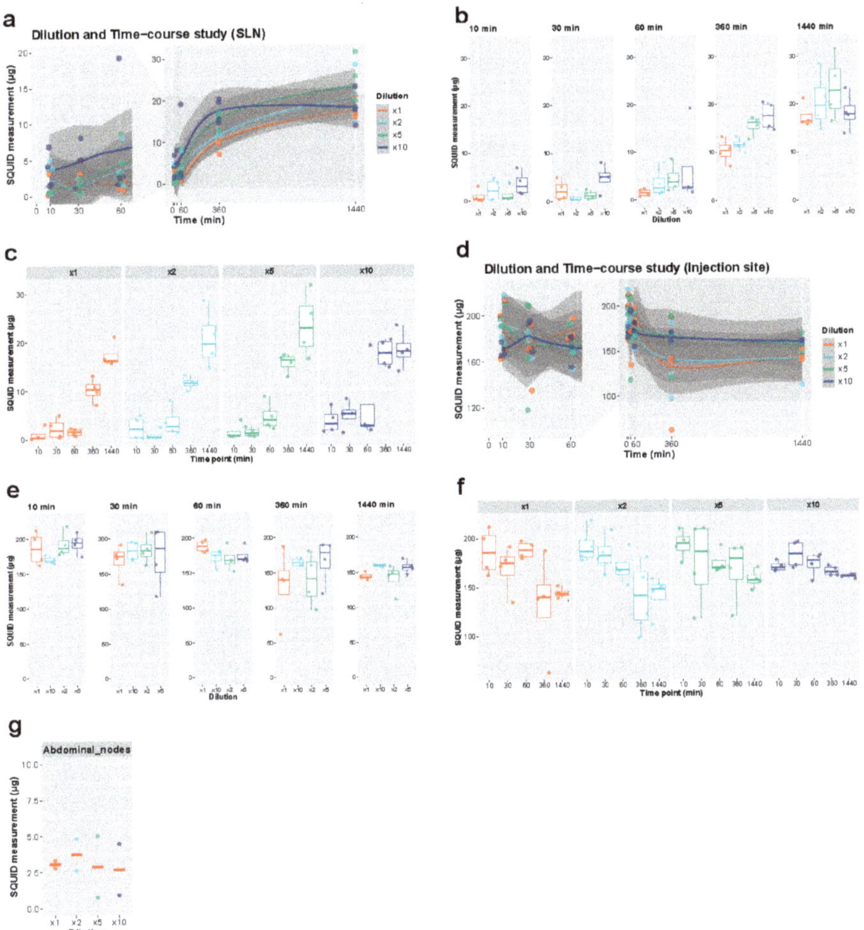

Figure 2. Dilution and time-course experiment. (**a–c**) Iron accumulation in the SLNs. (**d–f**) Iron accumulation at the injection sites. (**a,d**) All the measurements by SQUID were plotted along the time course. The solid lines represent regression curves in a LOESS model and the surrounding grey areas represent the confidence intervals. The time points between 0 and 60 min were magnified on the left. $n = 4$, each group. (**b,e**) Iron accumulation was compared between different dilutions at each time point. Each measurement (points) was overlaid on a box plot. (**c,f**) Iron accumulation was compared between different time points at each dilution. Each measurement (points) was overlaid on a box plot. (**g**) Iron accumulation in the secondary nodes 24 h after Resovist injection with varying dilution ($n = 2$). Corresponding results from SLN measurements can be found in **e** (1440 min). SLN: sentinel lymph node.

3.2.3. Statistical Analysis

To find the best single predictors of iron volume in the SLNs and at the injection site, Spearman correlations were calculated. Correlation coefficients of iron volume in the SLNs with dilution, time, and iron volume at the injection sites are 0.201 ($p = 0.07$), 0.810 ($p < 0.001$) and -0.624 ($p < 0.001$), respectively. This result showed that time is a strong predictor for iron volume in the SLNs with statistical significance. To study the uptake of iron in the SLNs, it was necessary to regress SLN iron accumulation on time, dilution, and iron concentration at the injection site. For this purpose, linear and squared terms of the predictors were used. In addition, the SLN iron concentration needed to be transformed. The Box–Cox transformation $y = (x^{\alpha} - 1)/\alpha$ was used, which reduces to a log transformation as $\alpha \rightarrow 0$. For the SLN regression, α had a 95% confidence interval of (0.3, 0.59). Hence, for simplicity, the square root of SLN iron volume was used in the regression. Detailed results are given in Table S3 and confirmed that dilution was not a significant predictor. Time was, however, and also time squared, with a negative coefficient, showing that iron accumulation increases sublinearly, i.e., it slows down with time. Turning to iron volume at the injection site, the correlation coefficients with dilution and time are 0.0108 ($p = 0.92$) and -0.554 ($p < 0.001$), respectively. Again, time is the best predictor, and dilution has no significant effect. For multiple regression, the iron volume at the injection site was fitted with $\alpha = 1$. As a result, dilution was not significant, whereas again the decrease with time was significant, and also slowed down significantly with time. Full results are given in Table S4. Consequently, time after injection is the most significant factor for both increasing iron accumulation in the SLNs and facilitating iron clearance at the injection site. Dilution does not have a significant effect on either.

Iron accumulation in the secondary nodes showed no difference between the different dilutions, 24 h after injection (Figure 2g). There was no apparent trend in injection pressure when compared between the different dilution groups, despite the varying injection volumes. In addition, no animal showed extremely high or low injection pressure, which would indicate inadequate injection (intradermal or intravenous injection) of SPIO. Averaged minimal pressure was 22.95 ± 5.78 mmHg and averaged maximal pressure was 97.34 ± 27.91 mmHg.

3.3. Massaging Experiment

Massaging of the injection site had no significant increasing effect on the amount of iron taken up by the SLNs (Mann–Whitney U = 5.0, $p = 0.15$) if the SLND was performed 30 min after injection of the magnetic tracer (Figure 3). The SLNs on the massage side contained a slightly higher amount of iron (median 3.25 µg, range 2.90–4.62 µg) compared to the untreated side (median 1.67 µg, range 1.30–4.28 µg). No statistical significance in iron at the injection site was found between the massage and no massage groups (Mann–Whitney U = 11.0, $p = 0.84$). The injected site on the massage side contained a median iron amount of 173.65 µg (range 108.95–338.96 µg), and that of the untreated side had a median amount of 174.14 µg (range 135.55–251.22 µg).

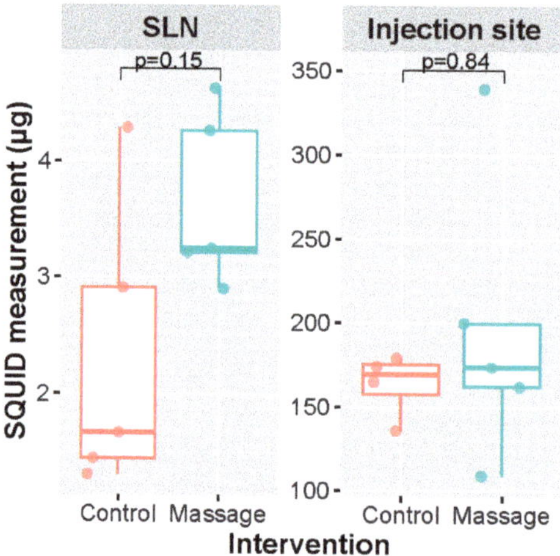

Figure 3. Results from massage experiment. $n = 4$, Mann–Whitney U test. SLN: sentinel lymph node.

3.4. MRI Experiment

The SLNs of 30 rats from the dose increase experiment and two controls were successfully scanned with MRI immediately after euthanasia. On T1-weighed images, SPIO accumulation in the SLNs was observed as a signal reduction of the entire lymph node in the 2-μg group and as a spherical artifact in the other five groups (Figure 4a). Control groups showed no signal reduction on MRI.

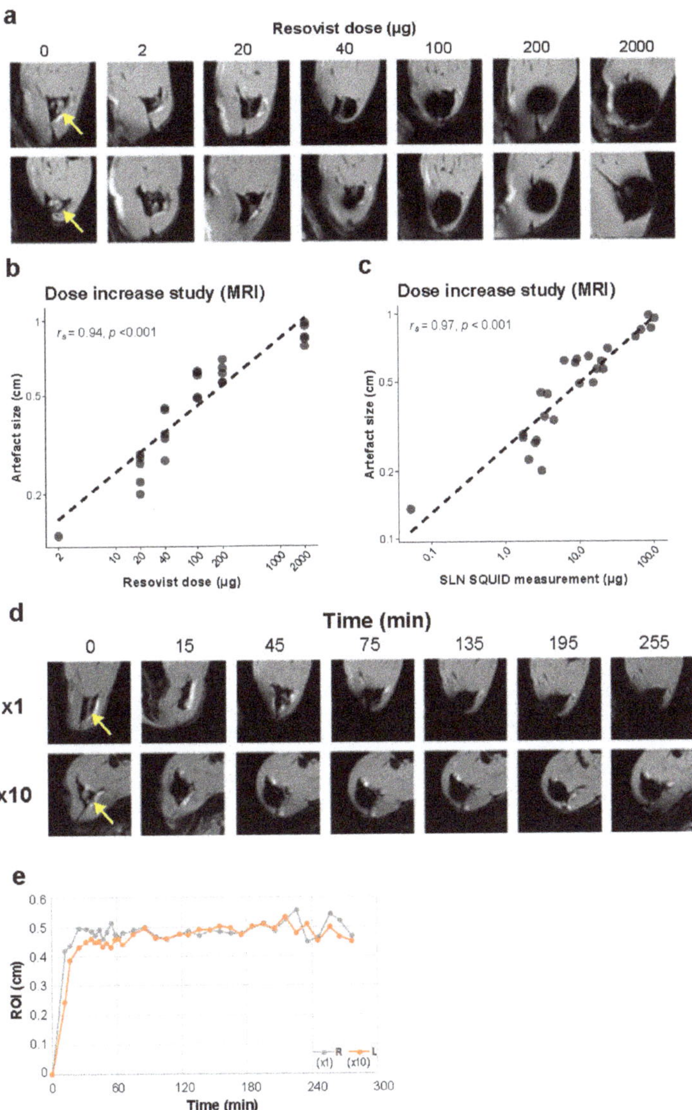

Figure 4. MRI experiment. (**a**) Representative transverse images of the rat hindlimbs at the popliteal lymph node level from magnetic resonance imaging (MRI). Yellow arrows indicate locations of the SLNs (popliteal lymph nodes) of control rats. (**b**) Relationship between the amount of Resovist injection and the size of artifacts measured on MRI. Both axes are on logarithmic scales, $n = 5$ for each group. (**c**) Relationship between iron accumulation measured by SQUID in SLNs and the size of artifact measured on MRI. Both axes are logarithmic scales. (**d**) Representative images from time-course MRI in a rat. (**e**) Time-course measurement of the size of artifact on MRI in a rat. MRI: magnetic resonance imaging.

The size of the artifact at the SLNs increased with the injected dose of iron (Spearman's r_s (30) = 0.94, $p < 0.001$, Figure 4b, Table S1). In addition, a statistically significant correlation was found between the amount of iron in the SLNs determined by the SQUID and the artifact size (Spearman's r_s (30) = 0.97, $p < 0.001$, Figure 4c, Table S1).

In the time-course MRI experiment, iron uptake in the SLNs was seen immediately after injection and a reduction in signal was immediately visible at both sides (Figure 4d). The artifact located at the SLN was on average 4.64 mm on the left ($\times 1$) and 4.84 mm on the right ($\times 10$) in maximum diameter, showing no major difference in artifact size between neat and 10-times diluted Resovist. No increase in artifact size over time was seen on both sides after reaching plateaus around 30–60 min (Figure 4e). The absolute amounts of iron in the left and the right SLNs were 5.48 and 8.70 µg after resection, respectively.

4. Discussion

The amount of iron in the SLN was proportional to the injection dose up to 200 µg. However, excessive injection of iron only causes an increase in residual iron at the injection site, as there was a plateau uptake of 80 µg in the SLNs. The total amount of iron at the injection site increased (with increasing injection dose) without any plateau. This suggests that each lymph node has a maximum capacity to accommodate magnetic particles. The maximum accommodation of iron in lymph nodes may depend on size of lymph node. Therefore, it may differ not only between different animal species, but also between individuals of the same species and the locations of the nodes in the body. Size of SPIO particles may matter as well. Injecting excessive iron would therefore not be of any additional benefit and would most likely cause larger MRI artifacts at the injection site. This indicates that it is important to determine the precise plateau point to optimize the injection volume and reduce the possibility of a susceptibility artifact. This study, by investigating wide range of SPIO injection volumes, revealed both a gradual increase in iron in the SLNs along with an increase in iron injection dose and the existence of a plateau in uptake. Proportional increase in iron in the SLNs was observed between 20 (0.1 mg Iron/BW in a 200 g rat) and 200 µg (1 mg Iron/kg BW) iron injection. For a person who weighs 60 kg, 2 mL of neat SPIO (either of Sienna+, Magtrace, or Resovist) injection is equal to 1 mg Iron/kg BW, approximately. Utilization rates were also low with injection of small iron volumes less than 10 µg. This could be caused by injected iron being trapped at the injection site such as phagocytosis by tissue macrophages, but further evaluation is required. Alternatively, this could be due to aggregation of the SPIO at lower iron concentrations, preventing uptake into the SLNs. This phenomenon, aggregation of the SPIO particles at low concentration, has not been identified so far. In most human clinical trials, SPIO agents were traditionally diluted at 2.5 times to facilitate SLN uptake. However, excessive dilution of SPIO may decrease iron uptake by the lymphatic system in human medicine as well.

During dilution experiments, longer time from injection increased iron uptake in the SLNs to a greater extent than increasing the dilution factor. This indicates that the lymphatic system may be able to take up more magnetic tracer if the injection is performed a day prior to surgery rather than injection just prior to surgery, as has already been shown in some studies [16,21–23]. For iron at the injection site, time is also a significant factor for washout. Dilution did not have a significant effect on either iron in the SLNs or at the injection site. Massaging may facilitate iron uptake by the SLNs, but the effect was small and not statistically significant in this study.

Although no human model was used for this study, it may be possible to suggest critical factors and conditions to optimize SPIO injection for breast cancer patients, together with the previous studies. The use of a magnetic tracer for SLND was first evaluated in a porcine model [18,25,27]. Anninga et al. [27] showed a significant correlation between magnetometer counts and iron in SLNs ($r = 0.86$, $p < 0.01$). Grading with both H&E and Perl's staining showed a correlation with iron content ($p = 0.001$, $p = 0.003$) and magnetometer counts ($p < 0.001$, $p = 0.004$). Pouw et al. [18] evaluated three different magnetic tracers and showed that Sienna+ had the highest number of detected transcutaneous hotspots and showed the highest ex-vivo magnetometer counts. Furthermore, ferumoxytol had the highest lymph node retrieval rate. The effect of concentration, volume and time of harvesting was initially evaluated in the same porcine model by Ahmed et al. [25]. This study showed a significant positive correlation between magnetometer counts and iron content of excised SLNs ($r = 0.82$; $p < 0.001$) and increasing time of injection ($p < 0.001$), plateauing at 60 min. The total number of excised SLNs showed a significant positive correlation with increasing magnetic nanoparticle volumes ($p < 0.001$), and iron content in SLNs with increasing concentration ($p = 0.006$). In addition, Ahmed et al. [28] evaluated the impact of site and timing of magnetic nanoparticle injection in a murine model, showing a rapid uptake on MRI, smaller "void artifacts" ($p < 0.001$) and a significant increase in iron content with time in the group receiving subcutaneous injection ($r = 0.94$; $p < 0.001$). Previous clinical studies also evaluated different protocols of the SPIO injection. Hersi et al. [16], performed a study comparing patients from the Nordic SentiMag trial (2.0 mL Sienna+ injected on the morning of surgery) with two new patient groups: one group that received an injection of 1.0 mL Magtrace 1–7 days prior to surgery and a second group that received 1.5 mL Magtrace on the day of surgery. This study showed as well that a lower dose is also non-inferior to the standard technique with similar detection rates between the three doses (97.5% versus 100% and 97.6%, $p = 0.11$) and more SLNs excised than higher doses (2.18, versus 1.85 and 1.83, $p = 0.003$). Rubio et al. [19] compared the injection of 1.0, 1.5 and 2.0 mL Magtrace diluted with saline on the morning of surgery and found each group of 45 patients to be non-inferior to the standard technique ($p = 0.654$). Our study further evaluated the magnetic nanoparticle injection based on these experiments, focusing on evaluation of the effect of iron dose, dilution, time, and massaging the injection site. Time was an important factor, and longer waiting time ended up in more iron accumulation in the SLN within the timeframe investigated. With injection a day prior to surgery, people could reduce required amount of SPIO injection, which would reduce MRI artifact and coloration at the injection site. The results also indicated that iron accumulation in the SLN is proportional to the amount of iron injection between 20 ug (0.1 mg Iron/kg) and 200 ug (1 mg/kg) in this model. Although there must be interspecies differences, researchers may be able to explore injection of the current SPIO agents ranging between 0.2 and 2 mL with expectation of proportional accumulation in the SLN.

Recently, the first randomized trial [21] was published evaluating the magnetic technique for SLND on its own. The use of the magnetic tracer in one hospital was compared to the use of the radioisotope in a second hospital. A total of 338 patients (343 SLNDs) were included and identification rates of 95.6% for the magnetic technique and 96.9% for the radioisotope were found. The lymph node retrieval rate was 1.35 nodes per patient for the magnetic technique and 1.89 nodes per patient for the radioisotope technique. This study also looked at the timing of the injection and found that SPIO injections performed at a median of 16 days prior to SLND obtained a better identification rate ($p = 0.031$) and higher lymph node retrieval rate ($p < 0.001$) compared to injections on the morning of surgery. Other studies also suggested feasibility of preoperative injection up to weeks or months prior to surgery [22,23,29]. Although this study had been designed before these publications and examined preoperative injection up to 24 h, a longer waiting period might further reduce required injection volume of magnetic particles. This speculation should be investigated in future studies.

The current drawback of the magnetic technique using liquid tracers are the residual MRI artifacts seen in patients at the injection site post-operatively, even for years [30–32]. Within the SentiMAG and MagSNOLL trials [4,33] all patients who underwent an MRI after their SLND were asked to participate in a sub-study. The MRIs of these patients were evaluated, and it was seen that no artifact was visible in patients who participated in the MagSNOLL trial, but an artifact was visible in patients of the SentiMAG trial. This is most likely due to the resection of the injected area in the MagSNOLL trial. Our results indicated that the amount of residual iron at the injection site is proportional to injection dose, and accumulation of very small amount of iron can cause a void artifact on MRI. Although reduction in the injection dose would lead to a less severe artifact, it might be advisable that the injection site should be included in resection to avoid any ambiguity on future diagnostic imaging.

A major limitation of this study is that we used a rat model to optimize SPIO injection. Although it enabled us to address the factors associated with the injection, the SLNs of rats are small, and ex vivo counts by a magnetometer were not taken in this study. Therefore, the theoretical basis provided in this study should be further evaluated in more relevant animal models, which may include dog and pig, or in clinical studies. Preoperative injection more than one day, up to months, prior to surgery needs to be investigated in the future as well, to see if a longer waiting period facilitates more accumulation of iron in the SLNs and/or leads to complicating accumulation in secondary and tertiary nodes. In a recent study, patients were injected with SPIO prior to neoadjuvant chemotherapy, and MRI lymphography was compared before and after chemotherapy with a median of 130 days interval [29]. As a result, SPIO accumulation was observed in the same lymph nodes. Therefore, it was suggested that SPIO does not migrate in higher nodes for months, which would support SPIO injection more than 1 day prior to surgery. Also, subsequent studies can include visual images and MRI scanning of the injection site to address artifact and coloration around the area. Another limitation is that this study lacks sample size calculation. Again, the findings may need to be validated with the appropriate statistical power considering the pilot results from this study.

Tumor microenvironments evolve dynamically and continuously, shaping a niche in favor of tumor cell proliferation and dissemination [34,35]. A tissue structure is edited and distorted compared to its normal counterpart, and the lymphatic system is not an exception. Vigorous lymphangiogenesis and expansion of tumor cell nests are reported to lead to enlarged peritumor lymphatic vessels as well as collapsed intratumoral vessels [35]. As such, peritumorally injected SPIO would be put in different lymphatic dynamics compared to subcutaneous injection into the normal tissue performed in this study. Different lymphatic system, such as the axillary versus neck networks, may have different draining machineries. A more relevant tumor-bearing animal model and clinical studies will reveal these points in the future.

5. Conclusions

Iron accumulation in the SLNs was proportional to injection doses within a certain range. Time was also a primary factor for SLN uptake of SPIO. Dilution and massaging did not have significant effects. From these results, preoperative injection may be advantageous for SLN detection, as has been shown in the pioneering clinical studies, as well as for reduction in residual iron at the injection site by the expected decrease in required injection dose. The result from this study provides a theoretical background that helps our understanding of the magnetic SLND.

Supplementary Materials: The following are available online at https://www.mdpi.com/article/10.3390/cancers13195031/s1. Figure S1: Flow diagram of subjects. Table S1: Descriptive statistics of the dose increase experiment and the MRI experiment. Table S2: Descriptive statistics of the dilution and time-course experiment. Table S3: Results of the regression analysis for square root of the SLN iron amount. Table S4: Results of the regression analysis for the iron amount at the injection site. Data S1: ARRIVE guideline checklist.

Author Contributions: Conceptualization, M.C.L.P., K.S., T.N., M.K., M.D. and M.S.; methodology, M.C.L.P., K.S., M.K., M.D. and M.S.; formal analysis, M.C.L.P., K.S., M.K., M.D., M.S., K.O., S.C., R.B. and T.N.; investigation, M.C.L.P., K.S., K.O., S.C., R.B. and T.N.; writing—original draft preparation, M.C.L.P. and K.S.; writing—review and editing, M.C.L.P., K.S., M.K., M.D. and R.B.; funding acquisition, T.N., M.K. and M.S. All authors have read and agreed to the published version of the manuscript.

Funding: This study was funded in part by the Japan Agency for Medical Research and Development (18he0902010h0004, T.N., M.K. and M.S.) and JSPS KAKENHI Grant Number 20H05759 (M.S.).

Institutional Review Board Statement: Animal studies were approved by the local Ethics Board (Accession Number P15-124) and the experiments were performed at the University of Tokyo according to the guidelines of the institution and ensured humane care of animals.

Informed Consent Statement: Not applicable.

Data Availability Statement: The data presented in this study are available on request from the corresponding author. The data are not publicly available due to ethical restrictions.

Acknowledgments: We would like to acknowledge the following people for assisting us in our research: Taeseong Woo (The University of Tokyo), Akihiro Kuwahata (The University of Tokyo) and Satoshi Sakamoto (Tokyo Institute of Technology).

Conflicts of Interest: The authors declare no conflict of interest. The funders had no role in the design of the study; in the collection, analyses, or interpretation of data; in the writing of the manuscript, or in the decision to publish the results.

References

1. Giuliano, A.E.; Kirgan, D.M.; Guenther, J.M.; Morton, D.L. Lymphatic Mapping and Sentinel Lymphadenectomy for Breast Cancer. *Ann. Surg.* **1994**, *220*, 391–398, discussion 398–401. [CrossRef]
2. Krag, D.N.; Anderson, S.J.; Julian, T.B.; Brown, A.M.; Harlow, S.P.; Costantino, J.P.; Ashikaga, T.; Weaver, D.L.; Mamounas, E.P.; Jalovec, L.M.; et al. Sentinel-Lymph-Node Resection Compared with Conventional Axillary-Lymph-Node Dissection in Clinically Node-Negative Patients with Breast Cancer: Overall Survival Findings from the NSABP B-32 Randomised Phase 3 Trial. *Lancet Oncol.* **2010**, *11*, 927–933. [CrossRef]
3. Kim, T.; Giuliano, A.E.; Lyman, G.H. Lymphatic Mapping and Sentinel Lymph Node Biopsy in Early-Stage Breast Carcinoma: A Metaanalysis. *Cancer* **2006**, *106*, 4–16. [CrossRef]
4. Douek, M.; Klaase, J.; Monypenny, I.; Kothari, A.; Zechmeister, K.; Brown, D.; Wyld, L.; Drew, P.; Garmo, H.; Agbaje, O.; et al. Sentinel Node Biopsy Using a Magnetic Tracer versus Standard Technique: The SentiMAG Multicentre Trial. *Ann. Surg. Oncol.* **2014**, *21*, 1237–1245. [CrossRef]
5. Thill, M.; Kurylcio, A.; Welter, R.; van Haasteren, V.; Grosse, B.; Berclaz, G.; Polkowski, W.; Hauser, N. The Central-European SentiMag Study: Sentinel Lymph Node Biopsy with Superparamagnetic Iron Oxide (SPIO) vs. Radioisotope. *Breast* **2014**, *23*, 175–179. [CrossRef]
6. Shiozawa, M.; Lefor, A.T.; Hozumi, Y.; Kurihara, K.; Sata, N.; Yasuda, Y.; Kusakabe, M. Sentinel Lymph Node Biopsy in Patients with Breast Cancer Using Superparamagnetic Iron Oxide and a Magnetometer. *Breast Cancer* **2013**, *20*, 223–229. [CrossRef] [PubMed]
7. Rubio, I.T.; Diaz-Botero, S.; Esgueva, A.; Rodriguez, R.; Cortadellas, T.; Cordoba, O.; Espinosa-Bravo, M. The Superparamagnetic Iron Oxide Is Equivalent to the Tc99 Radiotracer Method for Identifying the Sentinel Lymph Node in Breast Cancer. *Eur. J. Surg. Oncol.* **2015**, *41*, 46–51. [CrossRef] [PubMed]
8. Piñero-Madrona, A.; Torró-Richart, J.A.; de León-Carrillo, J.M.; de Castro-Parga, G.; Navarro-Cecilia, J.; Domínguez-Cunchillos, F.; Román-Santamaría, J.M.; Fuster-Diana, C.; Pardo-García, R. Grupo de Estudios Senológicos de la Sociedad Española de Patologia Mamaria (SESPM) Superparamagnetic Iron Oxide as a Tracer for Sentinel Node Biopsy in Breast Cancer: A Comparative Non-Inferiority Study. *Eur. J. Surg. Oncol.* **2015**, *41*, 991–997. [CrossRef] [PubMed]
9. Houpeau, J.-L.; Chauvet, M.-P.; Guillemin, F.; Bendavid-Athias, C.; Charitansky, H.; Kramar, A.; Giard, S. Sentinel Lymph Node Identification Using Superparamagnetic Iron Oxide Particles versus Radioisotope: The French Sentimag Feasibility Trial. *J. Surg. Oncol.* **2016**, *113*, 501–507. [CrossRef]
10. Ghilli, M.; Carretta, E.; Di Filippo, F.; Battaglia, C.; Fustaino, L.; Galanou, I.; Di Filippo, S.; Rucci, P.; Fantini, M.P.; Roncella, M. The Superparamagnetic Iron Oxide Tracer: A Valid Alternative in Sentinel Node Biopsy for Breast Cancer Treatment. *Eur. J. Cancer Care* **2017**, *26*, e12385. [CrossRef]
11. Ahmed, M.; Purushotham, A.D.; Douek, M. Novel Techniques for Sentinel Lymph Node Biopsy in Breast Cancer: A Systematic Review. *Lancet Oncol.* **2014**, *15*, e351–e362. [CrossRef]
12. Zada, A.; Peek, M.C.L.; Ahmed, M.; Anninga, B.; Baker, R.; Kusakabe, M.; Sekino, M.; Klaase, J.M.; Ten Haken, B.; Douek, M. Meta-Analysis of Sentinel Lymph Node Biopsy in Breast Cancer Using the Magnetic Technique. *Br. J. Surg.* **2016**, *103*, 1409–1419. [CrossRef] [PubMed]
13. Karakatsanis, A.; Christiansen, P.M.; Fischer, L.; Hedin, C.; Pistioli, L.; Sund, M.; Rasmussen, N.R.; Jørnsgård, H.; Tegnelius, D.; Eriksson, S.; et al. The Nordic SentiMag Trial: A Comparison of Super Paramagnetic Iron Oxide (SPIO) Nanoparticles versus Tc(99) and Patent Blue in the Detection of Sentinel Node (SN) in Patients with Breast Cancer and a Meta-Analysis of Earlier Studies. *Breast Cancer Res. Treat.* **2016**, *157*, 281–294. [CrossRef]

14. Teshome, M.; Wei, C.; Hunt, K.K.; Thompson, A.; Rodriguez, K.; Mittendorf, E.A. Use of a Magnetic Tracer for Sentinel Lymph Node Detection in Early-Stage Breast Cancer Patients: A Meta-Analysis. *Ann. Surg. Oncol.* **2016**, *23*, 1508–1514. [CrossRef]

15. Alvarado, M.D.; Mittendorf, E.A.; Teshome, M.; Thompson, A.M.; Bold, R.J.; Gittleman, M.A.; Beitsch, P.D.; Blair, S.L.; Kivilaid, K.; Harmer, Q.J.; et al. SentimagIC: A Non-Inferiority Trial Comparing Superparamagnetic Iron Oxide versus Technetium-99m and Blue Dye in the Detection of Axillary Sentinel Nodes in Patients with Early-Stage Breast Cancer. *Ann. Surg. Oncol.* **2019**, *26*, 3510–3516. [CrossRef]

16. Hersi, A.-F.; Pistiolis, L.; Dussan Luberth, C.; Vikhe-Patil, E.; Nilsson, F.; Mohammed, I.; Olofsson Bagge, R.; Wärnberg, F.; Eriksson, S.; Karakatsanis, A. Optimizing Dose and Timing in Magnetic Tracer Techniques for Sentinel Lymph Node Detection in Early Breast Cancers: The Prospective Multicenter SentiDose Trial. *Cancers* **2021**, *13*, 693. [CrossRef] [PubMed]

17. Taruno, K.; Kurita, T.; Kuwahata, A.; Yanagihara, K.; Enokido, K.; Katayose, Y.; Nakamura, S.; Takei, H.; Sekino, M.; Kusakabe, M. Multicenter Clinical Trial on Sentinel Lymph Node Biopsy Using Superparamagnetic Iron Oxide Nanoparticles and a Novel Handheld Magnetic Probe. *J. Surg. Oncol.* **2019**, *120*, 1391–1396. [CrossRef]

18. Pouw, J.J.; Ahmed, M.; Anninga, B.; Schuurman, K.; Pinder, S.E.; Van Hemelrijck, M.; Pankhurst, Q.A.; Douek, M.; ten Haken, B. Comparison of Three Magnetic Nanoparticle Tracers for Sentinel Lymph Node Biopsy in an In Vivo Porcine Model. *Int. J. Nanomed.* **2015**, *10*, 1235–1243. [CrossRef]

19. Rubio, I.T.; Rodriguez-Revuelto, R.; Espinosa-Bravo, M.; Siso, C.; Rivero, J.; Esgueva, A. A Randomized Study Comparing Different Doses of Superparamagnetic Iron Oxide Tracer for Sentinel Lymph Node Biopsy in Breast Cancer: The SUNRISE Study. *Eur. J. Surg. Oncol.* **2020**, *46*, 2195–2201. [CrossRef] [PubMed]

20. Huizing, E.; Anninga, B.; Young, P.; Monypenny, I.; Hall-Craggs, M.; Douek, M. 4. Analysis of Void Artefacts in Post-Operative Breast MRI Due to Residual SPIO after Magnetic SLNB in SentiMAG Trial Participants. *Eur. J. Surg. Oncol.* **2015**, *41*, S18. [CrossRef]

21. Karakatsanis, A.; Daskalakis, K.; Stålberg, P.; Olofsson, H.; Andersson, Y.; Eriksson, S.; Bergkvist, L.; Wärnberg, F. Superparamagnetic Iron Oxide Nanoparticles as the Sole Method for Sentinel Node Biopsy Detection in Patients with Breast Cancer. *Br. J. Surg.* **2017**, *104*, 1675–1685. [CrossRef] [PubMed]

22. Wärnberg, F.; Stigberg, E.; Obondo, C.; Olofsson, H.; Abdsaleh, S.; Wärnberg, M.; Karakatsanis, A. Long-Term Outcome After Retro-Areolar versus Peri-Tumoral Injection of Superparamagnetic Iron Oxide Nanoparticles (SPIO) for Sentinel Lymph Node Detection in Breast Cancer Surgery. *Ann. Surg. Oncol.* **2019**, *26*, 1247–1253. [CrossRef]

23. Karakatsanis, A.; Hersi, A.-F.; Pistiolis, L.; Olofsson Bagge, R.; Lykoudis, P.M.; Eriksson, S.; Wärnberg, F. SentiNot Trialists Group Effect of Preoperative Injection of Superparamagnetic Iron Oxide Particles on Rates of Sentinel Lymph Node Dissection in Women Undergoing Surgery for Ductal Carcinoma In Situ (SentiNot Study). *Br. J. Surg.* **2019**, *106*, 720–728. [CrossRef] [PubMed]

24. du Sert, N.P.; Hurst, V.; Ahluwalia, A.; Alam, S.; Avey, M.T.; Baker, M.; Browne, W.J.; Clark, A.; Cuthill, I.C.; Dirnagl, U.; et al. The ARRIVE Guidelines 2.0: Updated Guidelines for Reporting Animal Research. *PLoS Biol.* **2020**, *18*, e3000410. [CrossRef]

25. Ahmed, M.; Anninga, B.; Pouw, J.J.; Vreemann, S.; Peek, M.; Van Hemelrijck, M.; Pinder, S.; Ten Haken, B.; Pankhurst, Q.; Douek, M. Optimising Magnetic Sentinel Lymph Node Biopsy in an In Vivo Porcine Model. *Nanomedicine* **2015**, *11*, 993–1002. [CrossRef] [PubMed]

26. Lang, T.A.; Altman, D.G. Basic Statistical Reporting for Articles Published in Biomedical Journals: The "Statistical Analyses and Methods in the Published Literature" or the SAMPL Guidelines. *Int. J. Nurs. Stud.* **2015**, *52*, 5–9. [CrossRef]

27. Anninga, B.; Ahmed, M.; Van Hemelrijck, M.; Pouw, J.; Westbroek, D.; Pinder, S.; Ten Haken, B.; Pankhurst, Q.; Douek, M. Magnetic Sentinel Lymph Node Biopsy and Localization Properties of a Magnetic Tracer in an In Vivo Porcine Model. *Breast Cancer Res. Treat.* **2013**, *141*, 33–42. [CrossRef]

28. Ahmed, M.; Woo, T.; Ohashi, K.; Suzuki, T.; Kaneko, A.; Hoshino, A.; Zada, A.; Baker, R.; Douek, M.; Kusakabe, M.; et al. Magnetic Sentinel Lymph Node Biopsy in a Murine Tumour Model. *Nanomedicine* **2016**, *12*, 1045–1052. [CrossRef]

29. Jazrawi, A.; Pantiora, E.; Abdsaleh, S.; Bacovia, D.V.; Eriksson, S.; Leonhardt, H.; Wärnberg, F.; Karakatsanis, A. Magnetic-Guided Axillary UltraSound (MagUS) Sentinel Lymph Node Biopsy and Mapping in Patients with Early Breast Cancer. A Phase 2, Single-Arm Prospective Clinical Trial. *Cancers* **2021**, *13*, 4285. [CrossRef] [PubMed]

30. Arslan, G.; Yılmaz, C.; Çelik, L.; Çubuk, R.; Tasalı, N. Unexpected Finding on Mammography and MRI Due to Accumulation of Iron Oxide Particles Used for Sentinel Lymph Node Detection. *Eur. J. Breast Health* **2019**, *15*, 200–202. [CrossRef]

31. Aribal, E.; Çelik, L.; Yilmaz, C.; Demirkiran, C.; Guner, D.C. Effects of Iron Oxide Particles on MRI and Mammography in Breast Cancer Patients after a Sentinel Lymph Node Biopsy with Paramagnetic Tracers. *Clin. Imaging* **2021**, *75*, 22–26. [CrossRef] [PubMed]

32. Krischer, B.; Forte, S.; Niemann, T.; Kubik-Huch, R.A.; Leo, C. Feasibility of Breast MRI after Sentinel Procedure for Breast Cancer with Superparamagnetic Tracers. *Eur. J. Surg. Oncol.* **2018**, *44*, 74–79. [CrossRef]

33. Ahmed, M.; Anninga, B.; Goyal, S.; Young, P.; Pankhurst, Q.A.; Douek, M. MagSNOLL Trialists Group Magnetic Sentinel Node and Occult Lesion Localization in Breast Cancer (MagSNOLL Trial). *Br. J. Surg.* **2015**, *102*, 646–652. [CrossRef] [PubMed]

34. Stacker, S.; Williams, S.; Karnezis, T.; Shayan, R.; Fox, S.; Achen, M. Lymphangiogenesis and Lymphatic Vessel Remodelling in Cancer. *Nat. Rev. Cancer* **2014**, *14*, 159–172. [CrossRef] [PubMed]

35. Hinshaw, D.C.; Shevde, L.A. The Tumor Microenvironment Innately Modulates Cancer Progression. *Cancer Res.* **2019**, *79*, 4557–4566. [CrossRef]

Review

In Vitro Magnetic Techniques for Investigating Cancer Progression

Sarah Libring [1,2], Ángel Enríquez [1,2,3], Hyowon Lee [1,2,3,*] and Luis Solorio [1,4,*]

[1]   Weldon School of Biomedical Engineering, Purdue University, West Lafayette, IN 47907, USA; slibring@purdue.edu (S.L.); aenrique@purdue.edu (Á.E.)
[2]   Birck Nanotechnology Center, Purdue University, West Lafayette, IN 47907, USA
[3]   Center for Implantable Devices, Purdue University, West Lafayette, IN 47907, USA
[4]   Purdue Center for Cancer Research, Purdue University, West Lafayette, IN 47907, USA
*   Correspondence: hwlee@purdue.edu (H.L.); lsolorio@purdue.edu (L.S.)

Simple Summary: This review focuses on the advantages achieved by incorporating magnetic forces into culture platforms used to study cancer progression in the laboratory. Due to the complex interactions that occur between cancer cells and their environment throughout primary tumor growth and metastatic spread, benchtop techniques are essential for decoupling these factors at several stages of disease progression where traditional animal models would fail. Breakthroughs in our understanding of cancer biology and mechanics through these benchtop techniques can ultimately lead to better-designed precision medicine platforms and clinical therapeutics for patients.

Abstract: Worldwide, there are currently around 18.1 million new cancer cases and 9.6 million cancer deaths yearly. Although cancer diagnosis and treatment has improved greatly in the past several decades, a complete understanding of the complex interactions between cancer cells and the tumor microenvironment during primary tumor growth and metastatic expansion is still lacking. Several aspects of the metastatic cascade require in vitro investigation. This is because in vitro work allows for a reduced number of variables and an ability to gather real-time data of cell responses to precise stimuli, decoupling the complex environment surrounding in vivo experimentation. Breakthroughs in our understanding of cancer biology and mechanics through in vitro assays can lead to better-designed ex vivo precision medicine platforms and clinical therapeutics. Multiple techniques have been developed to imitate cancer cells in their primary or metastatic environments, such as spheroids in suspension, microfluidic systems, 3D bioprinting, and hydrogel embedding. Recently, magnetic-based in vitro platforms have been developed to improve the reproducibility of the cell geometries created, precisely move magnetized cell aggregates or fabricated scaffolding, and incorporate static or dynamic loading into the cell or its culture environment. Here, we will review the latest magnetic techniques utilized in these in vitro environments to improve our understanding of cancer cell interactions throughout the various stages of the metastatic cascade.

Keywords: magnetism; cancer; tumor; in vitro; metastatic cascade; review

Citation: Libring, S.; Enríquez, Á.; Lee, H.; Solorio, L. In Vitro Magnetic Techniques for Investigating Cancer Progression. *Cancers* **2021**, *13*, 4440. https://doi.org/10.3390/cancers 13174440

Academic Editors: Moriaki Kusakabe and Akihiro Kuwahata

Received: 30 July 2021
Accepted: 29 August 2021
Published: 3 September 2021

Publisher's Note: MDPI stays neutral with regard to jurisdictional claims in published maps and institutional affiliations.

1. Introduction

In 2021, there will be almost 1.9 million newly diagnosed cancer cases and over 600,000 cancer deaths in the United States [1]. Worldwide, there are currently around 18.1 million new cases and 9.6 million cancer deaths yearly [2]. It is clear that advances in the diagnosis and treatment of cancer remain a high priority in biological, healthcare, and engineering research disciplines. Several advances for both cancer treatment and basic research are achieved through the incorporation of magnetic technologies. For example, the untethered transmission of force attainable through magnetic force allows for remote access, facilitates targeted delivery and precise movement in vivo and in vitro, and enables the easy sorting of specific cell types. In this review, we will first give an overview of primary tumor growth and metastatic progression (Section 2) and the principles of magnetic transduction (Section 3). We will then briefly discuss recent advances using magnetic techniques for in vivo and ex vivo patient care (Section 4) before focusing on in vitro magnetic platforms as an invaluable supplement to in vivo animal models in improving our understanding of the complex interactions that occur in cancer biology (Sections 5 and 6).

2. Primary Tumor Growth and Metastatic Progression

In order for a solid carcinoma to grow and eventually metastasize, cells of a particular tissue must first acquire features that enable aberrant survival and rapid division and then must acquire additional hallmark features that enable local movement and systemic spread throughout the body. The growth of the primary tumor and the metastatic cascade can be broken into four main categories: primary tumor growth, invasion, survival in circulation, and overt metastasis [3]. In the first step, cancer cells evade antiproliferative and apoptotic signals typical of a tissue in homeostasis [4]. A permissive tumor microenvironment is orchestrated by the recruitment and reprogramming of cancer-associated fibroblasts and other stromal cells which foster angiogenesis and alter the extracellular matrix (ECM) content and architecture [5,6]. As the primary tumor develops, premetastatic niches also develop throughout the body, fueled by extracellular vesicle communication from the cancer cells. Changes to the ECM, suppression of the immune system, and an increase in vascularization to increase nutrient transport prime distant areas of the body to be amenable to the cancer cells upon arrival [7–12].

As epithelial cancer cells continue to lose their apical-basal polarity, weaken cell-cell junctions, and rearrange their cytoskeleton, invasive capacities are acquired in a subset of the population. This is referred to as the epithelial to mesenchymal transition (EMT). EMT is a common process utilized in embryonic development and tissue regeneration. This developmental pathway is reactivated, although typically incompletely, in many cancer types [13,14]. EMT is initialized by both internal transcription factors (e.g., the Snail, Twist, and Zeb families) and external microenvironmental cues, such as increased fibrosis and inflammation. EMT in cancer is characterized by the induction of these EMT-transcription factors, the loss of epithelial gene products and the gain of mesenchymal gene products (e.g., loss of E-cadherin and gain of vimentin), and the acquisition of a mesenchymal cell phenotype (i.e., spindle-shaped, migratory, loss of cell–cell cluster packing) [13]. Tumor cells with this mesenchymal phenotype migrate through the basement membrane and invade the stroma toward blood and lymphatic vessels by utilizing rearranged bundles of collagen and fibronectin that lead radially outward from the tumor edge [15].

After intravasation, cancer cells must withstand the shear forces of the vasculature. It is estimated that <0.02% of circulating tumor cells (CTCs) complete metastasis [16,17]. Tumor cells may circulate individually or they may form circulating emboli when small clusters of cells maintain intracellular junctions and intravasate together [15,18]. Clusters can provide a survival advantage by shielding internal CTCs from fluid shear stress and immune assault [19]. Additionally, heterotypic clusters, which include CTCs and additional cell types, such as neutrophils and/or cancer-associated fibroblasts, appear to be rare in peripheral circulation but seem to possess a significant metastatic advantage over other common CTC arrangements [19,20].

CTCs must eventually extravasate into new surrounding tissue. To do so, CTCs arrested in small capillaries adhere to the endothelium, transmigrate, and invade the stromal matrix [21]. Disseminated cells from various cancers preferentially reside in different organs, a feature known as organotropism. This non-random distribution is not accounted for by simple circulation patterns, and instead seems dictated by numerous additional factors including tumor-intrinsic factors and organ-specific niches [22–24]. Once at secondary sites, disseminated cancer cells may lay dormant until external stimuli are presented. Some stimuli found in recent research include further adjustments to the niche matrix, neutrophil extracellular traps induced by inflammation, and manipulation of tumor cell metabolic pathways [7,22,23,25] These mechanical and biochemical signals allow the cancer cells to undergo mesenchymal to epithelial transition, at which point a secondary tumor, fueled by rapidly proliferating epithelial cancer cells, develops in a similar manner to the primary tumor [25]. Although originally believed to be a linear process, wherein metastasis was the final product of a primary tumor outgrowing its original tissue, recent evidence has established that the dissemination of cancer cells occurs early in primary tumor growth and that cancer cells are shed continuously into the body for eventual colonization of secondary tumors [26,27]. This phenomenon has radical implications for the heterogeneity of CTCs and metastatic cells which must be carefully considered when designing techniques to analyze patient samples and suggest treatment regimens [27].

3. Magnetic Transduction

In magnetism, there are typically two poles, positive and negative. Although there is active research in observing a magnetic monopole in nature, the quest to find the particle continues to elude researchers [28]. In two-pole systems, like poles create a repellent force toward one another while opposite poles generate an attractive force. This force follows Coulomb's law where the magnitude of the force is dictated by how strong the poles themselves are and the distance between them.

$$\vec{F} = k_e \frac{q_1 q_2}{r^2} \hat{r} \tag{1}$$

Coulomb's law is shown in Equation (1), where $\vec{F}$ is the force vector, k_e is Coulomb's constant, q_1 and q_2 are the signed magnitudes of the two charges, r is the distance between the magnetic sources, and $\hat{r}$ is the unit vector directed along r [29].

3.1. Forces on Particles

The magnetic properties of a material are dictated primarily by the electrons which compose the atoms of the material. Most materials have atoms arranged in a random manner, where their respective electrons' magnetic states cancel each other. The force acting on a magnetic dipole when exposed to an external magnetic field is defined as

$$\vec{F}_m = \left(\vec{m} \cdot \nabla \right) \vec{B} \tag{2}$$

where $\vec{F}_m$ is the magnetic force, $\vec{m}$ is the magnetic dipole, and $\vec{B}$ is the magnetic flux density [29]. The magnetic dipole is defined by $\vec{m} = V_m \vec{M}$, where V_m is the volume of the particle and $\vec{M}$ is the magnetization of the material. The magnetic force is related to the differential of the magnetostatic field energy density. This can be further illustrated by defining

$$\vec{M} = \Delta \chi \, \vec{H} \tag{3}$$

$$\vec{B} = \mu_0 \vec{H}$$

where $\Delta \chi$ is the effective susceptibility of the magnetic nanoparticle relative to the environment it is placed in, $\vec{H}$ is the applied magnetic field, and μ_0 is the permeability constant of free space. When considering that there are no time-varying electric fields or currents in the medium, the equation for magnetic force transforms to

$$\vec{F}_m = V_m \Delta \chi \nabla \left(\frac{1}{2} \vec{B} \cdot \vec{H} \right) \tag{4}$$

where the magnetostatic field energy density, $\frac{1}{2} \vec{B} \cdot \vec{H}$, dictates that the resultant force for a particle in a magnetic field is proportional to the strength of the magnetic field, and to the field gradient that the particle experiences. The magnetic flux density gradient can apply a translational force at a distance whereas a uniform field can only apply a torque [30].

3.2. Magnetic Torque

A material is classified as ferromagnetic when it has a large number of unpaired electrons throughout its compositional atoms which, when aligned, create a strong unidirectional magnetic field. Because electrons can behave like magnets, a large number of electrons with the same pole orientation creates magnetic domains inside a material. If a ferromagnetic material is applied to the end of a beam fixed by a mechanical flexure on only one side, said cantilever will deflect out-of-plane when exposed to a uniform magnetic field [31]. Magnetic actuators can produce large out-of-plane deflections with high force without the need for onboard power or wiring. This force is generated as the magnetic element torques out-of-plane in the direction of the applied magnetic field. This force is defined as

$$T_{field} = v\vec{H}\vec{M}\sin(\gamma - \phi) \tag{5}$$

where v is the volume of the magnetic structure, $\vec{M}$ is the magnetization of the material, $\vec{H}$ is the applied magnetic field, and $\gamma - \phi$ is the angle between the magnetic field and the magnet.

As such, the deflection achieved is dependent on the strength of the magnetic components, as well as the length, width, thickness, moduli, and angular stiffness of the beam and mechanical flexure [31,32]. The flexure will remain suspended out-of-plane until the magnetic field is removed. Similarly, if the flexure is exposed to a magnetic field of cyclically changing strength, the magnitude of deflection will dynamically adjust as well. This is typically achieved by exposing the actuator device to an electromagnet powered by an alternating current or a rotating/translating permanent magnet.

A fast-moving cyclic magnetic actuator in a fluid will generate a shear force which can disrupt biological masses, such as thrombi [33,34]. However, if an elastic material is attached between the end of the cantilever and the adjacent outer frame of the device, this material will undergo uniaxial stretching as the distance from the cantilever edge and the device's outer frame changes due to the out-of-plane deflection. For example, Enríquez et al. recently used cantilever magnetic actuators to cyclically stretch fibronectin, a glycoprotein with elastic properties, in an effort to mimic the breathing cycle of the lungs as a platform to study changes in disseminated breast cancer cells upon arrival to the common metastatic location. The material experienced uniaxial stretching as the distance from the cantilever edge and the device's outer frame changed due to the out-of-plane deflection [35].

Permanent magnets will remain magnetized (remain aligned) after the external field is removed, while other ferromagnetic or ferrimagnetic materials do not have the ability to stay magnetized permanently. Common ferromagnetic materials include iron, cobalt, and nickel. Because the magnetic force is related to the number of electrons which can move, a magnetic field may also be produced by a current of electricity rather than by a magnet [29,36,37].

3.3. Thermal Energy

Based on their composition, iron oxide particles are ferromagnetic or ferrimagnetic by nature. In fact, most, but not all, iron oxide nanoparticles are magnetite (Fe_3O_4) or maghemite (γFe_2O_3), which are ferrimagnetic materials at room temperature [38]. Ferrimagnetic materials below a certain temperature threshold possess the same spontaneous magnetization as ferromagnetic materials. However, a non-uniform arrangement of atomic dipoles is created. Therefore, a lattice often forms of magnetic moments, with a strong magnetic moment directed parallel and a weaker magnetic moment directed antiparallel, leading to a magnetic field still being generated for the bulk material [39]. These materials can achieve magnetically induced heating due to their hysteretic properties when exposed to a time-varying magnetic field. Nevertheless, their heating efficiency is limited due to multiple magnetic domains present in these larger-sized particles.

However, if the diameter of the fabricated iron oxide particle is less than 200 nm, it will exhibit superparamagnetic properties instead. These particles are called superparamagnetic iron oxide particles (SPIOs/SPIONs), with particles under 50 nm being further classified as ultra-small SPIONs [40]. Paramagnetic materials have the same underlying principles as discussed above with ferromagnetic materials, but the coupling of the atomic magnetic moments is small. They do not exhibit a net magnetic moment without an external field and generate only a small magnetic moment when placed within a magnetic field [29]. Therefore, these materials have no magnetic remanence, meaning their magnetization relaxes to zero in a certain amount of time after the removal of the applied magnetic field. The relaxation time relates to either the Brownian relaxation time, the physical rotation of the particle dependent on the surrounding fluid, or the Néel relaxation time, the rotation of the atomic magnetic moments within each particle. The heating mechanisms in magnetic nanoparticles to induce hyperthermia include Neel and Brownian relaxation as well as hysteretic loss. SPION hyperthermia depends strongly on the particle size (<15 nm diameter), where Brownian relaxation exerts thermal energy as the particle rotates and applies shear force against the surrounding fluid, and Neel relaxation dissipates energy as the magnetic moment of the particle rotates before the physical particle [41]. Lastly, superparamagnetic nanoparticles are distinguished as such because they can generate larger field gradients than traditional paramagnetic materials, owning to the field being concentrated on a small particle area [42].

3.4. Benefits and Disadvantages of Using Magnetic Forces

A large appeal of generating forces through magnetism is the ability to apply an external magnetic field to direct the movement of a magnetized sample without direct contact or tethering [30,43–46]. In clinical settings, this has potential for less invasive, targeted therapeutic delivery, although attenuation of the field strength at deep tissue distances must still be overcome, as is similar with penetration of sound, light, and other external stimuli often proposed for non-invasive therapies. In vitro, the contactless, but precise, movement attainable with magnetic force can better preserve the sterility of biological samples and simplifies fabrication within the confines of traditional culturing equipment, such as commercial cell culture well plates, Petri dishes, and incubators, as compared to other force-generating apparatuses (e.g., pneumatics, electrostatic, piezoelectric, etc.) [35].

The equipment used to produce these wireless forces are simple and relatively inexpensive. Permanent magnets are commercially available and electromagnets consisting of wound wire around a high-permeability core can easily be made in the lab. Additionally, the size of the magnetic element is easily scalable, ranging from the size of a gene (single nanometer width) up to the macro-scale. Lastly, due to the diamagnetic nature of most biological materials, there is little interaction or sensitivity of the inorganic magnetic force component with existing cultures, as long as the researcher is mindful of the strength and frequency of the field that is required for a given application [47].

Conversely, the main disadvantage of using magnetic materials in culture is a pronounced cytotoxic effect. Permanent magnets and superparamagnetic particles typically must be coated in a biocompatible material before they can be utilized as a culture platform or a material for cellular uptake. For example, permanent magnets, such as neodymium (NdFeB), may be cytotoxic due to their corrosiveness [48]. Many common coatings are detailed below (see Sections 4 and 5 and Table 1). However, this is not always sufficient. Ketebo et al. and Shin et al. showed that silica-coated nanoparticles could damage a cell's cytoskeleton, impairing cell adhesion properties and reducing matrix rigidity/moduli sensing, due to the reactive oxygen species generated [49,50]. Beyond explicit cytotoxicity, all new magnetic particle formulations should undergo verification studies to ensure that they do not alter typical cell metabolism and function once taken up, as this could lead to untranslatable or misleading results [51].

Table 1. Magnetic-based 3D cell aggregation.

Magnetic Agent Used with Cells	Magnet Type	Type of Aggregation Assembly	Notes	Source
Cadolinium(III) chelates	Magnetized media	Dual magnet levitation	Multiple spheroids share media within a capillary tube	[52]
Gx				[53]
Paramagnetic metal halides				[54]
Gadopentatic acid (Gd-DTPA)		3D magnetic patterning		[55]
Magnetite nanoparticles isolated from magnetic bacteria	Internalized iron oxide nanoparticles	Single magnet levitation	Ring magnet [57,58]	[56]
Magnetite (Fe$_3$O$_4$), gold and bacteriophage nanoparticles (NanoShuttle)				[57,59]
NanoShuttle-PL		n3D magnetic drive system	Cell lumen formed [60,61]	[58]
				[60–66]
Magnetite nanoparticles with bovine serum albumin coating				[67]
Magnetite nanoparticles		3D magnetic patterning	Spheroids formed by hanging drop. Spheroids then patterned into lumens using magnetic patterning	[68]
Magnetite nanoparticles in liposomes	Internalized iron oxide cationic liposomes		Cells cultured in media and in collagen I Multiple spheroids share media	[69–71]

Each published system uses a combination of permanent magnets outside the culture area and a magnetic agent incubated with cells to improve the tunability, reproducibility, and precise patterning/movement of spheroids. Spheroids were composed of a variety of cell types, including non-transformed cells (e.g., fibroblasts and endothelial cells), various cancer cell lines (e.g., MCF7 (breast), MDA-MB-231 (breast), HCC827 (lung), DT66066) (pancreatic)), and co-cultured spheroids consisting of multiple cell types. These systems have been categorized based on the type of magnetic agent used and the type of levitation/patterning that the device achieves, with notes specifying if the system deviates from the standard practice of forcing cell aggregation into a single spheroid in a media bath.

4. Introduction to Magnetic Techniques in Cancer Treatment

Numerous magnetic techniques have been developed that show promise for in vivo and ex vivo clinical use. Generally, these techniques allow clinicians to sort, analyze, and/or treat primary and circulating tumor cells by functionalizing SPIONs, which then accumulate at the tumor site (in vivo) or isolate cancerous cells from bulk patient fluid samples (ex vivo). These SPIONs form the basis of many proposed targeted drug delivery treatments, which reduce a patient's chemotherapeutic burden as the nanoparticles accumulate at the tumor site, enabling a lower systemic dose and a higher local dose [72,73]. SPIONs also encompass a large proportion of proposed hyperthermia treatments, as described in Section 3, where heat is generated from the particles at the tumor site, causing apoptosis when tissue temperatures reach 42 °C and necrosis when temperatures exceed 46 °C up to 48 °C. [74–76].

For both treatments, the SPIONs must aggregate to the primary tumor. This is occasionally achieved through a direct local injection, but is more often achieved by external manipulation, where researchers guide the particles using an exterior magnetic field, or through self-aggregation using a tumor-specific antibody-coating on the nanoparticle [74,75,77]. This latter coating also forms the basis of ex vivo magnetic-associated sorting of circulating tumor cells or, occasionally, metastatic cells from patient fluid (e.g., blood, pleural effusions), which can then be analyzed based on cancer cell number isolated, marker expression, genetic profiling, or drug screening assays under the umbrella of precision medicine. Some common antibodies that are conjugated to SPIONs include anti-EpCAM

released by cancerous cells, and pulling down DNA, RNA, or specific proteins from lysed cells for quantification [105–111]. Magnetic particles have even been proposed as a method to improve transduction efficiency in conjunction with standard lentiviral particle use or instead of it [112–114]. These techniques often follow the particle principles outlined in Sections 3 and 4. In this section, we will instead focus on magnetic techniques that resolve the mechanical properties of cells and their environment or illustrate cells' biological response to dynamic mechanical stimulation.

6.1. Extracellular Matrix (ECM) Patterning and Detection of Remodeling

In contrast to the scaffold-free spheroid formation discussed in Section 5, scaffold-based 3D cell cultures are composed of cells seeded onto a supporting matrix. This matrix may be a gellable polymer, such as collagen, Matrigel (naturally derived), or polyethylene glycol (synthetic) [115]. The gelation process is typically simple and enables cells to be cultured on top of or throughout the gel. However, researchers may choose to undergo additional processing steps with their matrix of interest in order to create alternative scaffolding. One common process is called electrospinning. In this technique, a viscous liquid is drawn into a fiber which continuously builds on a collection plate until a complete mesh is fabricated [116]. In standard electrospinning, the fibers of the mesh are aligned and layered in a random orientation. However, using different mechanical, electrostatic, or magnetic interventions, the alignment of the electrospun fibers can be more precisely controlled. In magnetic-assisted electrospinning, an external magnetic field is generated through two parallel permanent magnets (Figure 2a). The fibers, drawn out of solution by standard electrospinning, will be driven to align parallel to the magnetic field lines as they travel toward the collection plate. This technique requires that the material being spun responds to the magnetic field generated. This can be achieved by incorporating one of the numerous nanoparticles into the matrix, such as silver nanorods, carbon nanotubes, or superparamagnetic nanoparticles [116–118].

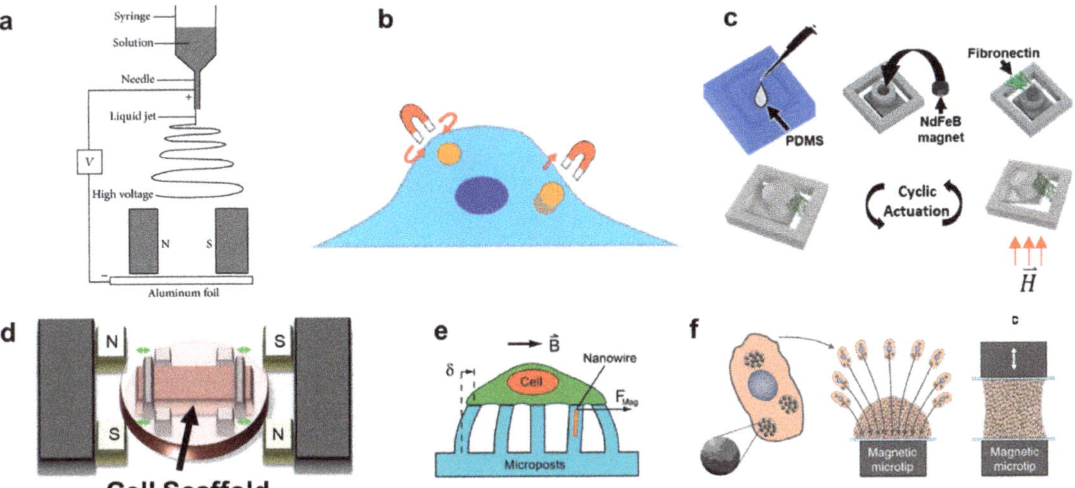

Figure 2. (**a**) Schematic depicting magnetically assisted electrospinning to create aligned matrix fibers [119]. (**b**) Schematic depicting how magnetic tweezers interact with the magnetic nanoparticles inside a cell [120]. (**c**) Fabrication process of fibronectin-coated polydimethylsiloxane (PDMS) cantilever with embedded permanent magnet. The cantilever deflects once exposed to a magnetic field enabling cyclic actuation [35]. (**d**) Uniaxial stretching platform using magnetic posts to stretch a cell scaffold [121]. (**e**) Magnetically active microposts to induce forces on cultured cells [46]. (**f**) Magnetic nanoparticle embedded cells for remote mechanical control. Magnetic microtips can aggregate the cells to create a tissue [45].

In addition to magnetic-assisted electrospinning, Kim et al. recently demonstrated that ECM proteins could be chemically crosslinked onto magnetic particles and self-assembled into numerous topographical patterns in a surrogate hydrogel by adjusting the external magnetic field applied [122]. Martin et al. incorporated iron oxide nanoparticles into nonmagnetic materials, such as silica and calcium phosphate, and developed an SLA-based 3D magnetic printing protocol. To demonstrate the technique, the researchers aligned an external magnetic field with respect to different axes and observed changing bulk mechanical properties of the printed device based on the orientation of the reinforcing elements of the discontinuous fiber composite [123,124]. Margolis et al. similarly demonstrated that the mechanical properties of alginate could be altered by loading the hydrogel with magnetic nanoparticles of various sizes and concentrations and exposing the system to an external magnetic field. The resulting gel formed an aligned microporous structure with anisotropic topical features and stiffnesses related to the direction of the magnetic field, which ultimately dictated the morphology of mouse myoblasts cultured within [125].

Lastly, a magnetic technique was developed to selectively identify ECM remodeled by cancerous cells in an in vitro tumor-like coculture environment [126]. Magnetic helical nanorobots were fabricated such that contactless forward or backward movement could be achieved when the devices were subjected to a rotating magnetic field. The nanorobots were made from silica with embedded iron particles. The robots were injected into a hydrogel of basement membrane protein with breast cancer cells and non-cancerous breast epithelial cells cocultured together. The robots were injected on one side of the hydrogel and driven across its length for approximately 30 min using an external magnetic field. The microrobots preferentially attached to ECM surrounding the cancer cells, with comparatively few microrobots found around the healthy cell type. This phenomenon seems driven by the difference in charge surrounding cancerous and healthy cells, where microrobots became irreversible stuck when passing up to 35 µm away from a cancer cell, but only up to 15 µm away from a healthy cell. This charge difference is driven by a sialic acid linkage aberrantly expressed in the cancer cells, which imparts a large negative charge on the surrounding ECM. At present, the experiment was only performed with different breast cancer cell lines in the reconstituted basement membrane protein [126]. Although sialylation changes have been reported in a number of different cancer types, it remains to be seen if the designed microrobots will retain the same preferential attachment for other cancer lines and/or in other matrices based on their unique ECM remodeling [126,127]. It is also of interest if the results would be replicated in environments with more cell complexity, as fibroblasts, not breast cancer cells, seem to be the main source of ECM remodeling in the tumor microenvironment [128].

6.2. Resolving Dynamic Mechanotransduction Behavior

Mechanotransduction is the process by which cells convert mechanical stimuli into biochemical signals [129]. This mechanical stimulation can include all aspects of the surrounding physical environment, including its moduli and topography, as well as dynamic compressive, tensile, or shear forces acting on the cells [120]. Cells have several mechanisms through which they detect mechanical stimulation. The surface of the cell itself has mechanical sensitive protein complexes, mechanosensitive ion channels, and transmembrane integrins which connect the intracellular cytoskeleton to the ECM [129,130]. Mechanoresponsive proteins also exist within the cell, typically as a downstream effect of a membrane mechanosensitive signal being triggered (e.g., integrin signal). The conformation of these protein complexes will transition in a mechanically stimulated environment (e.g., when the cell is under tension) which will alter the resulting binding properties or enzymatic function [129].

Using magnetic techniques to deliver mechanical stimulation to cells is very attractive due to its potential to be contactless, simple to fabricate, and easily integrated within current culturing equipment, as mentioned in Section 3.4. In addition, magnetic nanoparticles can be effectively utilized to deliver targeted, non-destructive mechanical stimulation to cells, or even directly to the surface mechanoreceptors of cells. In this way, some magnetic techniques offer the unique ability to mechanically load cells on soft or fragile biomaterials since no extracellular scaffold deformation is required [129]. In this section, we will highlight different magnetic devices that can provide dynamic mechanical stimulation either directly to cellular components or to the outside of the cell (i.e., scaffold deformation) and how these stimuli affect cell function.

6.2.1. Applying Mechanical Force Intracellularly

Magnetic pulling cytometry and magnetic twisting cytometry using magnetic tweezers are the main magnetic techniques available to measure intrinsic mechanical properties of cells [131] (Figure 2b). For these techniques, magnetic beads are typically incubated with cells in culture to allow for cellular uptake. In magnetic pulling cytometry, a magnetic needle is then positioned close to the magnetic bead in an isolated cell, such that the bead is pulled in-plane toward the needle with a known force. In magnetic twisting cytometry, a strong external magnetic field is first pulsed (usually $\geq$1000 G for <0.5 milliseconds) which magnetizes the bead in the direction of the magnetic field. Then, a weak twisting field is applied in the orthogonal direction to the magnetic moment of the bead, causing the bead to attempt to deflect out of plane, similar to the principles of magnetic actuation [120,132–134]. Magnetic manipulation of the internalized beads has revealed several intrinsic mechanical properties of cells, such as cytoplasm viscoelasticity [131,135]. However, this technique has also been used to generate intracellular force long-term ($\geq$hours) and observe the resulting cellular changes. For example, Qiu et al. demonstrated that prolonged intracellular force generated by internalized magnetic beads would result in the alignment of F-actin fibers in the direction of the magnetic force in endothelial cells [136].

Additionally, these techniques can also be used to measure the force of or dynamically manipulate specific cell receptors. For example, Boulter et al. used fibronectin-coated nanobeads and magnetic tweezers.to uncover crosstalk between integrin rigidity mechanosensing and cellular metabolism. They found that the integrin coreceptor, CD98hs, indirectly regulated sphingolipid synthesis after mechanical integrin disturbance through the prevention of several upstream regulators of RhoA, such as Src kinases [137]. Others have used numerous coatings, including for the RGD domain (to target integrin $\alpha_v\beta_3$), anti-β1 integrin, anti-PDGFRα, E-cadherin extracellular domains, and the extracellular loop of TREK-1 [133,135,138–144].

The main disadvantage to these techniques is that they are often extremely costly and technically challenging. These challenges are not unique to magnetic-based approaches (e.g., optical tweezers), but are a common issue associated with being able to stimulate and then observe minute intracellular forces [145]. Specifically for magnetic-based approaches, both finely calibrated magnetic field control equipment and extremely uniform magnetic beads are essential for accurate measurements [120]. Additionally, an overarching obstacle for the field of intracellular mechanotransduction is a lack of consensus between experimental techniques. Specifically, Wu et al. measured the MCF-7 breast cancer cell line using six techniques (atomic force microscopy, parallel-plate rheology, optical stretching, cell monolayer rheology, magnetic twisting cytometry, and particle tracking microrheology) and demonstrated that the obtained elastic and viscous moduli of the cells varied up to 1000- and 100-fold, respectively [135]. This inconsistency must be carefully considered as mechanosensitive cellular pathways become increasingly investigated by various techniques and research groups, such that our collective understanding of mechanotransduction throughout the metastatic cascade can continue to grow.

6.2.2. Extracellular Movement

Although static cultures have provided invaluable insights into metastatic progression, they fail to capture the response of cancer cells to dynamic extracellular forces, such as forces that are native to a distant organ on early disseminated cancer cells. Here, will we discuss magnetic techniques used to apply dynamic and static forces extracellularly, encompassing compressive, traction, and tensile forces.

The growth of a carcinoma is associated with tremendous compressive forces within the bulk of the tumor as cells, ECM, and fluid accumulate. Compressive forces also increase within the surrounding microenvironment as the growing tumor pushes against the boundaries of healthy tissue. In an unusual technique, Fernández-Sánchez et al. developed a method to generate compressive forces in vivo. Magnetic liposomes or ultra-magnetic liposomes (i.e., liposomes loaded with a magnetic aqueous fluid or loaded with superparamagnetic iron oxide nanocrystals, respectively) were fabricated in vitro and then intravenously injected in mice. Then a permanent disc magnet was inserted subcutaneously in front of the colon. After one week, the ultra-magnetic liposomes were concentrated in the stromal cells surrounding the distal colon crypts and remained with a stable concentration of iron per gram of tissue over a one-month span. The compressive force, and subsequent pathophysiological stress on the mesenchymal cells, was consistent with the mechanical pressure exerted by early tumor growth in adjacent crypts as measured in previous cancerous mouse models. Their subsequent biological investigation indicated that tumorigenic pathways may be activated in the non-cancerous stroma surrounding a carcinoma due to the compressive mechanical stimulation, which could contribute to an unstable positive feedback loop between oncogene expression and tumor induction [146].

In vitro, several devices have been built which can provide dynamic compressive and tensile forces onto cells on polymer substrates. In one system, magnetic actuators transmitted linear uniaxial compressive or tensile forces through positioning pins to a cell culture area, consisting of cell-seeded polyethylene glycol constructs. Although the entire device was sealed for sterility, it was built around one culture area, such that it cannot be easily upscaled for high-throughput assays [147,148]. In another system, a magnetic force was created between an external electromagnet and a permanent magnet embedded into a polydimethylsiloxane (PDMS) frame. The PDMS frame was fixed on one side and allowed to freely move toward the electromagnet on the other. In the middle of the PDMS frame was a mesh where a thin layer of Matrigel was deposited and cells were subsequently seeded to undergo cyclic stretching [149]. Enríquez et al. developed a magnetically actuating device along a similar principle (Figure 2c). Here, the matrix protein, fibronectin, was suspended between the cantilever and adjacent frame edge of a PDMS magnetic actuator, resulting in cyclic stretching of the matrix (and cells within) as described in Section 3. A main advantage of this recent design is the suspended matrix, which does not utilize any polymeric support mesh in the culture region of interest (Figure 2d). It also benefits from a high-throughput design, consisting of an array of permanent magnets in a linearly moving actuating platform that sits beneath a standard multi-well culture plate [35].

Lastly, similar devices have been developed to those described above, but which solely use PDMS gel as the culture substrate [150]. A unique subsection of this category is magnetic micropillars (Figure 2e). Cells are cultured on top of magnetized PDMS pillars, such that the bottom of the cell will undergo tractile forces as the pillars bend toward an external magnetic source. The pillars are typically magnetized by iron-coating over the PDMS or by embedding magnetic nanoparticles into the polymeric pillars [151–153]. Like magnetic tweezer techniques, these distinct pillars can be used to isolate and quantify forces at individual cell-matrix contact sites due to their free motion from one another. Additionally, this magnetic method is one of few that can detect traction forces that cells impose onto the substrate during migration [154]. However, the translation of these results to in situ cell response is likely to suffer due to the lack of ECM proteins used in the culture platform as well as the often unrealistic stiffness of the PDMS compared to the native tissue of the cell type. In contrast, Du et al. developed a magnetic tissue stretcher that utilizes no matrix or substrate at all. This technique begins identically to aggregation micropatterning, where magnetic nanoparticles are taken up by cells which are then forced to aggregate due to a magnet underneath the culture area (a glass slide) (Figure 2f). However, a second magnetic microtip-glass slide apparatus is then placed in contact with the top of the aggregate, and these cells are allowed to attach. The confluent band of tissue can then undergo cyclic tensile strain driven by the movement of one of the magnetic microtips. In this way, although the cells at the very edges are influenced by the topography and modulus of the glass slides, the bulk tissue sample is free of any substrate [45]. Because no in vitro system can replicate all aspects of the in situ tissue environment (nor is it meant to), we will likely find that it is a combination of scaffold and scaffold-free mechanotransduction techniques which together allow us to decouple different aspects of the metastatic cascade and eventually understand the complete and complex process.

7. Conclusions

The use of magnetic forces can greatly improve our ability to diagnose, treat, and fundamentally understand cancer. Magnetic nanoparticles are being widely explored for in vivo applications, including as MRI contrast agents, hyperthermic treatment agents, and targeted drug delivery vehicles [38,73]. They also form the basis of most ex vivo cell isolation techniques, such as for isolating circulating tumor cells, wherein antibody-conjugated SPIONs are used for negative or positive magnetophoresis. Here, we focused on in vitro culture platforms that utilize magnetic force to broadly (1) improve the reproducibility and ease of handling of spheroids and (2) impose mechanical force intracellularly and extracellularly. The latter allows researchers to understand the mechanotransduction of cancer cells throughout several unique steps of the metastatic cascade, such as during primary tumor growth or early dissemination. Although using magnets comes with several advantages, such as the untethered transmission of force, the potential effect of large magnetic fields on cells cannot be overlooked in these culture platforms. Both a prolonged strong magnetic field without nanobeads and the unstimulated magnetic nanobeads have been shown to affect proliferation rates, cell metabolism, ion channel activity, and the cytoskeletal organization in a small subset of experiments [49,50,155,156]. The inherent biological effect of using external magnetic forces must be especially accounted for when analyzing certain types of cancers, such as when using cancerous neural cells [157]. With the proper consideration and controls, magnetic force can be an invaluable addition to in vitro culture platforms for reliable probing of cancer cells throughout primary tumor growth and metastatic progression.

Author Contributions: Conceptualization, S.L., Á.E., H.L., L.S.; writing—original draft preparation, S.L.; writing—review and editing, S.L., Á.E., H.L., L.S.; visualization, S.L. and Á.E.; supervision, H.L. and L.S. All authors have read and agreed to the published version of the manuscript.

Funding: This publication was supported in part from the National Institutes of Health, National Center for Advancing Translational Sciences, Clinical and Translational Science Award (UL1TR002529) to S.L. This work was also supported in part by NIH/NCRR Indiana CTSI (UL1TR001108, UL/TR002529), and Clinical and Translational Science Award and The Advances in Medicine (AIM) grant from Cook Medical to H.L. This work received support from the National Science Foundation (United States) under grant ECCS-1944480 to H.L. This project received additional support from the Purdue University Women's Global Health Institute and from The Catherine Peachey Fund: A Member of the Heroes Foundation Family to H.L. and L.S. and received support from the Showalter Trust to L.S.

Acknowledgments: The authors would like to acknowledge the reuse of schematics from previously published works in Figure 2 to illustrate many of the common magnetic techniques developed to probe cells and their microenvironments in vitro. Each original article is cited in the figure legend. Sections 2a and 2f are licensed under the Creative Commons Attribution License (https: //creativecommons.org/licenses/by/4.0/ (accessed on 27 July 2021) with no changes made. Section 2b is reprinted from 16 Cancer Mechanobiology: Interaction of Biomaterials with Cancer Cells, In Biomaterials for Cancer Therapeutics (Second Edition), Libring, S., Solorio, L. Editor: Park, K., pp. 445–470, Woodhead Publishing (2020), with permission from Elsevier (license 5135691231943). Section 2c is adapted from a previously authored publication where the authors retained the right to re-use the content in a new publication from John Wiley and Sons. Section 2d is adapted from The Effect of Cyclic Stretch on Maturation and 3D Tissue Formation of Human Embryonic Stem

Cell-Derived Cardiomyocytes, 35(9), Mihic, A., Li, J., Miyahi, Y., Gagliardi, M., Li, S., Zu, J., Weisel, R.D., Keller, G., Li, R. pp. 2798–2808 (2014), with permission from Elsevier (license 5135700576730). The schematic is unaltered except for the "Cell Scaffold" annotation. Section 2e is reprinted under the exclusive PNAS License to Publish, Copyright 2007 National Academy of Sciences.

Conflicts of Interest: The authors declare no conflict of interest.

References

1. Siegel, R.L.; Miller, K.D.; Fuchs, H.E.; Jemal, A. Cancer Statistics, 2021. *CA Cancer J. Clin.* **2021**, *71*, 7–33. [CrossRef] [PubMed]
2. Bray, F.; Ferlay, J.; Soerjomataram, I.; Siegel, R.L.; Torre, L.A.; Jemal, A. Global cancer statistics 2018: GLOBOCAN estimates of incidence and mortality worldwide for 36 cancers in 185 countries. *CA Cancer J. Clin.* **2018**, *68*, 394–424. [CrossRef] [PubMed]
3. Eslami-S, Z.; Cortés-Hernández, L.E.; Alix-Panabières, C. The Metastatic Cascade as the Basis for Liquid Biopsy Development. *Front. Oncol.* **2020**, *10*, 1055. [CrossRef]
4. Hanahan, D.; Weinberg, R.A. The hallmarks of cancer. *Cell* **2000**, *100*, 57–70. [CrossRef]
5. Hanahan, D.; Weinberg, R.A. Hallmarks of cancer: The next generation. *Cell* **2011**, *144*, 646–674. [CrossRef] [PubMed]
6. Anderson, N.M.; Simon, M.C. The tumor microenvironment. *Curr. Biol.* **2020**, *30*, R921–R925. [CrossRef]
7. Doglioni, G.; Parik, S.; Fendt, S.M. Interactions in the (Pre)metastatic Niche Support Metastasis Formation. *Front. Oncol.* **2019**, *9*, 219. [CrossRef]
8. Kaplan, R.N.; Rafii, S.; Lyden, D. Preparing the "soil": The premetastatic niche. *Cancer Res.* **2006**, *66*, 11089–11093. [CrossRef]
9. Sleeman, J.P. The metastatic niche and stromal progression. *Cancer Metastasis Rev.* **2012**, *31*, 429–440. [CrossRef]
10. Wortzel, I.; Dror, S.; Kenific, C.M.; Lyden, D. Exosome-Mediated Metastasis: Communication from a Distance. *Dev. Cell* **2019**, *49*, 347–360. [CrossRef]
11. Isola, A.L.; Chen, S. Exosomes: The Messengers of Health and Disease. *Curr. Neuropharmacol.* **2017**, *15*, 157–165. [CrossRef] [PubMed]
12. Rajagopal, C.; Harikumar, K.B. The Origin and Functions of Exosomes in Cancer. *Front. Oncol.* **2018**, *8*, 66. [CrossRef]
13. Dudas, J.; Ladanyi, A.; Ingruber, J.; Steinbichler, T.B.; Riechelmann, H. Epithelial to Mesenchymal Transition: A Mechanism that Fuels Cancer Radio/Chemoresistance. *Cells* **2020**, *9*, 428. [CrossRef] [PubMed]
14. Yilmaz, M.; Christofori, G. EMT, the cytoskeleton, and cancer cell invasion. *Cancer Metastasis Rev.* **2009**, *28*, 15–33. [CrossRef] [PubMed]
15. Clark, A.G.; Vignjevic, D.M. Modes of cancer cell invasion and the role of the microenvironment. *Curr. Opin. Cell Biol.* **2015**, *36*, 13–22. [CrossRef]
16. Li, Y.; Wu, S.; Bai, F. Molecular characterization of circulating tumor cells-from bench to bedside. *Semin. Cell Dev. Biol.* **2018**, *75*, 88–97. [CrossRef]
17. Meng, S.; Tripathy, D.; Frenkel, E.P.; Shete, S.; Naftalis, E.Z.; Huth, J.F.; Beitsch, P.D.; Leitch, M.; Hoover, S.; Euhus, D.; et al. Circulating tumor cells in patients with breast cancer dormancy. *Clin. Cancer Res.* **2004**, *10*, 8152–8162. [CrossRef]
18. Scully, O.J.; Bay, B.H.; Yip, G.; Yu, Y. Breast cancer metastasis. *Cancer Genomi. Proteom.* **2012**, *9*, 311–320. [PubMed]
19. Perea Paizal, J.; Au, S.H.; Bakal, C. Squeezing through the microcirculation: Survival adaptations of circulating tumour cells to seed metastasis. *Br. J. Cancer* **2021**, *124*, 58–65. [CrossRef]
20. Castro-Giner, F.; Aceto, N. Tracking cancer progression: From circulating tumor cells to metastasis. *Genome Med.* **2020**, *12*, 31. [CrossRef]
21. Strilic, B.; Offermanns, S. Intravascular Survival and Extravasation of Tumor Cells. *Cancer Cell* **2017**, *32*, 282–293. [CrossRef] [PubMed]
22. Sai, B.; Xiang, J. Disseminated tumour cells in bone marrow are the source of cancer relapse after therapy. *J. Cell. Mol. Med.* **2018**, *22*, 5776–5786. [CrossRef] [PubMed]
23. Gao, Y.; Bado, I.; Wang, H.; Zhang, W.; Rosen, J.M.; Zhang, X.H. Metastasis Organotropism: Redefining the Congenial Soil. *Dev. Cell* **2019**, *49*, 375–391. [CrossRef]
24. Smith, H.A.; Kang, Y. Determinants of Organotropic Metastasis. *Ann. Rev. Cancer Biol.* **2017**, *1*, 403–423. [CrossRef]
25. Wei, S.C.; Yang, J. Forcing through Tumor Metastasis: The Interplay between Tissue Rigidity and Epithelial-Mesenchymal Transition. *Trends Cell Biol.* **2016**, *26*, 111–120. [CrossRef]
26. Luzzi, K.J.; MacDonald, I.C.; Schmidt, E.E.; Kerkvliet, N.; Morris, V.L.; Chambers, A.F.; Groom, A.C. Multistep nature of metastatic inefficiency: Dormancy of solitary cells after successful extravasation and limited survival of early micrometastases. *Am. J. Pathol.* **1998**, *153*, 865–873. [CrossRef]
27. Dasgupta, A.; Lim, A.R.; Ghajar, C.M. Circulating and disseminated tumor cells: Harbingers or initiators of metastasis? *Mol. Oncol.* **2017**, *11*, 40–61. [CrossRef]
28. Ray, M.W.; Ruokokoski, E.; Tiurev, K.; Möttönen, M.; Hall, D.S. Observation of isolated monopoles in a quantum field. *Science* **2015**, *348*, 544. [CrossRef] [PubMed]
29. Morrish, A.H. The Magnetic Field. In *The Physical Principles of Magnetism*; IEEE: Piscataway, NJ, USA, 2001; pp. 1–30.
30. Pankhurst, Q.A.; Thanh, N.T.K.; Jones, S.K.; Dobson, J. Progress in applications of magnetic nanoparticles in biomedicine. *J. Phys. D Appl. Phys.* **2009**, *42*, 224001. [CrossRef]

31. Judy, J.W.; Muller, R.S. Magnetic microactuation of torsional polysilicon structures. *Sens. Actuators A Phys.* **1996**, *53*, 392–397. [CrossRef]
32. Judy, J.W.; Muller, R.S. Magnetically actuated, addressable microstructures. *J. Microelectromech. Syst.* **1997**, *6*, 249–256. [CrossRef]
33. Xu, J.; Lee, H. Anti-Biofouling Strategies for Long-Term Continuous Use of Implantable Biosensors. *Chemosensors* **2020**, *8*, 66. [CrossRef]
34. Yang, Q.; Enriquez, A.; Devathasan, D.; Thompson, C.; Nayee, D.; Harris, R.; Satoski, D.; Obeng-Gyasi, B.; Lee, A.; Bentley, R.T. Magnetically actuated self-clearing catheter for rapid in situ blood clot clearance for improved hemorrhagic stroke treatment. *Res. Sq.* **2021**. [CrossRef]
35. Enríquez, Á.; Libring, S.; Field, T.C.; Jimenez, J.; Lee, T.; Park, H.; Satoski, D.; Wendt, M.K.; Calve, S.; Tepole, A.B.; et al. High-Throughput Magnetic Actuation Platform for Evaluating the Effect of Mechanical Force on 3D Tumor Microenvironment. *Adv. Funct. Mater.* **2020**, *31*, 2005021. [CrossRef]
36. Bozorth, R.M. Concepts of Ferromagnetism. In *Ferromagnetism*; IEEE: Piscataway, NJ, USA, 1978; pp. 1–13.
37. Morrish, A.H. Ferromagnetism. In *The Physical Principles of Magnetism*; IEEE: Piscataway, NJ, USA, 2001; pp. 259–331.
38. Stephen, Z.R.; Kievit, F.M.; Zhang, M. Magnetite Nanoparticles for Medical MR Imaging. *Mater. Today* **2011**, *14*, 330–338. [CrossRef]
39. Morrish, A.H. Ferrimagnetism. In *The Physical Principles of Magnetism*; IEEE: Piscataway, NJ, USA, 2001; pp. 486–538.
40. Yan, G.-P.; Robinson, L.; Hogg, P. Magnetic resonance imaging contrast agents: Overview and perspectives. *Radiography* **2007**, *13*, e5–e19. [CrossRef]
41. Deatsch, A.E.; Evans, B.A. Heating efficiency in magnetic nanoparticle hyperthermia. *J. Magn. Magn. Mater.* **2014**, *354*, 163–172. [CrossRef]
42. Orel, V.E.; Tselepi, M.; Mitrelias, T.; Zabolotny, M.; Shevchenko, A.; Rykhalskyi, A.; Romanov, A.; Orel, V.B.; Burlaka, A.; Lukin, S.; et al. The comparison between superparamagnetic and ferromagnetic iron oxide nanoparticles for cancer nanotherapy in the magnetic resonance system. *Nanotechnology* **2019**, *30*, 415701. [CrossRef] [PubMed]
43. Yang, Q.; Park, H.; Nguyen, T.N.H.; Rhoads, J.F.; Lee, A.; Bentley, R.T.; Judy, J.W.; Lee, H. Anti-biofouling implantable catheter using thin-film magnetic microactuators. *Sens. Actuators B Chem.* **2018**, *273*, 1694–1704. [CrossRef]
44. Park, H.; Raffiee, A.H.; John, S.W.M.; Ardekani, A.M.; Lee, H. Towards smart self-clearing glaucoma drainage device. *Microsyst. Nanoeng.* **2018**, *4*, 35. [CrossRef] [PubMed]
45. Du, V.; Luciani, N.; Richard, S.; Mary, G.; Gay, C.; Mazuel, F.; Reffay, M.; Menasché, P.; Agbulut, O.; Wilhelm, C. A 3D magnetic tissue stretcher for remote mechanical control of embryonic stem cell differentiation. *Nat. Commun.* **2017**, *8*, 400. [CrossRef]
46. Sniadecki, N.J.; Anguelouch, A.; Yang, M.T.; Lamb, C.M.; Liu, Z.; Kirschner, S.B.; Liu, Y.; Reich, D.H.; Chen, C.S. Magnetic microposts as an approach to apply forces to living cells. *Proc. Natl. Acad. Sci. USA* **2007**, *104*, 14553–14558. [CrossRef] [PubMed]
47. Blümler, P.; Friedrich, R.P.; Pereira, J.; Baun, O.; Alexiou, C.; Mailänder, V. Contactless Nanoparticle-Based Guiding of Cells by Controllable Magnetic Fields. *Nanotechnol. Sci. Appl.* **2021**, *14*, 91–100. [CrossRef]
48. Donohue, V.E.; McDonald, F.; Evans, R. In vitro cytotoxicity testing of neodymium-iron-boron magnets. *J. Appl. Biomater.* **1995**, *6*, 69–74. [CrossRef] [PubMed]
49. Shin, T.H.; Lee, D.Y.; Ketebo, A.A.; Lee, S.; Manavalan, B.; Basith, S.; Ahn, C.; Kang, S.H.; Park, S.; Lee, G. Silica-Coated Magnetic Nanoparticles Decrease Human Bone Marrow-Derived Mesenchymal Stem Cell Migratory Activity by Reducing Membrane Fluidity and Impairing Focal Adhesion. *Nanomaterials* **2019**, *9*, 1475. [CrossRef]
50. Ketebo, A.A.; Shin, T.H.; Jun, M.; Lee, G.; Park, S. Effect of silica-coated magnetic nanoparticles on rigidity sensing of human embryonic kidney cells. *J. Nanobiotechnol.* **2020**, *18*, 170. [CrossRef] [PubMed]
51. Naseer, S.M.; Manbachi, A.; Samandari, M.; Walch, P.; Gao, Y.; Zhang, Y.S.; Davoudi, F.; Wang, W.; Abrinia, K.; Cooper, J.M.; et al. Surface acoustic waves induced micropatterning of cells in gelatin methacryloyl (GelMA) hydrogels. *Biofabrication* **2017**, *9*, 015020. [CrossRef]
52. Türker, E.; Demirçak, N.; Arslan-Yildiz, A. Scaffold-free three-dimensional cell culturing using magnetic levitation. *Biomater. Sci.* **2018**, *6*, 1745–1753. [CrossRef] [PubMed]
53. Onbas, R.; Arslan Yildiz, A. Fabrication of Tunable 3D Cellular Structures in High Volume Using Magnetic Levitation Guided Assembly. *ACS Appl. Bio Mater.* **2021**, *4*, 1794–1802. [CrossRef]
54. Bwambok, D.K.; Thuo, M.M.; Atkinson, M.B.J.; Mirica, K.A.; Shapiro, N.D.; Whitesides, G.M. Paramagnetic Ionic Liquids for Measurements of Density Using Magnetic Levitation. *Anal. Chem.* **2013**, *85*, 8442–8447. [CrossRef]
55. Abdel Fattah, A.R.; Mishriki, S.; Kammann, T.; Sahu, R.P.; Geng, F.; Puri, I.K. 3D cellular structures and co-cultures formed through the contactless magnetic manipulation of cells on adherent surfaces. *Biomater. Sci.* **2018**, *6*, 683–694. [CrossRef]
56. Kim, J.A.; Choi, J.H.; Kim, M.; Rhee, W.J.; Son, B.; Jung, H.K.; Park, T.H. High-throughput generation of spheroids using magnetic nanoparticles for three-dimensional cell culture. *Biomaterials* **2013**, *34*, 8555–8563. [CrossRef] [PubMed]
57. Souza, G.R.; Molina, J.R.; Raphael, R.M.; Ozawa, M.G.; Stark, D.J.; Levin, C.S.; Bronk, L.F.; Ananta, J.S.; Mandelin, J.; Georgescu, M.-M.; et al. Three-dimensional tissue culture based on magnetic cell levitation. *Nat. Nanotechnol.* **2010**, *5*, 291–296. [CrossRef] [PubMed]
58. Timm, D.M.; Chen, J.; Sing, D.; Gage, J.A.; Haisler, W.L.; Neeley, S.K.; Raphael, R.M.; Dehghani, M.; Rosenblatt, K.P.; Killian, T.C.; et al. A high-throughput three-dimensional cell migration assay for toxicity screening with mobile device-based macroscopic image analysis. *Sci. Rep.* **2013**, *3*, 3000. [CrossRef]

59. Jaganathan, H.; Gage, J.; Leonard, F.; Srinivasan, S.; Souza, G.R.; Dave, B.; Godin, B. Three-dimensional in vitro co-culture model of breast tumor using magnetic levitation. *Sci. Rep.* **2014**, *4*, 6468. [CrossRef]

60. Souza, G.R.; Tseng, H.; Gage, J.A.; Mani, A.; Desai, P.; Leonard, F.; Liao, A.; Longo, M.; Refuerzo, J.S.; Godin, B. Magnetically Bioprinted Human Myometrial 3D Cell Rings as A Model for Uterine Contractility. *Int. J. Mol. Sci.* **2017**, *18*, 683. [CrossRef]

61. Tseng, H.; Gage, J.A.; Haisler, W.L.; Neeley, S.K.; Shen, T.; Hebel, C.; Barthlow, H.G.; Wagoner, M.; Souza, G.R. A high-throughput in vitro ring assay for vasoactivity using magnetic 3D bioprinting. *Sci. Rep.* **2016**, *6*, 30640. [CrossRef]

62. Baillargeon, P.; Shumate, J.; Hou, S.; Fernandez-Vega, V.; Marques, N.; Souza, G.; Seldin, J.; Spicer, T.P.; Scampavia, L. Automating a Magnetic 3D Spheroid Model Technology for High-Throughput Screening. *SLAS Technol.* **2019**, *24*, 420–428. [CrossRef]

63. Desai, P.K.; Tseng, H.; Souza, G.R. Assembly of Hepatocyte Spheroids Using Magnetic 3D Cell Culture for CYP450 Inhibition/Induction. *Int. J. Mol. Sci.* **2017**, *18*, 1085. [CrossRef] [PubMed]

64. Ferreira, J.N.; Hasan, R.; Urkasemsin, G.; Ng, K.K.; Adine, C.; Muthumariappan, S.; Souza, G.R. A magnetic three-dimensional levitated primary cell culture system for the development of secretory salivary gland-like organoids. *J. Tissue Eng. Regen. Med.* **2019**, *13*, 495–508. [CrossRef]

65. Abou Ali, E.; Bordacahar, B.; Mestas, J.L.; Batteux, F.; Lafon, C.; Camus, M.; Prat, F. Ultrasonic cavitation induces necrosis and impairs growth in three-dimensional models of pancreatic ductal adenocarcinoma. *PLoS ONE* **2018**, *13*, e0209094. [CrossRef] [PubMed]

66. Abdullah, M.; Bela, B.; Syam, A.F.; Simadibrata, M.; Meilany, S.; Annisa, F.; Amirulloh, D.; Makmun, D.; Rani, A.A. Establishment of primary 3D cell culture based on magnetic bioprinting for colorectal cancer cells from patients in Cipto Mangunkusumo National Hospital Indonesia. *AIP Conf. Proc.* **2019**, *2155*, 020025. [CrossRef]

67. Ghosh, S.; Kumar, S.R.; Puri, I.K.; Elankumaran, S. Magnetic assembly of 3D cell clusters: Visualizing the formation of an engineered tissue. *Cell Prolif.* **2016**, *49*, 134–144. [CrossRef] [PubMed]

68. Mattix, B.M.; Olsen, T.R.; Casco, M.; Reese, L.; Poole, J.T.; Zhang, J.; Visconti, R.P.; Simionescu, A.; Simionescu, D.T.; Alexis, F. Janus magnetic cellular spheroids for vascular tissue engineering. *Biomaterials* **2014**, *35*, 949–960. [CrossRef] [PubMed]

69. Okochi, M.; Matsumura, T.; Yamamoto, A.S.; Nakayama, E.; Jimbow, K.; Honda, H. Cell behavior observation and gene expression analysis of melanoma associated with stromal fibroblasts in a three-dimensional magnetic cell culture array. *Biotechnol. Prog.* **2013**, *29*, 135–142. [CrossRef]

70. Okochi, M.; Takano, S.; Isaji, Y.; Senga, T.; Hamaguchi, M.; Honda, H. Three-dimensional cell culture array using magnetic force-based cell patterning for analysis of invasive capacity of BALB/3T3/v-src. *Lab Chip* **2009**, *9*, 3378–3384. [CrossRef]

71. Yamamoto, S.; Hotta, M.M.; Okochi, M.; Honda, H. Effect of vascular formed endothelial cell network on the invasive capacity of melanoma using the in vitro 3D co-culture patterning model. *PLoS ONE* **2014**, *9*, e103502. [CrossRef]

72. Janko, C.; Dürr, S.; Munoz, L.E.; Lyer, S.; Chaurio, R.; Tietze, R.; Löhneysen, S.; Schorn, C.; Herrmann, M.; Alexiou, C. Magnetic drug targeting reduces the chemotherapeutic burden on circulating leukocytes. *Int. J. Mol. Sci.* **2013**, *14*, 7341–7355. [CrossRef] [PubMed]

73. Janko, C.; Ratschker, T.; Nguyen, K.; Zschiesche, L.; Tietze, R.; Lyer, S.; Alexiou, C. Functionalized Superparamagnetic Iron Oxide Nanoparticles (SPIONs) as Platform for the Targeted Multimodal Tumor Therapy. *Front. Oncol.* **2019**, *9*, 59. [CrossRef]

74. Su, Z.; Liu, D.; Chen, L.; Zhang, J.; Ru, L.; Chen, Z.; Gao, Z.; Wang, X. CD44-Targeted Magnetic Nanoparticles Kill Head and Neck Squamous Cell Carcinoma Stem Cells in an Alternating Magnetic Field. *Int. J. Nanomed.* **2019**, *14*, 7549–7560. [CrossRef]

75. Palanisamy, S.; Wang, Y.M. Superparamagnetic iron oxide nanoparticulate system: Synthesis, targeting, drug delivery and therapy in cancer. *Dalton Trans.* **2019**, *48*, 9490–9515. [CrossRef] [PubMed]

76. Musielak, M.; Piotrowski, I.; Suchorska, W.M. Superparamagnetic iron oxide nanoparticles (SPIONs) as a multifunctional tool in various cancer therapies. *Rep. Pract. Oncol. Radiother.* **2019**, *24*, 307–314. [CrossRef]

77. Dulińska-Litewka, J.; Łazarczyk, A.; Hałubiec, P.; Szafrański, O.; Karnas, K.; Karewicz, A. Superparamagnetic Iron Oxide Nanoparticles-Current and Prospective Medical Applications. *Materials* **2019**, *12*, 617. [CrossRef] [PubMed]

78. Yamamoto, S.; Shimizu, K.; Fei, J.; Iwata, H.; Okochi, M.; Nakanishi, H.; Honda, H. Ex vivo culture of circulating tumor cells using magnetic force-based coculture on a fibroblast feeder layer. *Biotechnol. J.* **2016**, *11*, 1433–1442. [CrossRef]

79. Wu, L.L.; Wen, C.Y.; Hu, J.; Tang, M.; Qi, C.B.; Li, N.; Liu, C.; Chen, L.; Pang, D.W.; Zhang, Z.L. Nanosphere-based one-step strategy for efficient and nondestructive detection of circulating tumor cells. *Biosens. Bioelectron.* **2017**, *94*, 219–226. [CrossRef]

80. Fina, E.; Callari, M.; Reduzzi, C.; D'Aiuto, F.; Mariani, G.; Generali, D.; Pierotti, M.A.; Daidone, M.G.; Cappelletti, V. Gene expression profiling of circulating tumor cells in breast cancer. *Clin. Chem.* **2015**, *61*, 278–289. [CrossRef] [PubMed]

81. Lu, N.N.; Xie, M.; Wang, J.; Lv, S.W.; Yi, J.S.; Dong, W.G.; Huang, W.H. Biotin-triggered decomposable immunomagnetic beads for capture and release of circulating tumor cells. *ACS Appl. Mater. Interfaces* **2015**, *7*, 8817–8826. [CrossRef]

82. Kang, J.H.; Krause, S.; Tobin, H.; Mammoto, A.; Kanapathipillai, M.; Ingber, D.E. A combined micromagnetic-microfluidic device for rapid capture and culture of rare circulating tumor cells. *Lab Chip* **2012**, *12*, 2175–2181. [CrossRef]

83. Wang, Z.; Wu, W.; Tang, Y.; Deng, Y.; Xu, L.; Tian, J.; Shi, Q. Ex vivo expansion of circulating lung tumor cells based on one-step microfluidics-based immunomagnetic isolation. *Analyst* **2016**, *141*, 3621–3625. [CrossRef]

84. Zhang, Y.; An, J.; Liu, M.; Li, N.; Wang, W.; Yao, H.; Yang, X.; Sun, Y.; Xu, N.; Wu, L. Efficient isolation, culture, purification, and stem cell expression profiles of primary tumor cells derived from uterine cervical squamous cell carcinoma. *Am. J. Reprod. Immunol.* **2020**, *84*, e13251. [CrossRef] [PubMed]

85. Ruiz, C.; Kustermann, S.; Pietilae, E.; Vlajnic, T.; Baschiera, B.; Arabi, L.; Lorber, T.; Oeggerli, M.; Savic, S.; Obermann, E.; et al. Culture and Drug Profiling of Patient Derived Malignant Pleural Effusions for Personalized Cancer Medicine. *PLoS ONE* **2016**, *11*, e0160807. [CrossRef]

86. A Phase I Clinical Trial of Neoadjuvant Chemotherapy with/without SPIONs/SMF for Patients with Osteosarcoma. Available online: https://clinicaltrials.gov/ct2/show/NCT04316091 (accessed on 30 June 2021).

87. SPIO-Enhanced MRI in Oral Cancer for Sentinel Lymph Node Identification (MAG-NODE). Available online: https://clinicaltrials.gov/ct2/show/NCT04803331 (accessed on 30 June 2021).

88. Radiotherapy with Iron Oxide Nanoparticles (SPION) on MR-Linac for Primary & Metastatic Hepatic Cancers. Available online: https://clinicaltrials.gov/ct2/show/NCT04682847 (accessed on 30 June 2021).

89. Delayed Sentinel Lymph Node Biopsy in Ductal Cancer in Situ (SENTINOT_2). Available online: https://clinicaltrials.gov/ct2/show/NCT04722692 (accessed on 30 June 2021).

90. Crețu, B.E.-B.; Dodi, G.; Shavandi, A.; Gardikiotis, I.; Șerban, I.L.; Balan, V. Imaging Constructs: The Rise of Iron Oxide Nanoparticles. *Molecules* **2021**, *26*, 3437. [CrossRef]

91. Jaroch, K.; Jaroch, A.; Bojko, B. Cell cultures in drug discovery and development: The need of reliable in vitro-in vivo extrapolation for pharmacodynamics and pharmacokinetics assessment. *J. Pharm. Biomed. Anal.* **2018**, *147*, 297–312. [CrossRef]

92. Brandon, E.F.; Raap, C.D.; Meijerman, I.; Beijnen, J.H.; Schellens, J.H. An update on in vitro test methods in human hepatic drug biotransformation research: Pros and cons. *Toxicol. Appl. Pharmacol.* **2003**, *189*, 233–246. [CrossRef]

93. Velasco, V.; Shariati, S.A.; Esfandyarpour, R. Microtechnology-based methods for organoid models. *Microsyst. Nanoeng.* **2020**, *6*, 76. [CrossRef]

94. Xie, B.; Teusch, N.; Mrsny, R. 14—Comparison of two- and three-dimensional cancer models for assessing potential cancer therapeutics. In *Biomaterials for Cancer Therapeutics*, 2nd ed.; Park, K., Ed.; Woodhead Publishing: Cambridge, UK, 2020; pp. 399–422.

95. Alghuwainem, A.; Alshareeda, A.T.; Alsowayan, B. Scaffold-Free 3-D Cell Sheet Technique Bridges the Gap between 2-D Cell Culture and Animal Models. *Int. J. Mol. Sci.* **2019**, *20*, 4926. [CrossRef] [PubMed]

96. Gunti, S.; Hoke, A.T.K.; Vu, K.P.; London, N.R., Jr. Organoid and Spheroid Tumor Models: Techniques and Applications. *Cancers* **2021**, *13*, 874. [CrossRef] [PubMed]

97. Han, S.J.; Kwon, S.; Kim, K.S. Challenges of applying multicellular tumor spheroids in preclinical phase. *Cancer Cell Int.* **2021**, *21*, 152. [CrossRef]

98. Hou, S.; Tiriac, H.; Sridharan, B.P.; Scampavia, L.; Madoux, F.; Seldin, J.; Souza, G.R.; Watson, D.; Tuveson, D.; Spicer, T.P. Advanced Development of Primary Pancreatic Organoid Tumor Models for High-Throughput Phenotypic Drug Screening. *SLAS Discov.* **2018**, *23*, 574–584. [CrossRef]

99. Paun, I.A.; Mustaciosu, C.C.; Mihailescu, M.; Calin, B.S.; Sandu, A.M. Magnetically-driven 2D cells organization on superparamagnetic micromagnets fabricated by laser direct writing. *Sci. Rep.* **2020**, *10*, 16418. [CrossRef]

100. Fu, C.Y.; Lin, C.Y.; Chu, W.C.; Chang, H.Y. A simple cell patterning method using magnetic particle-containing photosensitive poly (ethylene glycol) hydrogel blocks: A technical note. *Tissue Eng. Part C Methods* **2011**, *17*, 871–877. [CrossRef]

101. Ino, K.; Okochi, M.; Konishi, N.; Nakatochi, M.; Imai, R.; Shikida, M.; Ito, A.; Honda, H. Cell culture arrays using magnetic force-based cell patterning for dynamic single cell analysis. *Lab Chip* **2008**, *8*, 134–142. [CrossRef]

102. Ino, K.; Okochi, M.; Honda, H. Application of magnetic force-based cell patterning for controlling cell–cell interactions in angiogenesis. *Biotechnol. Bioeng.* **2009**, *102*, 882–890. [CrossRef]

103. Haisler, W.L.; Timm, D.M.; Gage, J.A.; Tseng, H.; Killian, T.C.; Souza, G.R. Three-dimensional cell culturing by magnetic levitation. *Nat. Protoc.* **2013**, *8*, 1940–1949. [CrossRef]

104. Greiner Bio-One International GmbH. 3D Cell Culture: Technology. Available online: https://3dcellculture.gbo.com/ (accessed on 18 June 2021).

105. Bronkhorst, A.J.; Ungerer, V.; Holdenrieder, S. Comparison of methods for the isolation of cell-free DNA from cell culture supernatant. *Tumour Biol.* **2020**, *42*, 1010428320916314. [CrossRef] [PubMed]

106. Ding, M.; Cavallin, A.; Hermansson, N.O.; Berntsson, P.; Jinton, L.; Rodrigo Blomqvist, S. Comparing Flow Cytometry QBeads PlexScreen Assays with Other Immunoassays for Determining Multiple Analytes. *SLAS Discov.* **2018**, *23*, 676–686. [CrossRef] [PubMed]

107. Kabe, Y.; Suematsu, M.; Sakamoto, S.; Hirai, M.; Koike, I.; Hishiki, T.; Matsuda, A.; Hasegawa, Y.; Tsujita, K.; Ono, M.; et al. Development of a Highly Sensitive Device for Counting the Number of Disease-Specific Exosomes in Human Sera. *Clin. Chem.* **2018**, *64*, 1463–1473. [CrossRef]

108. Díaz-Valdivia, N.I.; Díaz, J.; Contreras, P.; Campos, A.; Rojas-Celis, V.; Burgos-Ravanal, R.A.; Lobos-González, L.; Torres, V.A.; Perez, V.I.; Frei, B.; et al. The non-receptor tyrosine phosphatase type 14 blocks caveolin-1-enhanced cancer cell metastasis. *Oncogene* **2020**, *39*, 3693–3709. [CrossRef]

109. Filipova, D.; Walter, A.M.; Gaspar, J.A.; Brunn, A.; Linde, N.F.; Ardestani, M.A.; Deckert, M.; Hescheler, J.; Pfitzer, G.; Sachinidis, A.; et al. Gene profiling of embryonic skeletal muscle lacking type I ryanodine receptor $Ca^{(2+)}$ release channel. *Sci. Rep.* **2016**, *6*, 20050. [CrossRef] [PubMed]

110. Kimura, M.; Horie, T.; Baba, O.; Ide, Y.; Tsuji, S.; Ruiz Rodriguez, R.; Watanabe, T.; Yamasaki, T.; Otani, C.; Xu, S.; et al. Homeobox A4 suppresses vascular remodeling by repressing YAP/TEAD transcriptional activity. *EMBO Rep.* **2020**, *21*, e48389. [CrossRef]

111. Rupp, M.; Hagenbuchner, J.; Rass, B.; Fiegl, H.; Kiechl-Kohlendorfer, U.; Obexer, P.; Ausserlechner, M.J. FOXO3-mediated chemo-protection in high-stage neuroblastoma depends on wild-type TP53 and SESN3. *Oncogene* **2017**, *36*, 6190–6203. [CrossRef]

112. Padmanaban, V.; Grasset, E.M.; Neumann, N.M.; Fraser, A.K.; Henriet, E.; Matsui, W.; Tran, P.T.; Cheung, K.J.; Georgess, D.; Ewald, A.J. Organotypic culture assays for murine and human primary and metastatic-site tumors. *Nat. Protoc.* **2020**, *15*, 2413–2442. [CrossRef]

113. Cheung, K.J.; Gabrielson, E.; Werb, Z.; Ewald, A.J. Collective invasion in breast cancer requires a conserved basal epithelial program. *Cell* **2013**, *155*, 1639–1651. [CrossRef] [PubMed]

114. Yiu, H.H.; McBain, S.C.; Lethbridge, Z.A.; Lees, M.R.; Dobson, J. Preparation and characterization of polyethylenimine-coated Fe_3O_4-MCM-48 nanocomposite particles as a novel agent for magnet-assisted transfection. *J. Biomed. Mater. Res. A* **2010**, *92*, 386–392. [CrossRef]

115. Caliari, S.R.; Burdick, J.A. A practical guide to hydrogels for cell culture. *Nat. Methods* **2016**, *13*, 405–414. [CrossRef] [PubMed]

116. Xue, J.; Wu, T.; Dai, Y.; Xia, Y. Electrospinning and Electrospun Nanofibers: Methods, Materials, and Applications. *Chem. Rev.* **2019**, *119*, 5298–5415. [CrossRef]

117. Zhang, C.L.; Lv, K.P.; Hu, N.Y.; Yu, L.; Ren, X.F.; Liu, S.L.; Yu, S.H. Macroscopic-scale alignment of ultralong Ag nanowires in polymer nanofiber mat and their hierarchical structures by magnetic-field-assisted electrospinning. *Small* **2012**, *8*, 2936–2940. [CrossRef]

118. Yarin, A.L.; Zussman, E. Upward needleless electrospinning of multiple nanofibers. *Polymer* **2004**, *45*, 2977–2980. [CrossRef]

119. Lee, J.W. 3D Nanoprinting Technologies for Tissue Engineering Applications. *J. Nanomater.* **2015**, *2015*, 213521. [CrossRef]

120. Libring, S.; Solorio, L. 16—Cancer mechanobiology: Interaction of biomaterials with cancer cells. In *Biomaterials for Cancer Therapeutics*, 2nd ed.; Park, K., Ed.; Woodhead Publishing: Cambridge, UK, 2020; pp. 445–470.

121. Mihic, A.; Li, J.; Miyagi, Y.; Gagliardi, M.; Li, S.H.; Zu, J.; Weisel, R.D.; Keller, G.; Li, R.K. The effect of cyclic stretch on maturation and 3D tissue formation of human embryonic stem cell-derived cardiomyocytes. *Biomaterials* **2014**, *35*, 2798–2808. [CrossRef] [PubMed]

122. Kim, J.; Tanner, K. Three-Dimensional Patterning of the ECM Microenvironment Using Magnetic Nanoparticle Self Assembly. *Curr. Protoc. Cell Biol.* **2016**, *70*. [CrossRef] [PubMed]

123. Martin, J.J.; Fiore, B.E.; Erb, R.M. Designing bioinspired composite reinforcement architectures via 3D magnetic printing. *Nat. Commun.* **2015**, *6*, 8641. [CrossRef] [PubMed]

124. Erb, R.M.; Libanori, R.; Rothfuchs, N.; Studart, A.R. Composites reinforced in three dimensions by using low magnetic fields. *Science* **2012**, *335*, 199–204. [CrossRef]

125. Margolis, G.; Polyak, B.; Cohen, S. Magnetic Induction of Multiscale Anisotropy in Macroporous Alginate Scaffolds. *Nano Lett.* **2018**, *18*, 7314–7322. [CrossRef]

126. Dasgupta, D.; Pally, D.; Saini, D.K.; Bhat, R.; Ghosh, A. Nanomotors Sense Local Physicochemical Heterogeneities in Tumor Microenvironments. *Angew. Chem. Int. Ed. Engl.* **2020**, *59*, 23690–23696. [CrossRef] [PubMed]

127. Efremov, Y.M.; Cartagena-Rivera, A.X.; Athamneh, A.I.M.; Suter, D.M.; Raman, A. Mapping heterogeneity of cellular mechanics by multi-harmonic atomic force microscopy. *Nat. Protoc.* **2018**, *13*, 2200–2216. [CrossRef]

128. Libring, S.; Shinde, A.; Chanda, K.M.; Nuru, M.; George, H.; Saleh, M.A.; Abdullah, A.; Kinzer-Ursem, L.T.; Calve, S.; Wendt, K.M.; et al. The Dynamic Relationship of Breast Cancer Cells and Fibroblasts in Fibronectin Accumulation at Primary and Metastatic Tumor Sites. *Cancers* **2020**, *12*, 1270. [CrossRef] [PubMed]

129. Henstock, J.R.; Markides, H.; Bin, H.; El Haj, A.J.; Dobson, J. Applications of magnetic nanoparticles in tissue engineering and regenerative medicine. In *Nanomagnetic Actuation in Biomedicine*; CRC Press: Boca Raton, FL, USA, 2018; pp. 205–228.

130. Ridone, P.; Vassalli, M.; Martinac, B. Piezo1 mechanosensitive channels: What are they and why are they important. *Biophys. Rev.* **2019**, *11*, 795–805. [CrossRef]

131. Bausch, A.R.; Möller, W.; Sackmann, E. Measurement of local viscoelasticity and forces in living cells by magnetic tweezers. *Biophys. J.* **1999**, *76*, 573–579. [CrossRef]

132. Muhamed, I.; Chowdhury, F.; Maruthamuthu, V. Biophysical Tools to Study Cellular Mechanotransduction. *Bioengineering* **2017**, *4*, 12. [CrossRef]

133. Zhang, Y.; Wei, F.; Poh, Y.C.; Jia, Q.; Chen, J.; Luo, J.; Yao, W.; Zhou, W.; Huang, W.; Yang, F.; et al. Interfacing 3D magnetic twisting cytometry with confocal fluorescence microscopy to image force responses in living cells. *Nat. Protoc.* **2017**, *12*, 1437–1450. [CrossRef]

134. Bush, J.; Maruthamuthu, V. In situ determination of exerted forces in magnetic pulling cytometry. *AIP Adv.* **2019**, *9*, 035221. [CrossRef]

135. Wu, P.H.; Aroush, D.R.; Asnacios, A.; Chen, W.C.; Dokukin, M.E.; Doss, B.L.; Durand-Smet, P.; Ekpenyong, A.; Guck, J.; Guz, N.V.; et al. A comparison of methods to assess cell mechanical properties. *Nat. Methods* **2018**, *15*, 491–498. [CrossRef]

136. Qiu, Y.; Tong, S.; Zhang, L.; Sakurai, Y.; Myers, D.R.; Hong, L.; Lam, W.A.; Bao, G. Magnetic forces enable controlled drug delivery by disrupting endothelial cell-cell junctions. *Nat. Commun.* **2017**, *8*, 15594. [CrossRef] [PubMed]

137. Boulter, E.; Estrach, S.; Tissot, F.S.; Hennrich, M.L.; Tosello, L.; Cailleteau, L.; de la Ballina, L.R.; Pisano, S.; Gavin, A.C.; Féral, C.C. Cell metabolism regulates integrin mechanosensing via an SLC3A2-dependent sphingolipid biosynthesis pathway. *Nat. Commun.* **2018**, *9*, 4862. [CrossRef]

138. Matthews, B.D.; Overby, D.R.; Mannix, R.; Ingber, D.E. Cellular adaptation to mechanical stress: Role of integrins, Rho, cytoskeletal tension and mechanosensitive ion channels. *J. Cell Sci.* **2006**, *119*, 508–518. [CrossRef]

139. Hu, B.; El Haj, A.J.; Dobson, J. Receptor-targeted, magneto-mechanical stimulation of osteogenic differentiation of human bone marrow-derived mesenchymal stem cells. *Int. J. Mol. Sci.* **2013**, *14*, 19276–19293. [CrossRef] [PubMed]

140. Yoon, A.R.; Stasinopoulos, I.; Kim, J.H.; Yong, H.M.; Kilic, O.; Wirtz, D.; Bhujwalla, Z.M.; An, S.S. COX-2 dependent regulation of mechanotransduction in human breast cancer cells. *Cancer Biol. Ther.* **2015**, *16*, 430–437. [CrossRef]

141. Park, J.; Kim, D.H.; Kim, H.N.; Wang, C.J.; Kwak, M.K.; Hur, E.; Suh, K.Y.; An, S.S.; Levchenko, A. Directed migration of cancer cells guided by the graded texture of the underlying matrix. *Nat. Mater.* **2016**, *15*, 792–801. [CrossRef]

142. Mousavizadeh, R.; Hojabrpour, P.; Eltit, F.; McDonald, P.C.; Dedhar, S.; McCormack, R.G.; Duronio, V.; Jafarnejad, S.M.; Scott, A. β1 integrin, ILK and mTOR regulate collagen synthesis in mechanically loaded tendon cells. *Sci. Rep.* **2020**, *10*, 12644. [CrossRef]

143. Bays, J.L.; Campbell, H.K.; Heidema, C.; Sebbagh, M.; DeMali, K.A. Linking E-cadherin mechanotransduction to cell metabolism through force-mediated activation of AMPK. *Nat. Cell Biol.* **2017**, *19*, 724–731. [CrossRef]

144. Hughes, S.; McBain, S.; Dobson, J.; El Haj, A.J. Selective activation of mechanosensitive ion channels using magnetic particles. *J. R. Soc. Interface* **2008**, *5*, 855–863. [CrossRef]

145. Ciobanasu, C.; Faivre, B.; Le Clainche, C. Reconstituting actomyosin-dependent mechanosensitive protein complexes in vitro. *Nat. Protoc.* **2015**, *10*, 75–89. [CrossRef]

146. Fernández-Sánchez, M.E.; Barbier, S.; Whitehead, J.; Béalle, G.; Michel, A.; Latorre-Ossa, H.; Rey, C.; Fouassier, L.; Claperon, A.; Brullé, L.; et al. Mechanical induction of the tumorigenic β-catenin pathway by tumour growth pressure. *Nature* **2015**, *523*, 92–95. [CrossRef]

147. Raeber, G.P.; Mayer, J.; Hubbell, J.A. Part I: A novel in-vitro system for simultaneous mechanical stimulation and time-lapse microscopy in 3D. *Biomech. Model. Mechanobiol.* **2008**, *7*, 203–214. [CrossRef] [PubMed]

148. Raeber, G.P.; Lutolf, M.P.; Hubbell, J.A. Part II: Fibroblasts preferentially migrate in the direction of principal strain. *Biomech. Model. Mechanobiol.* **2008**, *7*, 215–225. [CrossRef] [PubMed]

149. Harshad, K.; Jun, M.; Park, S.; Barton, M.J.; Vadivelu, R.K.; St John, J.; Nguyen, N.T. An electromagnetic cell-stretching device for mechanotransduction studies of olfactory ensheathing cells. *Biomed. Microdevices* **2016**, *18*, 45. [CrossRef] [PubMed]

150. Seriani, S.; Del Favero, G.; Mahaffey, J.; Marko, D.; Gallina, P.; Long, C.S.; Mestroni, L.; Sbaizero, O. The cell-stretcher: A novel device for the mechanical stimulation of cell populations. *Rev. Sci. Instrum.* **2016**, *87*, 084301. [CrossRef]

151. Bidan, C.M.; Fratzl, M.; Coullomb, A.; Moreau, P.; Lombard, A.H.; Wang, I.; Balland, M.; Boudou, T.; Dempsey, N.M.; Devillers, T.; et al. Magneto-active substrates for local mechanical stimulation of living cells. *Sci. Rep.* **2018**, *8*, 1464. [CrossRef]

152. Nagayama, K.; Inoue, T.; Hamada, Y.; Sugita, S.; Matsumoto, T. Direct application of mechanical stimulation to cell adhesion sites using a novel magnetic-driven micropillar substrate. *Biomed. Microdevices* **2018**, *20*, 85. [CrossRef] [PubMed]

153. Monticelli, M.; Jokhun, D.S.; Petti, D.; Shivashankar, G.V.; Bertacco, R. Localized mechanical stimulation of single cells with engineered spatio-temporal profile. *Lab Chip* **2018**, *18*, 2955–2965. [CrossRef] [PubMed]

154. Geng, Y.; Wang, Z. Review of cellular mechanotransduction on micropost substrates. *Med. Biol. Eng. Comput.* **2016**, *54*, 249–271. [CrossRef] [PubMed]

155. Hughes, S.; El Haj, A.J.; Dobson, J.; Martinac, B. The influence of static magnetic fields on mechanosensitive ion channel activity in artificial liposomes. *Eur. Biophys. J.* **2005**, *34*, 461–468. [CrossRef]

156. Blyakhman, F.A.; Melnikov, G.Y.; Makarova, E.B.; Fadeyev, F.A.; Sedneva-Lugovets, D.V.; Shabadrov, P.A.; Volchkov, S.O.; Mekhdieva, K.R.; Safronov, A.P.; Fernández Armas, S.; et al. Effects of Constant Magnetic Field to the Proliferation Rate of Human Fibroblasts Grown onto Different Substrates: Tissue Culture Polystyrene, Polyacrylamide Hydrogel and Ferrogels γ-Fe. *Nanomaterials* **2020**, *10*, 1697. [CrossRef] [PubMed]

157. Kruszewski, M.; Sikorska, K.; Meczynska-Wielgosz, S.; Grzelak, A.; Sramkova, M.; Gabelova, A.; Kapka-Skrzypczak, L. Comet assay in neural cells as a tool to monitor DNA damage induced by chemical or physical factors relevant to environmental and occupational exposure. *Mutat. Res.* **2019**, *845*, 402990. [CrossRef] [PubMed]

cancers

MDPI

Article

A Comprehensive Grading System for a Magnetic Sentinel Lymph Node Biopsy Procedure in Head and Neck Cancer Patients

Eliane R. Nieuwenhuis [1,2], Barry Kolenaar [2], Jurrit J. Hof [3], Joop van Baarlen [4], Alexander J. M. van Bemmel [5], Anke Christenhusz [1,6], Tom W. J. Scheenen [7], Bernard ten Haken [1], Remco de Bree [8] and Lejla Alic [1,*]

[1] Magnetic Detection and Imaging Group, Technical Medical Centre, University of Twente, 7522 NB Enschede, The Netherlands; e.r.nieuwenhuis@utwente.nl (E.R.N.); a.christenhusz@utwente.nl (A.C.); b.tenhaken@utwente.nl (B.t.H.)

[2] Department of Maxillofacial Surgery—Head and Neck Surgical Oncology, Medisch Spectrum Twente, 7512 KZ Enschede, The Netherlands; b.kolenaar@utwente.nl

[3] Department of Radiology, Medisch Spectrum Twente, 7512 KZ Enschede, The Netherlands; J.Hof@mst.nl

[4] Laboratorium Pathologie Oost Nederland, 7555 BB Hengelo, The Netherlands; J.vanBaarlen@labpon.nl

[5] Department of Otorhinolaryngology—Head and Neck Surgical Oncology, Medisch Spectrum Twente, 7512 KZ Enschede, The Netherlands; A.vanBemmel@mst.nl

[6] Department of Surgery, Medisch Spectrum Twente, 7512 KZ Enschede, The Netherlands

[7] Department of Medical Imaging, Radboud University Medical Center, 6525 GA Nijmegen, The Netherlands; Tom.Scheenen@radboudumc.nl

[8] Department of Head and Neck Surgical Oncology, University Medical Center Utrecht, 3584 CX Utrecht, The Netherlands; R.deBree@umcutrecht.nl

* Correspondence: l.alic@utwente.nl; Tel.: +31-534-898-731

Citation: Nieuwenhuis, E.R.; Kolenaar, B.; Hof, J.J.; van Baarlen, J.; van Bemmel, A.J.M.; Christenhusz, A.; Scheenen, T.W.J.; ten Haken, B.; de Bree, R.; Alic, L. A Comprehensive Grading System for a Magnetic Sentinel Lymph Node Biopsy Procedure in Head and Neck Cancer Patients. *Cancers* **2022**, *14*, 678. https://doi.org/10.3390/cancers14030678

Academic Editors: Moriaki Kusakabe and Akihiro Kuwahata

Received: 10 December 2021
Accepted: 26 January 2022
Published: 28 January 2022

Publisher's Note: MDPI stays neutral with regard to jurisdictional claims in published maps and institutional affiliations.

Simple Summary: With 30% of clinically negative early-stage oral cancer patients harboring occult metastasis, an accurate staging of metastatic lymph nodes (LN) is of utmost importance for treatment planning. A magnetic sentinel lymph node biopsy (SLNB) procedure is offered as an alternative to conventional SLNB in oral oncology, however, a grading system is missing. A proper grading system is preferred to connect the different components of the magnetic SLNB: preoperative imaging, intraoperative detection, and histopathological examination of sentinel lymph nodes (SLNs). This study aims to provide a first grading system based on the distribution of a magnetic tracer, by means of preoperative magnetic resonance imaging (MRI), intraoperative estimation of iron content, and histopathological assessment of resected nodes. Pre- and post-operative MRI and harvested SLNs of eight tongue cancer patients with successful magnetic SLNB procedure were used for analyses.

Abstract: A magnetic sentinel lymph node biopsy ((SLN)B) procedure has recently been shown feasible in oral cancer patients. However, a grading system is absent for proper identification and classification, and thus for clinical reporting. Based on data from eight complete magnetic SLNB procedures, we propose a provisional grading system. This grading system includes: (1) a qualitative five-point grading scale for MRI evaluation to describe iron uptake by LNs; (2) an ex vivo count of resected SLN with a magnetic probe to quantify iron amount; and (3) a qualitative five-point grading scale for histopathologic examination of excised magnetic SLNs. Most SLNs with iron uptake were identified and detected in level II. In this level, most variance in grading was seen for MRI and histopathology; MRI and medullar sinus were especially highly graded, and cortical sinus was mainly low graded. On average 82 ± 58 µg iron accumulated in harvested SLNs, and there were no significant differences in injected tracer dose (22.4 mg or 11.2 mg iron). In conclusion, a first step was taken in defining a comprehensive grading system to gain more insight into the lymphatic draining system during a magnetic SLNB procedure.

Keywords: oral cancer; sentinel lymph node; magnetic tracer; superparamagnetic iron oxide (SPIO); tracer distribution; MRI; lymphography; histopathology; grading system

---

1. Introduction

In oral cancer, the presence of cervical lymph node (LN) metastases is one of the most important factors for prognosis [1]. Therefore, the detection of metastatic LN is important for treatment planning [2,3]. However, in 30% of early oral cancer (cT1/T2) patients, metastases are not identified during clinical examination or by diagnostic imaging modalities [4–6]. To identify these occult metastases, a sentinel lymph node biopsy (SLNB) procedure can be performed [7,8]. The sentinel lymph node (SLN), which is the first draining lymph node, represents the lymphatic status of regional LNs; if the SLN is found to be tumor negative then regional lymph nodes are considered negative too. The conventional SLNB procedure utilizes peritumorally administered radioisotopes for preoperative localization of SLNs by lymphoscintigraphy and for intraoperative detection of SLNs by a gamma probe. After SLNs are harvested, the nodal status is assessed by histopathological examination. The conventional procedure requires strict regulation concerning radioisotope production, transport, and usage [9]. It would therefore be beneficial to have a non-ionizing alternative.

A magnetic SLNB is proposed as an alternative for a radioactive SLNB in, e.g., breast cancer [10,11], prostate cancer [12], and, more recently, for thyroid carcinoma [13] and oral cancer [14,15]. The magnetic SLNB utilizes a magnetic tracer (superparamagnetic iron oxide (SPIO) nanoparticles) and a magnetometer. Similar to the conventional SLNB procedure, a magnetic tracer is peritumorally administered and drains via lymphatic vessels, freely or by macrophages, to the SLNs. In contrast to a radioactive tracer, a magnetic tracer does not have a short half-life time which is advantageous for an SLNB as it provides great freedom in surgical planning. Injected tracer can be used for preoperative SPIO-enhanced magnetic resonance imaging (MRI) to visualize and identify the location of LNs with iron uptake [16,17]. Simultaneously, SPIOs are also used for intraoperative detection of SLNs by a magnetometer [10–15]. What should be considered is the difference in SPIO-sensitivity between MRI and the magnetometer [18]. Consequently, the dose of the injected magnetic tracer should lead to a sufficient signal to be detected by a magnetometer without large signal voids on MRI affecting their examination [19].

Since a magnetic SLNB procedure in oral oncology was recently shown to be feasible [14,15], it raises the question of whether it is possible to develop a comprehensive grading system. This grading system could be based on a combination of preoperative MRI, intraoperative LN detection and post-operative histopathology. In the complex anatomy of the head and neck area, a grading system should be of help in discriminating SLNs from higher echelon nodes and determining the effect of signal voids in SPIO-enhanced MRI [15]. Furthermore, determination of the amount of iron within excised magnetic SLNs gives insight into what is still detectable and how much of the injected dose drains to SLN(s). Lastly, the iron content in SLNs, based on counts of the magnetometer, can be related to visual histopathology grading of the iron content.

With the abovementioned components of a magnetic SLNB procedure, we propose a first step towards a comprehensive grading system. Therefore, this study reports the iron distribution to (S)LNs and grading based on preoperative SPIO-enhanced MRI, intraoperative detection by a magnetometer, and histopathology of resected LNs in oral cancer patients who successfully underwent a magnetic SLNB procedure.

2. Materials and Methods

Our grading system is based on a selection of patients ($n = 8/10$) with oral squamous cell carcinoma (OSCC), recruited for a feasibility study on a magnetic SLNB procedure (NL6656, Netherlands Trial Register) in the period February 2018–December 2019 at Medisch Spectrum Twente [15]. To assess iron distribution in SLNs, based on SPIO-enhanced MRI, magnetometer counts and histopathological analysis, only patients with a successful magnetic SLNB procedure were selected. All patients were clinically diagnosed with T1-T2N0M0 and scheduled for resection of the primary tumor (border of the tongue) and consecutive elective neck dissection (END) level I-III. Patients were three females and five males, with a mean age of 67 years (range 43–77 years). The SPIO nanoparticle tracer used was Sienna+® [28 mg iron/mL] (Endomagnetics Ltd., Cambridge, UK), which was submucosally injected around the tumor within 24 h prior to surgery. A total volume of 0.4 mL in three patients and 0.8 mL in five patients (corresponding to an iron dose of 11.2 mg and 22.4 mg, respectively) was administered in 4 aliquots. Prior to entering the feasibility study, all patients provided oral and written informed consent. Ethical approval was given by the local medical ethics committee, METC Twente.

2.1. Magnetic Resonance Imaging Grading

To identify iron-containing LNs, a preoperative SPIO-enhanced MRI (1.5T, Ingenia, Philips Medical Systems, Best, the Netherlands) was acquired shortly after tracer injection, as described above (range: 22–101 min). Additionally, a postoperative SPIO-enhanced MRI was performed (range 23–49 days post injection) to observe the influence and distribution of SPIOs on MRI during follow-up. The following MRI sequences were acquired in transversal plane of the entire neck using a 16-channel dedicated head and neck coil:

- T1-weighted (T1w) 3D fast field echo (TR/TE = 25/4.6 ms, flip angle 30°, voxel size 0.75 mm × 0.75 mm × 1.6 mm, FOV = 251–269);
- T2*-weighted (T2*w) fast field echo (TR/TE = 1700/18.41 ms, flip angle 18°, voxel size 0.62 mm × 0.62 mm × 3 mm, FOV = 247–267).

To evaluate the effect of SPIOs in LNs on SPIO-enhanced MRI, a qualitative five-point grading scale was developed based upon T1w MRI. T2*-weighted MRI was used to confirm the presence of iron scored at T1w sequence. This scale is defined below and illustrated in Figure 1:

- 1: Internal spots of signal void
- 2: Confluence of internal spots of signal void, <25%
- 3: Partial (peripheral) signal void of the node, 25–75%
- 4: Complete signal void of the node, >75%
- 5: Blooming beyond border of the gland

Scale	Illustration	T1	T1 with marked LN border	T2*
1				
2				
3				
4				
5				

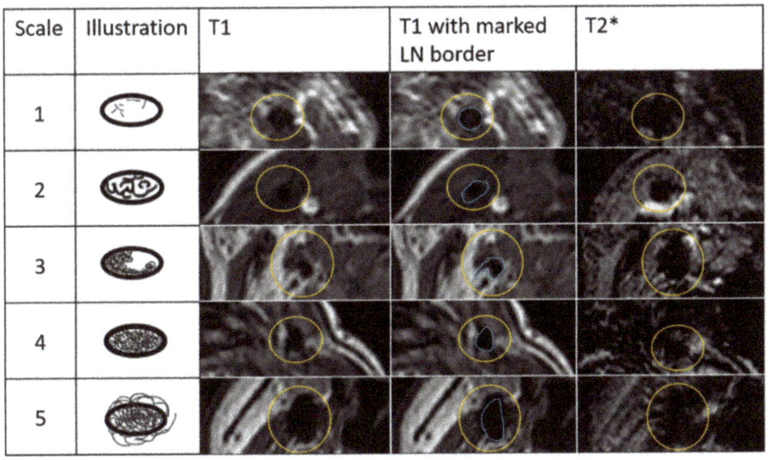

Figure 1. Qualitative five-point grading scale to assess effect of iron uptake in lymph nodes (LNs). Scale: 1—Internal spots of signal void; 2—Confluence of internal spots of signal void, <25%; 3—Partial (peripheral) signal void of the node, 25–75%; 4—Complete signal void of the node >75%; 5—Blooming beyond border of the gland. Column T1 shows examples of LNs with iron uptake. The yellow circle shows corresponding areas in T1 and T2* of the LN with iron uptake. In column T1 with marked LN border, blue line represents LN border.

All LNs with iron uptake in level I–III were graded following application of the five-point grading scale by a radiologist experienced in head and neck oncology (J.J.H.).

2.2. Amount of Iron in Sentinel Lymph Nodes

Before the SLNB procedure the Sentimag® system was checked for correct operation using a reference sample. SLNs were intraoperatively detected using a Sentimag (Endomagnetics Ltd., Cambridge, UK) magnetometer, within 24 h post injection (range 04 h 06 min–21 h 14 min). One or more magnetically detected SLNs per patient were resected and the amount of iron present in SLNs was determined by ex vivo magnetic readout of, in total, 24 SLNs and the use of a predefined look-up table (LUT). Ex vivo measurements of SLNs were performed, with a handheld probe vertically aligned and the probe tip in an upward position. The Sentimag system was balanced in the air without metal materials in the direct surroundings. When balanced, the SLN was placed at the middle of the probe tip. The following data was recorded per SLN: sensitivity setting of Sentimag, ex vivo Sentimag counts, neck level at which SLN was harvested.

The LUT was generated by evaluating the magnetic readout of ten Sienna+® samples. For each sample, a glass tube with an outer diameter of 8 mm and an inner diameter of 6.5 mm was filled with a Sienna+® dilution up to a volume of 35 µL. Sienna+® was diluted using 0.9% saline and samples contained the following iron doses: 1, 5, 10, 28, 50, 101, 140, 280, 420 and 504 µg. Correct operation of the handheld probe was checked by measuring a reference sample before Sienna+® samples were measured. Prior to each Sienna+® sample measurement, the handheld probe was balanced without metallic materials in the direct surrounding. Measurements were acquired six times per sample, with the sample placed in the middle of and directly next to the probe tip. The handheld probe was held in the same orientation as for the ex vivo SLN measurements. A Styrofoam cubic guaranteed a similar position of each sample to the probe tip. A relation between the amount of iron and corresponding counts was experimentally established and modeled as a first order polynomial by in-house developed software (Matlab environment, R2020a, Mathworks, Inc., Natick, MA, USA), using the six measurements per sample for each sensitivity setting as input. The created LUT is available through 4TU.ResearchData [20]. The first column shows the iron content from 1 µg–500 µg in steps of 0.01 µg. Consecutive columns show corresponding counts for sensitivity settings 1, 2 and 3 of the magnetometer, respectively. For analysis, the iron dose corresponding to magnetic readout per SLN was noted to one decimal place.

2.3. Histopathology Grading

SLNs were paraffin embedded in slices of 2 mm, which were histopathologically analyzed using step serial sectioning (five levels with 200 µm interval) and hematoxylin and eosin (H&E) staining. Cytokeratin AE1/3 staining was used to detect metastasis [15]. Iron content in cortical and medullar sinuses was assessed based on H&E stained coupes, using a qualitative five-point grading scale in cortical and medullar sinuses: no iron (grade 0), 1–25% iron (grade 1), 25–50% (grade 2), 50–75% (grade 3), and 75–100% (grade 4) was estimated in the region of interest; an example is given in Figure 2. Each harvested SLN was graded following the five-point grading scale, for one slide, by an experienced pathologist (J.v.B.).

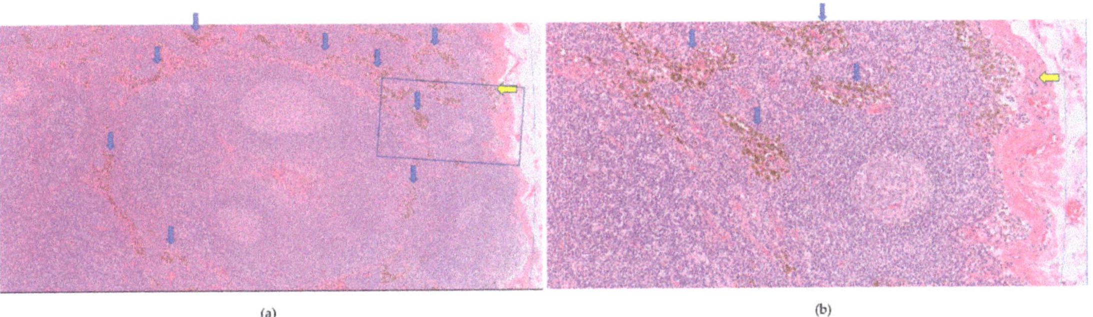

Figure 2. Hematoxylin and eosin-stained coupe of magnetic sentinel lymph node (**a**) 50× and (**b**) 200× magnification, (**b**) is the squared blue box in (**a**). The cortical sinus is shown at the right side (thin blue line) in (**a,b**). The iron content was graded as 1 (0–25%), see yellow arrow for focally located iron in cortical sinus. The medullar sinus was scored with grade 4 (75–100%), see blue arrows pointing at iron uptake.

2.4. Data Analysis

Grading of SPIO-enhanced MRI and histopathology of LNs was represented as the median (range) for each neck level. The calculated iron content in SLNs based upon LUT was represented as mean $\pm$ standard deviation. To compare groups of 22.4 mg and 11.2 mg iron dose injected, a Mann-Whitney U test was performed with confidence interval of 95% (IBM SPSS statistics, version 27). The correlation coefficient (Spearman's rho SPSS) was determined for iron content and histopathology grading.

3. Results

3.1. Magnetic Resonance Imaging Grading

For preoperative MRI the total number of identified SPIO-enhanced LNs in level I-III was 71 (90% at ipsilateral site). Eleven of these were identified as most likely SLNs, based on location relative to the tumor and iron susceptibility extent [15]. The highest number of SPIO-enhanced LNs was observed in level II at the ipsilateral side: 54% (38/71), 2(a). These LNs were mainly graded with a higher value compared to LNs in levels I and III, Figure 3b. Level I, II and III were graded (median (range)), respectively: 3.5 (1–5), 4 (1–5) and 3.5 (1–5) at the ipsilateral side and 1 (1–5), 5 (4–5) and 1 (1–1) at the contralateral side. Within preoperative MRI, no significant difference was found at the ipsilateral side, comparing groups with iron doses 11.2 mg and 22.4mg (p = 0.24). No LNs with iron uptake were seen at the contralateral side when a dose of 11.2mg iron was administered.

As a result of SLNB and END procedures, most of the identified SPIO-enhanced LNs were removed, however, postoperative SPIO-enhanced MRI still showed areas of susceptibility as a result of iron. These remaining iron deposits were minimally visible on MRI sequences used in daily clinical practice; an example is shown in Figure 4.

When possible, correlation of excised SLNs and preoperative MRI grades of SLNs in most cases showed an MRI-grade 4 or higher. Two of the three identified magnetic SLNs containing metastasis were given an MRI-grade 5 and were assigned as most likely SLNs on MRI. The remaining one was graded 4 or 5, but since three SPIO-enhanced LNs were seen in this level, it was not certain to which the SLN was related.

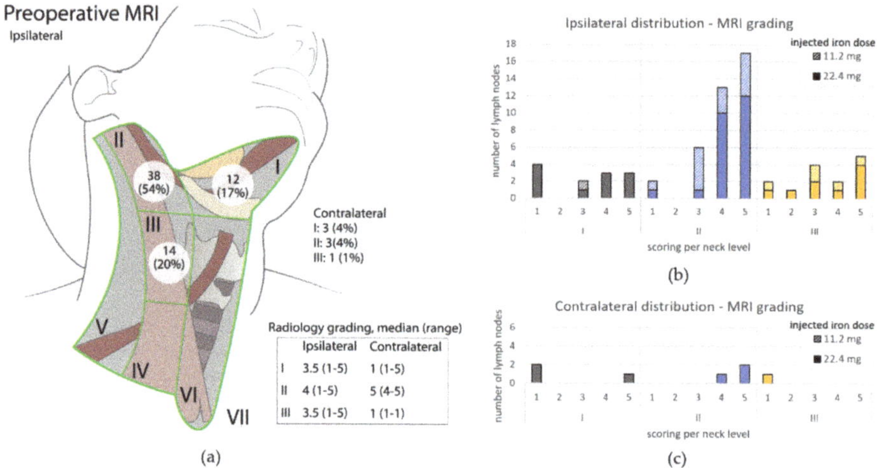

Figure 3. (a) Schematic overview of ipsilateral neck. Per neck level the amount of identified SPIO-enhanced LNs and grading on preoperative MRI, median (range) is given. Figure adapted with permission of original authors from [15]; (b) Overview of graded LNs at ipsilateral side of SPIO-enhanced preoperative MRI. Per neck level (I = grey, II = blue and III = yellow), the total number of LNs is given based upon their MRI grading score. Fully colored part of the bar represents the identified number of LNs when an iron dose of 22.4 mg was administered; diagonally striped part of the bar includes the LNs graded for patients who received an iron dose of 11.2 mg; (c) Overview of graded LNs distributed at contralateral site of SPIO-enhanced preoperative MRI. Description as for (b).

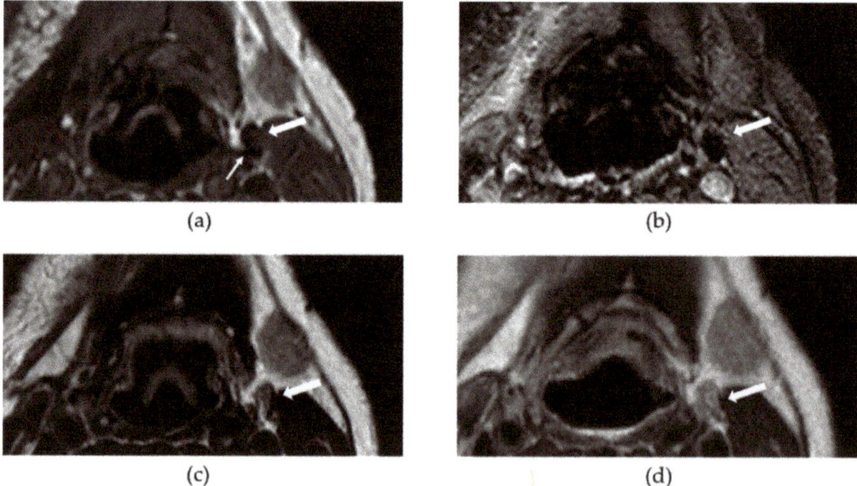

Figure 4. This figure shows an example of an iron containing lymph node (LN) at postoperative MRI for two superparamagnetic iron oxide nanoparticle (SPIO-) enhanced sequences (a,b) and two sequences for daily clinical practice (c) and (d). The big white arrow in each subfigure points to a LN (level IIa, left side) with SPIO uptake; the small white arrow in (a) points at part of LN with SPIO accumulation; (a) T1w 3D FFE (T1-weighted 3-dimensional fast field echo); (b) T2*w FFE (T2*-weighted fast field echo); (c) T2w SE (T2-weighted spin echo); (d) T2w Dixon–in-phase.

3.2. Amount of Iron in Sentinel Lymph Nodes

The LUT demonstrated, for all three sensitivity settings of Sentimag, a linear relationship between the magnetic readout and the iron content of Sienna+® samples [20]. Of 24 harvested SLNs, 18 were harvested after an injection dose of 22.4 mg iron and six SLNs as a result of an administered dose of 11.2 mg iron. On average, an SLN contained 82 ± 58 µg iron. Based upon iron content, no significant difference was found between the groups with 22.4 mg (81 ± 58 µg) and 11.2 mg (86 ± 64 µg) injected iron dose ($p = 0.87$) (see Figure 5a). On average, SLNs contained 0.36% (22.4 mg) and 0.77% (11.2 mg) iron of the total injected dose. Figure 5b shows, per patient, the number of nodes identified on preoperative SPIO-enhanced MRI, and detected by magnetic detector over time.

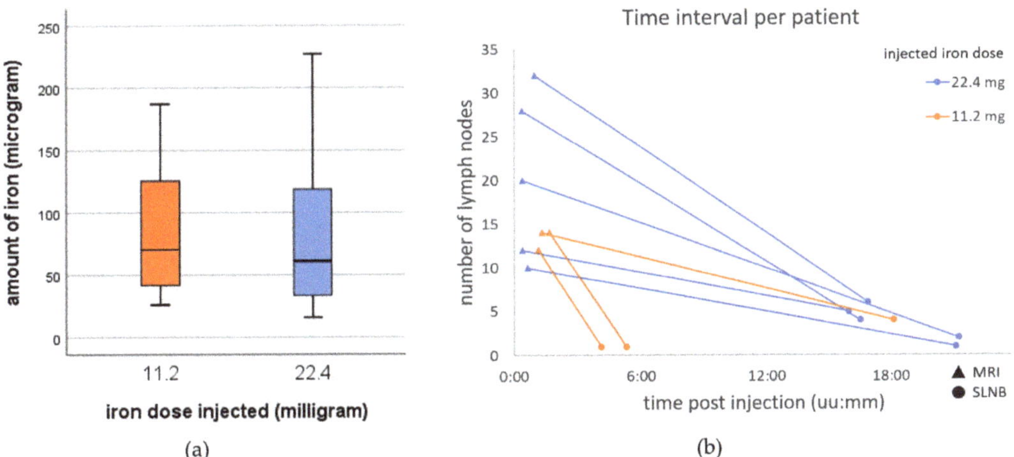

Figure 5. (**a**) Boxplot representing iron content of all detected SLNs for 11.2 mg ($n = 6$) and 22.4 mg ($n = 18$) iron dose injected, no significant difference; (**b**) Per patient the number of lymph nodes showing iron uptake on preoperative magnetic nanoparticle enhanced (triangles) and detected during sentinel lymph node biopsy (SLNB, circles); post magnetic tracer injection is shown in hours:minutes (uu:mm). Orange and blue, respectively, represent patients with 11.2 mg and 22.4 mg iron dose injected.

In level I, seven (30%) and one (4%) SLN(s) were harvested from the ipsilateral and contralateral sides, respectively. For level II and III, SLNs originated from the ipsilateral side only; these levels contained, respectively, 13 (54%) and three (13%) SLNs. The average iron content of SLNs per level found was: level I: 57 ± 42 µg, level II: 91 ± 61 µg and level III: 114 ± 73 µg. Table A1 illustrates patient characteristics including LN number(s), neck level and iron content.

3.3. Histopathology Grading

Of 24 SLNs, three contained metastasis (Appendix A). Regarding iron content grading: level I cortical and medullar sinus were graded as (median) 1.5 (range 0–3) and 3 (0–4), respectively; level II: 2 (1–4) and 4 (2–4); and level III: 4 (2–4) and 4 (3–4). The number of LNs per grade are shown in Figure 6. The three positive SLNs were graded as 2–4, 1–4 and 2–2 for cortical and medullar sinus, respectively. For individual SLNs, the medullar sinus was graded higher in 17 cases, equal in 6 cases and lower in 1 case compared to grading of the cortical sinus (Appendix A). The correlation coefficient of medullar grade and iron content was 0.50 ($p = 0.013$). No significant correlation was found for cortical grade and iron content (0.374, $p = 0.072$).

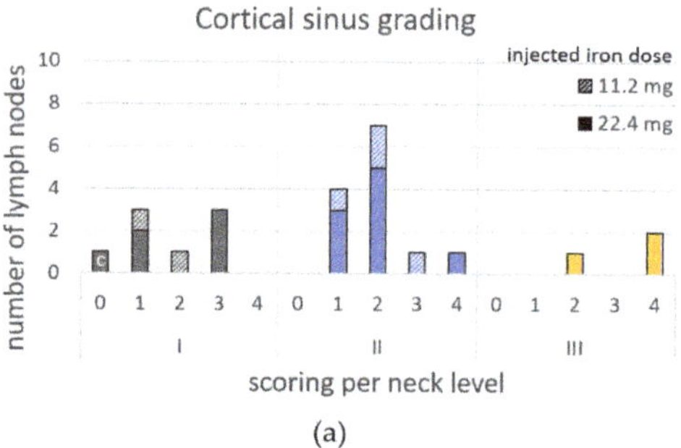
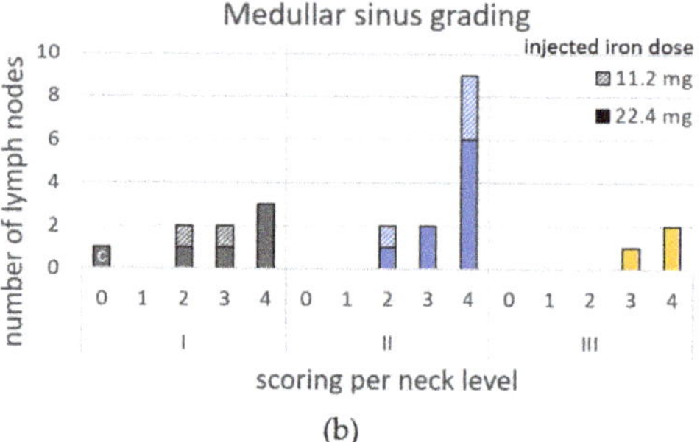

(a)   (b)

Figure 6. The total number of SLNs per histological iron content grade for each level (I: grey, II: blue and III: yellow), (**a**) cortical sinus and (**b**) medullar sinus. The columns represent the total group, whereas subgroups are identified by diagonally striped part (11.2 mg iron dose injected) and full colored part (22.4 mg iron dose injected).

4. Discussion

A comprehensive grading system was proposed in this study, based on iron distribution of (S)LNs in the head and neck area, as a result of magnetic tracer injection for an SLNB procedure. It included analysis of iron-containing LNs on SPIO-enhanced MRI, with SLNs detected by magnetometer and histopathology. Of eight clinically T1-T2N0M0 tongue cancer patients, most draining LNs were located in level II on preoperative MRI and detected by magnetometer. This is in accordance with the preferred draining level of tongue cancers and thus of their potential metastasis [21,22]. Mainly high grades were scored in this level for SPIO-enhanced MRI and medullar sinus on histopathology coupes. A high grade refers to a high iron uptake, which is expected for SLN as the first draining node.

On average 82 ± 58 µg iron was detected by a handheld magnetic probe in SLNs, regardless of the injected iron dose. This makes the percentage accumulated iron in the SLNs, 0.36% for a 22.4 mg injected iron dose and 0.77% for a dose of 11.2 mg. A study comparing two radioisotopes in oral oncology reported 1.95% and 3.16% as radioactive uptake in SLNs based on lymphoscintigraphy [23]. For breast cancer, tracer drainage to SLN was reported for a magnetic tracer to be 0.3% of the injected iron dose [24], and for radioisotopes this was 0.96% (0.0038–5.14%) of the injected dose [25]. Information on the tracer uptake in SLNs can be used for dose optimization. No significant difference was seen for the different iron doses injected, suggesting that an injected iron dose of 11.2 mg should be enough for the detection of SLNs with potential for less disturbing MRI artefacts [26]. Moreover, no additional clinical benefit is achieved when magnetic tracer deposits are sufficient. It is therefore recommended to not overload SLNs with increased tracer volume [27]. When we take the iron amount of SLNs containing metastasis into account (all found in level II and, in total, only three [15]), the iron amount of these SLNs was, in each case, not the highest among excised SLNs of individual patients. It is known that iron does not locate at sites of tumor cells and fat. In the case of a significant metastasis, this will lead to a reduction in iron capacity or even not show iron at all. However, in the case of clinically node negative, it has been shown that the uptake of SPIO is not influenced by occult metastases [28]. Although the three nodes harboring metastasis did not contain the highest iron content, they met the 10% rule (i.e., at least 10% of the hottest node excised) [7] and belonged to the top three hottest nodes, which is considered clinically sufficient for an SLNB procedure using radioisotopes [29]. One should be aware that this was determined ex vivo, while in vivo magnetic SLNs were searched and might be more difficult to determine if an SLN belongs to the top three hottest nodes. The magnetic readout of the handheld magentic probe is not only determined by the iron content, but by the distance of the source to the probe tip as well. Therefore, preoperative imaging, to indicate the amount of iron, or which LN is most likely to be an SLN, would be beneficial to the surgeon during a magnetic SLNB.

Regarding the SPIO-enhanced preoperative MRI, it was intended to identify LNs with iron uptake. The sequence and timing of MRI and the dose of SPIO are open for optimization, but with available data a grading system was initiated in this study, since a high number of LNs were seen with iron uptake on preoperative MRI [15]. It is expected that not all these LNs showing iron uptake are SLNs, especially since one third of the number was detected by a magnetometer and assigned as SLN. For lymphoscintigraphy, discrimination of SLNs and higher echelons is clear; the number of SLNs is half of the LNs seen with radiotracer uptake [23]. Differentiating SLNs from higher echelon LNs is of utmost importance in a complex anatomy, such as the head and neck [3]. Therefore, a grading system for SPIO-enhanced MRI should evaluate the amount of iron and help in identifying most likely SLNs, so SLNs can be distinguished from other LNs taking up iron. A one-on-one comparison of SPIO-enhanced MRI grading to iron content based on magnetometer readout was not completely possible due to the difference in time and following an END. The influence of time on tracer drainage can be limited by having the SLNB planned directly after SPIO-enhanced MRI. From another perspective, as long as the maximum capacity of iron content in SLN is not reached, the SLN is expected to contain the higher amount of accumulated iron, since the hydrodynamic diameter of SPIO is found to be optimal for an SLNB procedure [30,31]. In cases where a one-on-one comparison of most likely SLNs (including corresponding grade) and SLNs detected by magnetometer (including corresponding iron content) is preferred, it is advised to perform an MRI before and after SLNB (without performing an END). Furthermore, the T2*w sequence can be optimized to support the T1w sequence to identify most likely SLNs. The parameters for T2* were chosen to maximize visualization of the effect of SPIOs. Adjustment of the parameters (including shorter repetition time) may reduce this effect; ideally at T2* only iron uptake in SLNs would be visualized. Postoperative SPIO-enhanced MRI showed iron remainders in LNs resulting from either the time between injection and resection, or from iron remaining at the injection site after tumor resection. Sequences used in daily clinical practice seem to be less influenced by these iron remainders in LNs, due to the use of spin echo sequences which correct field inhomogeneities. Regarding the injected iron dose, a melanoma study showed a significant correlation ($r = 0.71$, $p = 0.009$) between the injected iron dose and graded MRI artefacts [26]. It also showed that a low tracer dose (0.02 mL Magtrace®, 0.6 mg iron) produced intraoperatively detectable SLNs [26].

Histopathology grading of the medullar sinus showed comparable distribution over levels of iron grading when compared to SPIO-enhanced MRI; most LNs in level II were graded high, in contrast to the cortical sinus grading in this same level, which included mainly low grades. This means that more iron accumulated within the medullar sinus, confirming descriptions of iron uptake in SLNs of breast cancer patients and in a porcine model study [27,32]. This suggests that the detectability of SLNs (i.e., iron quantity) is primarily determined by the presence of iron in the medullary sinus. Approaching an LN in vivo or the orientation of an LN during ex vivo measurements on the probe tip should then matter to a lesser degree. This is supportive in detecting SLNs, by determination of the top three LNs with highest iron content. It should be noted, however, that surrounding tissue can influence the magnetic readout, due to its diamagnetism.

The main limitations of this study are the small cohort, the differences in time between procedural steps, and the differences in injected magnetic tracer amount. Consequently, a larger (multi-center) cohort is recommended. This would also enable evaluation of (1) iron dose injected, and (2) time between tracer injection and intraoperative SLN detection on the number of nodes detected and graded. Based on information from this study, an administered iron dose of 11.2 mg found fewer identified (S)LNs per patient and (S)LNs were limited to the ipsilateral side on MRI and for SLNB; however, the iron content and iron grading distribution did not appear different. Furthermore, it was difficult in retrospect to directly link LNs visualized by SPIO-enhanced MRI to harvested and histopathologically examined SLNs. It would therefore be recommended to limit the time between preoperative MRI and an SLNB procedure for research purposes.

5. Conclusions

In conclusion, a first step was taken in defining a comprehensive grading system, where data for SPIO-enhanced MRI, ex vivo magnetometer readouts of harvested SLNs, and histopathologic assessment were used to gain more insight into the lymphatic draining system during magnetic SLNB. A direct link between harvested SLNs to preoperative visualized LNs on MRI was difficult to make. SLNs contained, on average, 82 ± 58 µg iron based upon ex vivo magnetometer readouts. No significant differences were found between injection groups (11.2 mg and 22.4 mg iron dose) regarding grading of LN on preoperative SPIO-enhanced MRI, iron content and histopathologic grading.

Author Contributions: Conceptualization, E.R.N., B.K. and L.A.; methodology, E.R.N., J.J.H., J.v.B. and L.A.; validation, R.d.B., A.C., T.W.J.S., B.K. and B.t.H.; formal analysis, E.R.N.; investigation, E.R.N., B.K., A.J.M.v.B., J.J.H., J.v.B. and A.C.; resources, B.t.H., B.K. and A.J.M.v.B.; data curation, E.R.N. and L.A.; writing—original draft preparation, E.R.N. and L.A.; writing—review and editing, B.K., J.J.H., J.v.B., A.J.M.v.B., A.C., B.t.H., T.W.J.S. and R.d.B.; visualization, E.R.N.; supervision, L.A., B.t.H., T.W.J.S. and R.d.B.; project administration, E.R.N.; funding acquisition, B.t.H., T.W.J.S. and R.d.B. All authors have read and agreed to the published version of the manuscript.

Funding: This research was funded by the KWF Dutch Cancer Society and NWO Domain Applied and Engineering Sciences (AES), as part of their joint strategic research program 'Technology for Oncology' (grant number: 15194).

Institutional Review Board Statement: The study was conducted according to the guidelines of the Declaration of Helsinki and approved by the local Medical Ethics Committee of METC Twente, (pp. 17–23, 19 December 2017). The study was registered in the Netherlands Trial Register (ID: NL6656).

Informed Consent Statement: Informed consent was obtained from all subjects involved in the study.

Data Availability Statement: The look-up table is available as a dataset via 4TU.ResearchData. Doi: 10.4121/15169674.

Appendix A

Table A1. Overview of study characteristics per patient.

Pt nr.	Tumor Stage	Total LNs	Injected Iron Dose (mg)	Interval			SLN nr.	Neck Level	Iron Content SLN (µg)	Cortical Grade	Medullar Grade
				I-MRI1 (hh:mm)	I-SLNB (hh:mm)	I-MRI2 (hh)					
1	cT2	29	22.4	00:24	15:57	672	1	II	116	1	3
							2	I	51	1	4
							3	I	124	3	3
							4	III	37	2	3
							5 & 6[1]	I[2]	16	0	0
2	cT1	32	22.4	00:22	16:31	672	1	II	95	2	4
							2	II	80	2	4
							3	III	121	4	4
							4	III	183	4	4
3	cT2	19	22.4	00:58	16:53	1009	1	II	30	1	4
							2	II	227	2	4
							3	II	22	2	3
							4	II	35	4	4
							5[3]	II	59	2	4
							6	I	64	3	4
4	cT2	11	22.4	00:24	21:14	1176	1	I	32	3	2
							2	I	115	1	4
5	cT1	11	22.4	00:38	21:07	841	1	II	57	1	2
6	cT1	4	11.2	01:41	05:17	561	1	II	136	3	4
7	cT2	20	11.2	01:20	18:07	841	1[3]	II	59	1	4
							2	II	187	2	4
							3	I	26	1	2
							4	I	26	2	3
8	cT1	25	11.2	01:09	04:06	656	1[3]	II	82	2	2

Abbreviations and notes: pt, patient; LNs, lymph nodes (includes all excised (S)LNs as result of surgery); I, injection; MRI1, preoperative superparamagnetic iron oxide (SPIO) nanoparticle enhanced magnetic resonance lymphography (MRI); SLN(B), sentinel lymph node (biopsy); MRI2, postoperative SPIO-enhanced MRI; [1] 5&6 as one resected, at pathology it seems to be two separate lymph nodes, for analysis processed as one. [2] SLN harvested level I contralateral. [3] SLNs containing metastasis.

References

1. Gourin, C.; Conger, B.; Porubsky, E.; Sheils, W.; Bilodeau, P.; Coleman, T. The effect of occult nodal metastases on survival and regional control in patients with head and neck squamous cell carcinoma. *Laryngoscope* **2008**, *118*, 1191–1194. [CrossRef] [PubMed]
2. De Zinis, L.O.R.; Bolzoni, A.; Piazza, C.; Nicolai, P. Prevalence and localization of nodal metastases in squamous cell carcinoma of the oral cavity: Role and extension of neck dissection. *Eur. Arch. Oto-Rhino-Laryngol. Head Neck* **2006**, *263*, 1131–1135. [CrossRef] [PubMed]
3. Giammarile, F.; Schilling, C.; Gnanasegaran, G.; Bal, C.; Oyen, W.J.G.; Rubello, D.; Schwarz, T.; Tartaglione, G.; Miller, R.N.; Paez, D.; et al. The EANM practical guidelines for sentinel lymph node localisation in oral cavity squamous cell carcinoma. *Eur. J. Nucl. Med. Mol. Imaging* **2019**, *46*, 623–637. [CrossRef] [PubMed]
4. Shah, J.; Candela, F.; Poddar, A. The patterns of cervical lymph node metastases from squamous carcinoma of the oral cavity. *Cancer* **1990**, *66*, 109–113. [CrossRef]
5. Thompson, C.F.; St. John, M.A.; Lawson, G.; Grogan, T.; Elashoff, D.; Mendelsohn, A.H. Diagnostic value of sentinel lymph node biopsy in head and neck cancer: A meta-analysis. *Eur. Arch. Oto-Rhino-Laryngol.* **2013**, *270*, 2115–2122. [CrossRef]
6. Liu, M.; Wang, S.J.; Yang, X.; Peng, H. Diagnostic efficacy of sentinel lymph node biopsy in early oral squamous cell carcinoma: A meta-analysis of 66 studies. *PLoS ONE* **2017**, *12*, 1–18. [CrossRef]
7. Schilling, C.; Stoeckli, S.J.; Vigili, M.G.; de Bree, R.; Lai, S.Y.; Alvarez, J.; Christensen, A.; Cognetti, D.M.; D'Kruz, A.K.; Frerich, B.; et al. Surgical consensus guidelines on sentinel node biopsy (SNB) in patients with oral cancer. *Head Neck* **2019**, *41*, 2655–2664. [CrossRef]
8. de Bree, R.; de Keizer, B.; Civantos, F.J.; Takes, R.P.; Rodrigo, J.P.; Hernandez-Prera, J.C.; Halmos, G.B.; Rinaldo, A.; Ferlito, A. What is the role of sentinel lymph node biopsy in the management of oral cancer in 2020? *Eur. Arch. Oto-Rhino-Laryngol.* **2021**, *278*, 3181–3191. [CrossRef]
9. Johnson, C.B.; Boneti, C.; Korourian, S.; Adkins, L.; Klimberg, V.S. Intraoperative injection of subareolar or dermal radioisotope results in predictable identification of sentinel lymph nodes in breast cancer. *Ann. Surg.* **2011**, *254*, 612–618. [CrossRef]
10. Zada, A.; Peek, M.C.L.; Ahmed, M.; Anninga, B.; Baker, R.; Kusakabe, M.; Sekino, M.; Klaase, J.M.; ten Haken, B.; Douek, M. Meta-analysis of sentinel lymph node biopsy in breast cancer using the magnetic technique. *Br. J. Surg.* **2016**, *103*, 1409–1419. [CrossRef]
11. Alvarado, M.D.; Mittendorf, E.A.; Teshome, M.; Thompson, A.M.; Bold, R.J.; Gittleman, M.A.; Beitsch, P.D.; Blair, A.L.; Kivilaid, K.; Harmer, Q.J.; et al. SentimagIC: A Non-inferiority Trial Comparing Superparamagnetic Iron Oxide Versus Technetium-99m and Blue Dye in the Detection of Axillary Sentinel Nodes in Patients with Early-Stage Breast Cancer. *Ann. Surg. Oncol.* **2019**, *26*, 3510–3516. [CrossRef] [PubMed]
12. Winter, A.; Woenkhaus, J.; Wawroschek, F. A novel method for intraoperative sentinel lymph node detection in prostate cancer patients using superparamagnetic iron oxide nanoparticles and a handheld magnetometer: The initial clinical experience. *Ann. Surg. Oncol.* **2014**, *21*, 4390–4396. [CrossRef] [PubMed]
13. Baena Fustegueras, J.A.; Gonzalez, F.H.; Caldero, S.G.; de la Fuente Juarez, M.C.; Lopez, S.R.; Riu, F.R.; Mestres Petit, N.; Álvarez, P.M.; Lecube Torelló, A.; Matias-Guiu, X.; et al. Magnetic detection of sentinel lymph node in papillary thyroid carcinoma: The MAGIC-PAT study results. *Eur. J. Surg. Oncol.* **2019**, *45*, 1175–1181. [CrossRef] [PubMed]
14. Hernando, J.; Aguirre, P.; Aguilar-Salvatierra, A.; Leizaola-Cardesa, I.O.; Bidaguren, A.; Gomez-Moreno, G. Magnetic detection of sentinel nodes in oral squamous cell carcinoma by means of superparamagnetic iron oxide contrast. *J. Surg. Oncol.* **2019**, *121*, 244–248. [CrossRef]
15. Nieuwenhuis, E.R.; Kolenaar, B.; van Bemmel, A.J.M.; Hof, J.J.; van Baarlen, J.; Christenhusz, A.; Pouw, J.J.; ten Haken, B.; Alic, L.; de Bree, R. A complete magnetic sentinel lymph node biopsy procedure in oral cancer patients: A pilot study. *Oral Oncol.* **2021**, *121*, 105464. [CrossRef]
16. Mizokami, D.; Kosuda, S.; Tomifuji, M.; Araki, K.; Yamashita, T.; Shinmoto, H.; Shiotani, A. Superparamagnetic iron oxide-enhanced interstitial magnetic resonance lymphography to detect a sentinel lymph node in tongue cancer patients. *Acta Otolaryngol.* **2013**, *133*, 418–423. [CrossRef]
17. Mahieu, R.; de Maar, J.S.; Nieuwenhuis, E.R.; Deckers, R.; Moonen, C.; Alic, L.; ten Haken, B.; de Keizer, B.; de Bree, R. New developments in imaging for sentinel lymph node biopsy in early-stage oral cavity squamous cell carcinoma. *Cancers* **2020**, *12*, 3055. [CrossRef]
18. Winter, A.; Kowald, T.; Engels, S.; Wawroschek, F. Magnetic resonance sentinel lymph node imaging and magnetometer-guided intraoperative detection in prostate cancer using superparamagnetic iron oxide nanoparticles. *Int. J. Nanomed.* **2018**, *13*, 6689–6698. [CrossRef]
19. Aribal, E.; Çelik, L.; Yilmaz, C.; Demirkiran, C.; Guner, D.C. Effects of iron oxide particles on MRI and mammography in breast cancer patients after a sentinel lymph node biopsy with paramagnetic tracers. *Clin. Imaging* **2021**, *75*, 22–26. [CrossRef]
20. Nieuwenhuis, E.R.; Christenhusz, A.; Alic, L. Look-up table (LUT) for Sentimag and Sienna, relation between magnetic readout (counts) and iron content. *4TU.ResearchData* **2021**. [CrossRef]
21. Byers, R.M.; Wolf, P.F.; Ballantyne, A.J. Rationale for Elective Modified. *Head Neck Surg.* **1988**, *10*, 160–167. [CrossRef] [PubMed]

22. Tartaglione, G.; Stoeckli, S.J.; de Bree, R.; Schilling, C.; Flach, G.B.; Bakholdt, V.; Sorensen, J.A.; Bilde, A.; von Buchwald, C.; Lawson, G.; et al. Sentinel node in oral cancer: The nuclear medicine aspects. A survey from the sentinel European node trial. *Clin. Nucl. Med.* **2016**, *41*, 534–542. [CrossRef] [PubMed]

23. Den Toom, I.J.; Mahieu, R.; van Rooij, R.; van Es, R.J.J.; Hobbelink, M.G.G.; Krijger, G.C.; Tijink, B.M.; de Keizer, B.; de Bree, R. Sentinel lymph node detection in oral cancer: A within-patient comparison between [99mTc]Tc-tilmanocept and [99mTc]Tc-nanocolloid. *Eur. J. Nucl. Med. Mol. Imaging* **2021**, *48*, 851–858. [CrossRef] [PubMed]

24. Sekino, M.; Kuwahata, A.; Ookubo, T.; Shiozawa, M.; Ohashi, K.; Kaneko, M.; Saito, I.; Inoue, Y.; Ohsaki, H.; Takei, H.; et al. Handheld magnetic probe with permanent magnet and Hall sensor for identifying sentinel lymph nodes in breast cancer patients. *Sci. Rep.* **2018**, *8*, 1–9. [CrossRef]

25. Waddington, W.A.; Keshtgar, M.R.S.; Taylor, I.; Lakhani, S.R.; Short, M.D.; Ell, P.J. Original article Radiation safety of the sentinel lymph node technique in breast cancer. *Eur. J. Nucl. Med.* **2000**, *27*, 377–391. [CrossRef]

26. Mirzaei, N.; Katsarelias, D.; Zaar, P.; Johansson, I.; Leonhardt, H.; Wärnberg, F.; Olofsson Bagge, R. Sentinel lymph node localization and staging with a low-dose of superparamagnetic iron oxide (SPIO) enhanced MRI and magnetometer in patients with cutaneous melanoma of the extremity-The MAGMEN feasibility study. *Eur. J. Surg. Oncol.* **2022**. Epub ahead of print. [CrossRef] [PubMed]

27. Anninga, B.; Ahmed, M.; Van Hemelrijck, M.; Pouw, J.J.; Westbroek, D.; Pinder, S.; ten Haken, B.; Pankhurst, Q.; Douek, M. Magnetic sentinel lymph node biopsy and localization properties of a magnetic tracer in an in vivo porcine model. *Breast Cancer Res. Treat.* **2013**, *141*, 33–42. [CrossRef] [PubMed]

28. Houpeau, J.-L.; Chauvet, M.-P.; Guillemin, F.; Bendavid-Athias, C.; Charitansky, H.; Kramar, A.; Giard, S. Sentinel lymph node identification using superparamagnetic iron oxide particles versus radioisotope: The French Sentimag feasibility trial. *J. Surg. Oncol* **2016**, *113*, 501–507. [CrossRef]

29. Atula, T.; Shoaib, T.; Ross, G.L.; Gray, H.W.; Soutar, D.S. How many sentinel nodes should be harvested in oral squamous cell carcinoma? *Eur. Arch. Oto-Rhino-Laryngol.* **2008**, *265*, 19–23. [CrossRef]

30. Pouw, J.J.; Ahmed, M.; Anninga, B.; Schuurman, K.; Pinder, S.E.; van Hemelrijck, M.; Pankhurst, Q.A.; Douek, M.; ten Haken, B. Comparison of three magnetic nanoparticle tracers for sentinel lymph node biopsy in an in vivo porcine model. *Int. J. Nanomed.* **2015**, *10*, 1235–1243. [CrossRef]

31. Schauer, A.J.; Becker, W.; Reiser, M.; Possinger, K. Specific developments in sentinel node labeling using 99mTc-colloids. In *The Sentinel Lymph Node Concept*; Heilmann, U., McHugh, W., Pfaff, U., Eds.; Springer: Berlin/Heidelberg, Germany, 2005; pp. 59–69.

32. Johnson, L.; Pinder, S.E.; Douek, M. Deposition of superparamagnetic iron-oxide nanoparticles in axillary sentinel lymph nodes following subcutaneous injection. *Histopathology* **2013**, *62*, 481–486. [CrossRef] [PubMed]

Article

Magnetometer-Guided Sentinel Lymph Node Dissection in Prostate Cancer: Rate of Lymph Node Involvement Compared with Radioisotope Marking

Svenja Engels †, Bianca Michalik †, Luca-Marie Meyer, Lena Nemitz, Friedhelm Wawroschek and Alexander Winter *

University Hospital for Urology, Klinikum Oldenburg, Department of Human Medicine, School of Medicine and Health Sciences, Carl von Ossietzky University Oldenburg, 26122 Oldenburg, Germany;
engels.svenja@klinikum-oldenburg.de (S.E.); michalik.bianca@klinikum-oldenburg.de (B.M.);
luca-marie.meyer@uol.de (L.-M.M.); nemitz.lena@klinikum-oldenburg.de (L.N.);
wawroschek.friedhelm@klinikum-oldenburg.de (F.W.)
* Correspondence: winter.alexander@klinikum-oldenburg.de
† These authors contributed equally to this work.

Simple Summary: Pelvic lymph node dissection is recommended in prostate cancer according to the patients' individual risk for nodal metastases. Targeted removal of sentinel lymph nodes increases the number of detected lymph node metastases in patients with prostate cancer. We previously established magnetometer-guided sentinel lymph node dissection in patients with prostate cancer to overcome logistical and technical disadvantages associated with the standard radioisotope-guided technique. This retrospective study compared the magnetometer-guided and standard techniques in terms of their ability to detect lymph node metastases. Using the magnetometer-guided technique, more sentinel lymph nodes were detected per patient. The detected rates of lymph node involvement matched the predictions in both techniques equally well. Our findings confirm the reliability of magnetometer-guided sentinel lymph node dissection and highlight the importance of the sentinel technique for detecting lymph node metastases in prostate cancer.

Abstract: Sentinel pelvic lymph node dissection (sPLND) enables the targeted removal of lymph nodes (LNs) bearing the highest metastasis risk. In prostate cancer (PCa), sPLND alone or combined with extended PLND (ePLND) reveals more LN metastases along with detecting sentinel LNs (SLNs) outside the conventional ePLND template. To overcome the disadvantages of radioisotope-guided sPLND in PCa treatment, magnetometer-guided sPLND applying superparamagnetic iron oxide nanoparticles as a tracer was established. This retrospective study compared the nodal staging ability between magnetometer- and radioisotope-guided sPLNDs. We analyzed data of PCa patients undergoing radical prostatectomy and magnetometer- (848 patients, 2015–2021) or radioisotope-guided (2092 patients, 2006–2015) sPLND. To reduce heterogeneity among cohorts, we performed propensity score matching and compared data considering sentinel nomogram-based probabilities for LN involvement (LNI). Magnetometer- and radioisotope-guided sPLNDs had SLN detection rates of 98.12% and 98.09%, respectively; the former detected more SLNs per patient. The LNI rates matched nomogram-based predictions in both techniques equally well. Approximately 7% of LN metastases were detected outside the conventional ePLND template. Thus, we confirmed the reliability of magnetometer-guided sPLND in nodal staging, with results comparable with or better than radioisotope-guided sPLND. Our findings highlight the importance of the sentinel technique for detecting LN metastases in PCa.

Keywords: prostate cancer; sentinel lymph node; lymphadenectomy; metastases; superparamagnetic iron oxide nanoparticles; radioisotopes

Citation: Engels, S.; Michalik, B.; Meyer, L.-M.; Nemitz, L.; Wawroschek, F.; Winter, A. Magnetometer-Guided Sentinel Lymph Node Dissection in Prostate Cancer: Rate of Lymph Node Involvement Compared with Radioisotope Marking. *Cancers* **2021**, *13*, 5821. https://doi.org/10.3390/cancers13225821

Academic Editors: Moriaki Kusakabe and Akihiro Kuwahata

Received: 21 September 2021
Accepted: 17 November 2021
Published: 20 November 2021

1. Introduction

Pelvic lymph node dissection (PLND) is the most reliable technique for lymph node (LN) staging in clinically localized prostate cancer (PCa) [1]. LN status is a therapeutically crucial prognostic factor in PCa because the presence and extent of LN involvement (LNI) are related to an increased risk of systemic dissemination and progression of the disease [2–4]. Moreover, PLND or resection of LN metastases has been indicated to have therapeutic benefits, particularly in patients with minimal LNI [1,5–8].

The detection of LNI directly correlates with the number of dissected LNs as well as with the anatomical limits of PLND [9,10]. The European Association of Urology (EAU) guidelines, therefore, recommend an extended PLND (ePLND) for LN staging in patients with >5% risk of LNI as diagnosed by systematic random biopsy [11,12] or in those with >7% risk of LNI as diagnosed by multiparametric magnetic resonance imaging (MRI) and MRI-targeted biopsy [12,13].

The complication rate, however, also increases alongside the increase in the number of LNs removed [14–16]. Therefore, Wawroschek et al. [17] adopted the techniques and concepts of radioisotope-guided sentinel LN (SLN) identification from other tumor entities for use in PCa, and this has subsequently been independently confirmed in several studies [18,19]. Sentinel PLND (sPLND) enables the targeted removal of clinically negative LNs, which bear a high probability of containing metastases [20] because SLNs are the first lymphatic drainage stations of their primary organs or the respective tumor [21]. In PCa, sPLND alone or in combination with ePLND increases the number of detected LN metastases [19,22,23]. Moreover, during sPLND, SLNs occurring outside the conventional ePLND template can be removed [22–24]. Thus, sPLND provides additional diagnostic value by adjusting the degree and anatomical extent of PLND to the patient's individual lymphatic drainage situation [19,22,23,25].

Unfortunately, the use of radioisotope tracers for SLN marking is associated with several technical and logistical disadvantages [26,27]. For example, the practicality of the radioisotope-guided sPLND technique depends on the accessibility to radioisotope tracers and nuclear medicine facilities. Thus, this technique is used only in more developed countries or hospitals with access to such technology. Furthermore, this technique exposes patients and surgical staff to radiation, an aspect that is strongly controlled by legislation. In patients with breast cancer, superparamagnetic iron oxide nanoparticles (SPIONs) have been successfully developed as an equivalent, easy-to-use, and radiation-free alternative for SLN marking and intraoperative detection [26,28]. Our group subsequently adopted this technique of intraoperative magnetic detection of SLNs for use in patients with PCa; we use a system that comprises a magnetic tracer and a handheld magnetometer [23,24,27,29].

This retrospective study primarily aimed to compare magnetometer- and radioisotope-guided sPLND techniques in terms of their ability to detect LN metastases in patients with PCa. The secondary aim was to evaluate the anatomical distribution of dissected SLNs and detected LN metastases. We analyzed two large data sets from patients with PCa who underwent radical prostatectomy in combination with either of the two sPLND techniques at a tertiary referral hospital and performed a matched-pair analysis. The rates of LNI were compared between the two techniques while considering the patients' individual probabilities for LNI as inferred from our sentinel nomogram [30].

2. Materials and Methods

2.1. Patient Populations

This retrospective study considered two cohorts of patients with PCa consecutively documented in the database of the University Hospital for Urology Oldenburg. We initially enrolled 2186 patients with PCa who underwent open retropubic radical prostatectomy combined with radioisotope-guided sPLND between January 2006 and February 2015. Then, we excluded the data of 11 patients who received only one-sided sPLND and of one patient in whom the time between tracer injection and surgery extended the manufacturer-guaranteed tracer detectability period. Furthermore, 45 patients who underwent hormonal treatment and 34 who underwent transurethral prostate surgeries before prostatectomy were also excluded. LN dissection data were incomplete in three additional patients. The final sample for the analysis of radioisotope-guided sPLND included 2092 patients.

Initially, we included 881 patients with PCa who underwent open retropubic radical prostatectomy combined with magnetometer-guided sPLND between February 2015 and May 2021. Then, we excluded the data of seven patients who received only one-sided sPLND, three patients with limited tracer detectability because of metal implants, and two patients in whom the time between tracer injection and surgery extended the manufacturer-guaranteed detectability period. Furthermore, we also excluded 16 patients who underwent hormonal treatment and 4 who underwent transurethral prostate surgeries prior to prostatectomy. The LN dissection data were incomplete in one additional patient. The final sample for the analysis of magnetometer-guided sPLND included 848 patients.

All patients were informed verbally and in writing about the open retropubic radical prostatectomy and sPLND; all signed a consent form before surgery.

2.2. sPLND Technique and Histopathological Examination

All patients were administered with transrectal tracer injection, either 99mTechnetium nanocolloid (160 MBq Nanocoll®, Nycomed Amersham Sorin, Milan, Italy) or superparamagnetic iron oxide nanoparticles (02/2015–01/2019 Sienna+®, Sysmex Europe GmbH, Norderstedt, Germany; 01/2019–05/2021 Magtrace®, Sysmex Europe GmbH, Norderstedt, Germany, into the prostate under ultrasonic guidance a day before surgery. SPLND was performed as described by Wawroschek et al. [31] and Winter et al. [32] for radioisotope-guided surgery and as described by Winter et al. [23] for magnetometer-guided surgery. All SLNs detected by a gamma probe (C-Trak System, Care Wise, Morgan Hill, CA, USA, or Crystal Probe SG04, Crystal Photonics GmbH, Berlin, Germany) or a magnetometer (Sentimag®, Sysmex Europe GmbH, Norderstedt, Germany) together with lymphatic fatty tissue directly adjoining or adhering to the identified SLNs were removed surgically. After sentinel-guided surgery, risk-adapted or if no SLN was detected at all, PLND was completed by ePLND using the anatomic template as described by Weingärtner et al. [33], which included all lymphatic fatty tissue along the external and internal iliac vessels and the obturator fossa as well as within the area dorsal to the obturator nerve, from the bifurcation of the common iliac artery (proximal limit) to the femoral canal (distal limit) and from the pelvic sidewall (lateral limit) to the perivesical fatty tissue (medial limit).

After surgery, LNs were cut into 3 mm transverse sections, routinely processed, and embedded into paraffin. Then, 4–5 μm sections were stained with hematoxylin-eosin (HE; Figure 1). In rare cases of inconclusive conventional histology, samples were immunohistochemically stained with AE1/AE3 pancytokeratin antibodies to check for (micro-)metastases.

2.3. Data Analyses

All data analyses were performed using R 4.1.0 [34]. For each sPLND technique, we calculated the median numbers of SLNs detected per patient, SLN detection rate (proportion of patients with detected SLNs), rate of LNI (proportion of patients with pathologic nodal stage 1; pN1), and the false-negative rate (proportion of LN-positive but SLN-negative cases). Numbers and proportions were compared statistically using the Wilcoxon rank-sum test and χ^2 proportions test, respectively. Tests were repeated after 1:1 optimal pair matching according to clinical information (i.e., age, prostate-specific antigen, clinical tumor stage, Gleason score, and percentage of positive biopsy cores) based on the results of a propensity score analysis. A summary of the propensity score-adjusted data is presented in Table A1 of Appendix A. We estimated the probability of LNI for each patient according to our nomogram [30]. This nomogram predicts a patient's individual probability for the presence of lymphogenic metastases based on clinical information such as prostate-specific antigen (PSA) value, clinical tumor stage, biopsy Gleason score, and percentage of positive biopsy cores [30]. In our data set, clinical tumor stage and/or biopsy data (biopsy Gleason score and/or the percentage of tumor-positive biopsy cores) were unavailable for 24 patients. To evaluate the diagnostic accuracy of each sPLND technique, we plotted the observed LNI rates against the nomogram-predicted probabilities of LNI in steps of approximately 3.33% (30 bins). Curves were smoothed, and 95% confidence intervals were calculated using the Loess method (local polynomial regression fitting). A χ^2 goodness-of-fit test was computed for each technique to estimate the deviation of the curves from the ideal curve wherein observed, and predicted rates of LNI match perfectly. Moreover, the numbers of dissected SLNs and LN metastases were counted for each anatomical region of sPLND. The region data were unavailable for 22 SLNs detected during magnetometer-guided sPLND.

3. Results

In total, we dissected 12,331 LNs in 848 patients who underwent magnetometer-guided sPLND and 22,565 LNs in 2092 patients who underwent radioisotope-guided sPLND. Among these dissected LNs, 6478 and 12,981 were SLNs. Table 1 summarizes clinical and histopathological tumor characteristics as well as LN dissection data of the two patient cohorts. A summary of these data after propensity score matching can be found in Table A1. The SLN detection rates did not differ between the magnetometer- and radioisotope-guided sPLND techniques (98.11%, n = 832 patients vs. 98.18%, n = 2054 patients, respectively; Table 2). However, we detected significantly more SLNs per patient using magnetometer-guided sPLND (median = 7, IQR: 4–10) than that using radioisotope-guided sPLND (median = 6, IQR: 4–8; Table 2).

Table 1. Patient characteristics.

Method	Magnetometer-Guided sPLND			Radioisotope-Guided sPLND		
	Overall	**pN0**	**pN1**	**Overall**	**pN0**	**pN1**
n (%)	848	655 (77.24)	193 (22.76)	2092	1696 (81.07)	396 (18.93)
Age (IQR)	67 (62–71)	67 (61–71)	68 (64–73)	67 (62–71)	67 (61–71)	68 (63–71)
Total PSA ng/mL (IQR)	8.7 (6.1–13.5)	8 (5.8–11.8)	12.8 (8.6–27.7)	7.8 (5.5–12.5)	7.2 (5.3–10.9)	12.0 (7.9–20.6)
Dissected LNs (IQR)	14 (10–18)	13 (10–17)	16 (12–21)	10 (7–14)	10 (7–13)	12 (9–15)
Positive LNs (IQR)	0 (-)	0 (-)	2 (1–4)	0 (-)	0 (-)	2 (1-3)
Dissected SLNs (IQR)	7 (4–10)	7 (5–10)	6 (4–10)	6 (4–8)	6 (4–8)	6 (3–8)
Positive SLNs (IQR)	0 (-)	0 (-)	1 (1-2)	0 (-)	0 (-)	1 (1-2)
Clinical tumor stage (%)	*		*	**	**	**
cT1	436 (51.42)	397 (60.61)	39 (20.21)	1129 (53.97)	1027 (60.55)	102 (25.76)
cT2	368 (43.40)	244 (37.25)	124 (64.25)	919 (43.93)	658 (38.80)	261 (65.91)
cT3	41 (4.83)	14 (2.14)	27 (13.99)	36 (1.72)	6 (0.35)	30 (7.58)
cT4	2 (0.24)	0	2 (1.04)	2 (0.10)	0	2 (0.51)
Biopsy Gleason sum (%)				***	***	
$\leq$6	162 (19.10)	150 (22.90)	12 (6.22)	998 (47.71)	938 (55.31)	60 (15.15)
=7 (3 + 4)	402 (47.41)	354 (54.05)	48 (24.87)	724 (34.61)	570 (33.61)	154 (38.89)
=7 (4 + 3)	129 (15.21)	88 (13.44)	41 (21.24)	191 (9.13)	109 (6.43)	82 (20.71)
$\geq$8	155 (18.28)	63 (9.62)	92 (47.67)	176 (8.41)	76 (4.48)	100 (25.25)
Postoperative Gleason sum (%)						
$\leq$6	30 (3.54)	30 (4.58)	0	349 (16.68)	348 (20.52)	1 (0.25)
=7 (3 + 4)	443 (52.24)	423 (64.58)	20 (10.36)	1122 (53.63)	1052 (62.03)	70 (17.68)
=7 (4 + 3)	216 (25.47)	147 (22.44)	69 (35.75)	420 (20.08)	230 (13.56)	190 (47.98)
$\geq$8	159 (18.75)	55 (8.40)	104 (53.89)	201 (9.61)	66 (3.89)	135 (34.09)

Table 1. *Cont.*

Method	Magnetometer-Guided sPLND			Radioisotope-Guided sPLND		
	Overall	pN0	pN1	Overall	pN0	pN1
Pathologic tumor stage (%)						
pT1c	2 (0.24)	2 (0.31)	0	1 (0.05)	1 (0.06)	0
pT2a	41 (4.83)	41 (6.26)	0	184 (8.80)	180 (10.61)	4 (1.01)
pT2b	21 (2.48)	19 (2.90)	2 (1.04)	40 (1.91)	39 (2.30)	1 (0.25)
pT2c	399 (47.05)	390 (59.54)	9 (4.66)	1086 (51.91)	1048 (61.79)	38 (9.60)
pT3a	178 (20.99)	141 (21.53)	37 (19.17)	407 (19.46)	300 (17.69)	107 (27.02)
pT3b	197 (23.23)	61 (9.31)	136 (70.47)	318 (15.20)	113 (6.66)	205 (51.77)
pT4	10 (1.18)	1 (0.15)	9 (4.66)	56 (2.68)	15 (0.88)	41 (10.35)

Data are presented as median (interquartile range) or frequency (percentage). sPLND: sentinel pelvic lymph node dissection; pN: pathologic nodal stage; IQR: interquartile range; (S)LN: (sentinel) lymph node; PSA: prostate-specific antigen; * clinical T-category could not be assessed in one patient (pN1); ** clinical T-category could not be assessed in six patients (pN0: $n = 5$, pN1: $n = 1$); *** incomplete biopsy data in three patients (pN0).

Table 2. Comparison between magnetometer-guided sPLND and radioisotope-guided sPLND either with original data or with propensity score-adjusted data.

Comparison	Original Results	Test Statistic	Adjusted Results	Test Statistic
SLN detection rate	98.11% (832) vs. 98.18% (2054)	$\chi^2 < 0.001$, df $= 1$, $p = 1$	98.11% (832) vs. 95.87% (812)	$\chi^2 = 6.55$, df $= 1$, $p = 0.011$ *
Number of dissected SLNs	7 (4–10) vs. 6 (4–8)	W $= 1{,}059{,}411$, $p < 0.001$ ***	7 (4–10) vs. 5 (3–7)	W $= 471{,}031$, $p < 0.001$ ***
Rate of LNI	22.76% (193) vs. 18.93% (396)	$\chi^2 = 5.29$, df $= 1$, $p = 0.021$ *	22.76% (192) vs. 25.97% (220)	$\chi^2 = 2.34$, df $= 1$, $p = 0.126$
Rate of LN+ but SLN-	7.25% (14) vs. 9.85% (39)	$\chi^2 = 0.77$, df $= 1$, $p = 0.379$	7.29% (14) vs. 14.55% (32)	$\chi^2 = 4.73$, df $= 1$, $p = 0.030$ *
False-negative rate	3.63% (7) vs. 5.05% (20)	$\chi^2 = 0.32$, df $= 1$, $p = 0.572$	3.65% (7) vs. 5.91% (13)	$\chi^2 = 0.700$, df $= 1$, $p = 0.403$

Data are presented as percentage (n) or median (interquartile range). (S)LN: (sentinel) lymph node; LNI: lymph node involvement; LN+: lymph node positivity; SLN-: sentinel lymph node negativity. * 5% significance level, *** 0.1% significance level.

We found metastases in 621 LNs of 193 patients who underwent magnetometer-guided sPLND and in 1010 LNs of 396 patients who underwent radioisotope-guided sPLND (Table 1). Figure 1 shows a representative example of a HE staining of an SLN metastasis as revealed by the magnetic tracer. On the one hand, there were significantly more patients with LN positivity among those who underwent magnetometer-guided sPLND than among those who underwent radioisotope-guided sPLND (22.76% vs. 18.93%, respectively; Table 2), which leveled off after propensity score matching (Table 2). On the other hand, no difference was noted between the two techniques in the proportion of patients who had metastases only in non-SLNs (7.25%, $n = 14$ vs. 9.85%, $n = 39$, respectively; Table 2). Of these, in five patients who underwent magnetometer-guided sPLND and in 15 patients who underwent radioisotope-guided sPLND, respectively, no SLNs could be detected at all, and in two and four patients, respectively, macroscopically visible metastases were surgically removed without measuring tracer activity. Excluding these cases from the sample of only patients with non-SLN positivity, the resulting false-negative rates were 3.63% ($n = 7$) for magnetometer-guided sPLND and 5.05% ($n = 20$) for radioisotope-guided sPLND (Table 2).

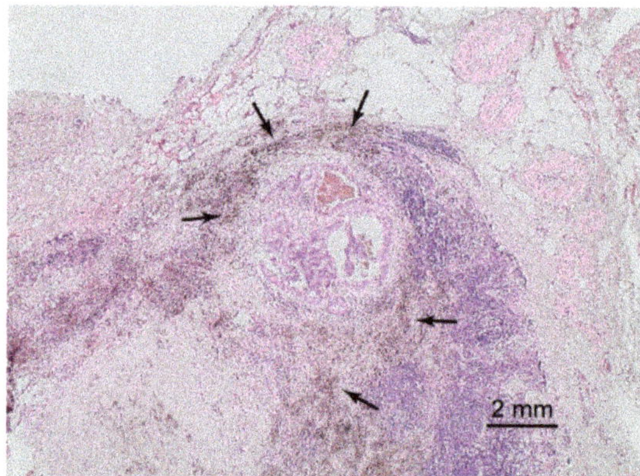

Figure 1. Hematoxylin-eosin staining of a magnetically traced sentinel lymph node. In the center of the image, a 4 mm metastasis of a Gleason 8 adenocarcinoma of the prostate is present. Note the brownish discoloration (arrows) of macrophages containing the magnetic tracer (superparamagnetic iron oxide nanoparticles). Total magnification 40×.

The observed proportions of patients with LN positivity did not differ from the proportions predicted by the nomogram in both magnetometer- (goodness-of-fit test: $\chi^2 = 0.73$, df = 29, $p = 1$) and radioisotope-guided (goodness-of-fit test: $\chi^2 = 0.75$, df = 29, $p = 1$; Figure 2) sPLND techniques.

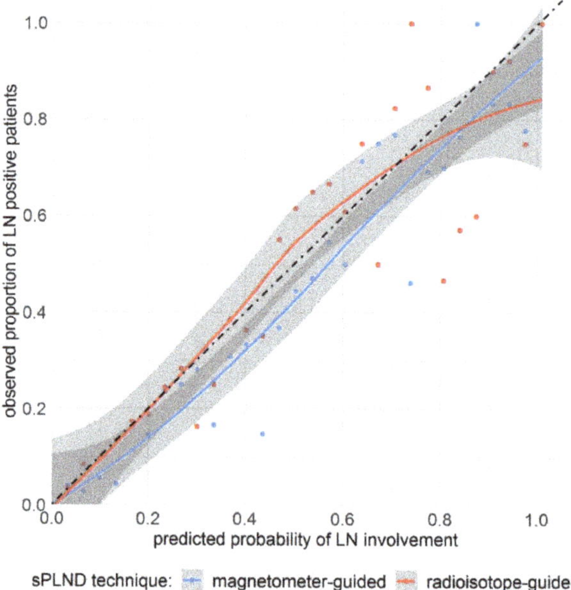

Figure 2. Observed proportion of patients with lymph node (LN) positivity in relation to the probability of LN involvement as predicted by the nomogram for magnetometer (blue)- and radioisotope (red)-guided sentinel pelvic lymph node dissection (sPLND). Gray shaded areas represent the 95% confidence intervals of the smoothed blue and red curves, respectively. The black dot-dashed line represents the ideal curve wherein the predicted probabilities and observed proportions match perfectly.

Anatomic regions outside the standard template for ePLND accounted for 4.5% of SLNs detected by a magnetometer and 2.9% of SLNs detected by a radioprobe, respectively (Figure 3a). Accordingly, approximately 7.6% and 5.7% of LN metastases detected by magnetometer-guided and radioisotope-guided sPLND, respectively, occurred in LNs outside the standard ePLND template (Figure 3b).

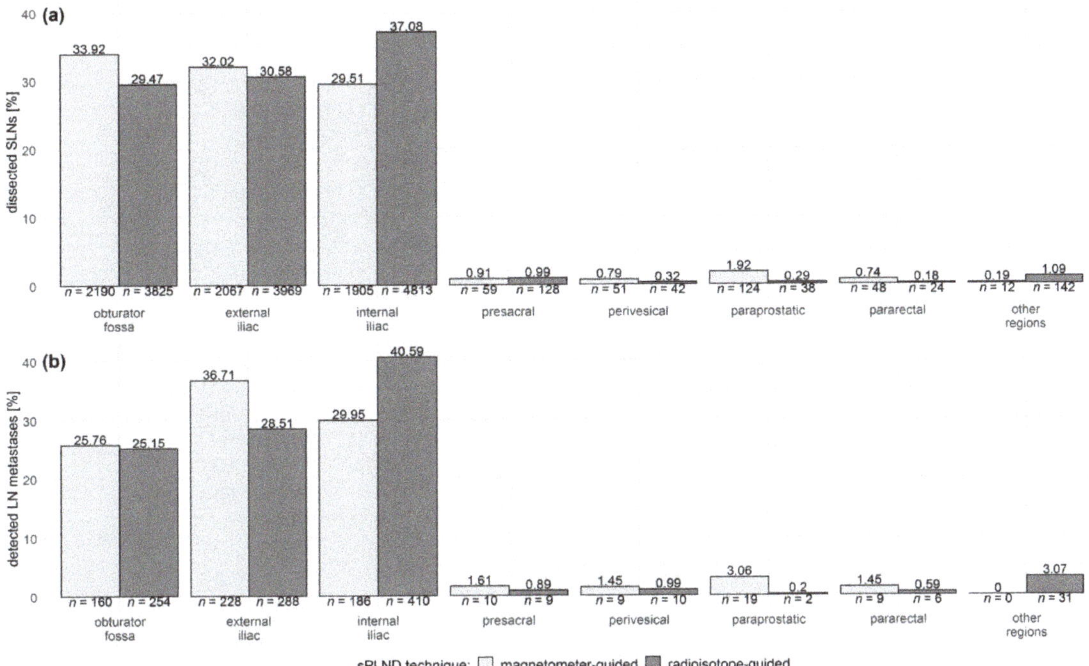

Figure 3. Anatomical distribution of (**a**) the dissected sentinel lymph nodes (SLNs) and (**b**) the lymph node (LN) metastases detected using magnetometer (light gray bars)- and radioisotope (dark gray bars)-guided sentinel pelvic lymph node dissection (sPLND) techniques, respectively. Data of the anatomic region were not available for 22 SLNs dissected using magnetometer-guided sPLND.

4. Discussion

In this retrospective study of pelvic sentinel lymphadenectomy in patients with PCa, SLN detection rates were equally high for both magnetometer- and radioisotope-guided sPLND techniques; however, the former revealed a greater number of SLNs per patient. The accordance between the observed rates of LNI and nomogram-based predictions was equally high in both sPLND techniques. Both sPLND techniques revealed a considerable proportion of lymphogenic metastases outside the conventional ePLND template.

The SLN detection rates found in our study closely match those reported in other studies on sPLND in PCa [19,22,35]. Our observed rates of LNI were accordingly high [19,32,36] and even higher than expected from the ePLND data [37–40]. We resect a comparatively large number of SLN with both procedures. This can probably be explained by the fact that we do not perform dynamic imaging in clinical routine and remove all active LNs regardless of their activity level. Therefore, strictly speaking, our procedure could rather be called "lymphatic mapping".

To predict the individual probability of LNI in patients with PCa, Winter et al. [30] developed a nomogram, which is based on clinical information and radioisotope-guided sPLND data. This nomogram was subsequently validated externally [41]. Our analyses revealed high accordance between nomogram-based predictions and LNI rates in both sPLND techniques. These results suggest that magnetometer-guided sPLND is a reliable and promising alternative that can be used in PCa to overcome the disadvantages of radioisotope marking [26,27], as observed in other tumor entities [26,28,42–45]. Furthermore, an improved flow of the magnetic tracer may lead to an increased number of detected SLNs in magnetometer-guided sPLND, as observed in the present study. The size of the magnetic tracer particles is approximately 60 nm, which is slightly smaller and more homogeneous than the radioisotope tracer [26]. The radioisotope tracer particles were 80 nm (95%), 80–100 nm (4%), and >100 nm (1%) [27]. Thus, the particle size might have influenced the drain of the two different tracers through the lymphatic pathways and/or trapping in the LNs, leading to more SLNs being detected using magnetometer-guided sPLND.

At our hospital, the magnetometer-guided sPLND has largely replaced radioisotope-guided sPLND, which is an obvious limitation of our retrospective study design as the two techniques were used in different study periods. Consequently, the change in surgeons, as well as their increasing experience over time, could be the reason why more SLNs per patient were detected when using the magnetometer technique. Routine histopathological examination of the dissected LNs and the distinct use of supplementary immunohistochemistry have also improved over the years at our hospital. Thus, the higher prevalence of LNI found in patients being treated with magnetometer-guided sPLND could have partly resulted from these developments. Furthermore, the advancements in PCa diagnostics as well as treatment options for low risk PCa (e.g., active surveillance) have led to updates in guidelines (e.g., EAU) for PCa over time with implications for clinical practice and a stage shift [46]. Therefore, it is unsurprising that the two patient cohorts in our retrospective study also differed in their clinical tumor properties. For example, the patient group treated with magnetometer-guided sPLND had more advanced tumors according to biopsy Gleason score (Table 1). As reflected in our results, the Gleason score may serve as a predictor for LNI [30,39], and this might partly explain the differences in the rates of LNI between the two sPLND techniques. In agreement with this, the observed differences in the rates of LNI are no longer present after propensity score analysis through optimal pair matching of the data according to clinical information. Nevertheless, the rates of LNI in both sPLND techniques accurately match the rates predicted by our sentinel-based nomogram [30]. Thus, our results show that both magnetometer- and radioisotope-guided sPLND techniques are reliable tools for LN staging in PCa. Since the performance of the magnetic method seems to be even slightly better and is also associated with a simpler clinical workflow, we will continue to use it to replace the radioactive sentinel procedure if there are no contraindications.

The magnetic SLN marking technique has some limitations [23,24,29]. The magnetic approach is not applicable in patients with pacemakers or other implanted electronic devices in the chest wall as well as those with hypersensitivity to iron or with iron overload disease. Furthermore, the detectability of the magnetic tracer is reduced in patients with metal implants such as hip prostheses or other metallic pelvic implants. In these cases, patients would still benefit from radioisotope-guided sPLND.

The sentinel approach bears one significant drawback. When LNs are fully metastasized, or lymphatic pathways are blocked, the afferent lymph could be directed to other LNs, which might not necessarily be SLNs [47,48]; this has been observed, for example, in inguinal SLNs in penile carcinoma [49,50]. Performing sPLND alone in these cases would yield false-negative results. The false-negative rates observed in our study were as low as those reported in other studies [19,22,35,48]. Thus, the diagnostic value of sPLND for LN staging in patients with PCa is similar to that when using a combination of sPLND and ePLND; this would have a higher prevalence of LNI than using ePLND alone [19,22,32,41].

In addition to the selective removal of LNs bearing the highest risk of containing metastases, the sentinel approach also enables surgical treatment, which is adjusted to the patient's individual lymphatic drainage situation [19,25,48]. Consequently, we identified a considerable proportion of LN metastases outside the conventional ePLND template [33] using both sPLND techniques. These findings are well in line with those of other studies and highlight the diagnostic value of sPLND for LN staging in patients with PCa [19,22,25,36,51].

5. Conclusions

Our study confirms the diagnostic accuracy of magnetometer-guided sPLND in nodal staging in patients with PCa. We suggest magnetometer-guided sPLND as a reliable and promising alternative sPLND technique in PCa treatment to overcome the technical and logistic disadvantages of radioisotope marking. Furthermore, our results highlight the additional diagnostic value of the sentinel technique in PCa because it allows not only the selective removal of LNs bearing the highest risk of containing metastases but also surgical treatment that is adapted to the patient's individual lymphatic drainage situation.

Author Contributions: Conceptualization, A.W. and S.E.; methodology, S.E., B.M., and A.W.; validation, S.E. and B.M.; formal analysis, B.M.; resources, F.W.; data curation, S.E., L.-M.M., A.W., and L.N.; writing—original draft preparation, S.E. and B.M.; writing—review and editing, A.W., S.E., B.M., L.-M.M., L.N., and F.W.; visualization, B.M.; supervision, A.W. and F.W.; project administration, S.E. and A.W.; funding acquisition, A.W. and F.W. All authors have read and approved the final version of the manuscript.

Funding: This research received no external funding.

Institutional Review Board Statement: This retrospective study was conducted in accordance with the Declaration of Helsinki and received ethical approval from the Medical Ethics Committee of the University of Oldenburg (02/06/2021, reference: 2018-140). The study is registered in an international clinical trials register (Research Registry, researchregistry7004).

Informed Consent Statement: All patients provided their written informed consent for the documentation and scientific evaluation of their data.

Data Availability Statement: The data presented in this study are available upon reasonable request from the corresponding author.

Conflicts of Interest: The authors declare no conflict of interest. The funders had no role in the design of the study; collection, analyses, or interpretation of data; writing of the manuscript; or decision to publish the results.

Appendix A

Table A1. Patient characteristics after 1:1 optimal pair matching according to clinical information (i.e., age, prostate-specific antigen, clinical tumor stage, Gleason score, and percentage of positive biopsy cores) and adjusted to propensity scores.

Method	Magnetometer-Guided sPLND			Radioisotope-Guided sPLND		
	Overall	pN0	pN1	Overall	pN0	pN1
n (%)	847 *	655 (77.33)	192 (22.67)	847	627 (74.03)	220 (25.97)
Age (IQR)	67 (62–71)	67 (61–71)	68 (64–73)	67 (62–71)	67 (62–71)	68 (63–72)
Total PSA ng/mL (IQR)	8.7 (6.1–13.4)	8.0 (5.8–11.8)	12.7 (8.6–27.3)	8.6 (6.0–14.3)	7.8 (5.6–12.4)	12.2 (8.1–20.2)
Dissected LNs (IQR)	14 (10–18)	13 (10–17)	16 (12–20)	10 (8–14)	10 (7–13)	12 (9–15)
Positive LNs (IQR)	0 (-)	0 (-)	2 (1–4)	0 (0–1)	0 (-)	2 (1–3)
Dissected SLNs (IQR)	7 (4–10)	7 (5–10)	6 (4–10)	5 (3–7)	5 (3–7)	5 (3–8)
Positive SLNs (IQR)	0 (-)	0 (-)	1 (1–2)	0 (-)	0 (-)	1 (1–2)
Clinical tumor stage (%)						
cT1	436 (51.48)	397 (60.61)	39 (20.31)	464 (54.78)	401 (63.96)	63 (28.64)
cT2	368 (43.45)	244 (37.25)	124 (64.58)	357 (42.15)	222 (35.41)	135 (61.36)
cT3	41 (4.84)	14 (2.14)	27 (14.06)	24 (2.83)	4 (0.64)	20 (9.09)
cT4	2 (0.24)	0	2 (1.04)	2 (0.24)	0	2 (0.91)

Table A1. *Cont.*

Method	Magnetometer-Guided sPLND			Radioisotope-Guided sPLND		
	Overall	pN0	pN1	Overall	pN0	pN1
Biopsy Gleason sum (%)						
≤6	162 (19.13)	150 (22.90)	12 (6.25)	162 (19.13)	153 (24.40)	9 (4.09)
=7 (3 + 4)	401 (47.34)	354 (54.05)	47 (24.48)	422 (49.82)	349 (55.66)	73 (33.18)
=7 (4 + 3)	129 (15.23)	88 (13.44)	41 (21.35)	115 (13.58)	64 (10.21)	51 (23.18)
≥8	155 (18.30)	63 (9.62)	92 (47.92)	148 (17.47)	61 (9.73)	87 (39.55)
Postoperative Gleason sum (%)						
≤6	30 (3.54)	30 (4.58)	0	58 (6.85)	58 (9.25)	0
=7 (3 + 4)	443 (52.30)	423 (64.58)	20 (10.42)	438 (51.71)	406 (64.75)	32 (14.55)
=7 (4 + 3)	215 (25.38)	147 (22.44)	68 (35.42)	204 (24.09)	111 (17.7)	93 (42.27)
≥8	159 (18.77)	55 (8.4)	104 (54.17)	147 (17.36)	52 (8.29)	95 (43.18)
Pathologic tumor stage (%)						
pT1c	2 (0.24)	2 (0.31)	0	1 (0.12)	1 (0.16)	0
pT2a	41 (4.84)	41 (6.26)	0	64 (7.56)	61 (9.73)	3 (1.36)
pT2b	21 (2.48)	19 (2.90)	2 (1.04)	8 (0.94)	8 (1.28)	0
pT2c	399 (47.11)	390 (59.54)	9 (4.69)	378 (44.63)	362 (57.74)	16 (7.27)
pT3a	177 (20.90)	141 (21.53)	36 (18.75)	180 (21.25)	127 (20.26)	53 (24.09)
pT3b	197 (23.26)	61 (9.31)	136 (70.83)	188 (22.20)	62 (9.89)	126 (57.27)
pT4	10 (1.18)	1 (0.15)	9 (4.69)	28 (3.31)	6 (0.96)	22 (10)

Data are presented as median (interquartile range) or frequency (percentage). sPLND: sentinel pelvic lymph node dissection; pN: pathologic nodal stage; IQR: interquartile range; (S)LN: (sentinel) lymph node; PSA: prostate-specific antigen; * clinical T-category could not be assessed in one patient (pN1).

References

1. Withrow, D.R.; DeGroot, J.M.; Siemens, D.R.; Groome, P.A. Therapeutic value of lymph node dissection at radical prostatectomy: A population-based case-cohort study. *BJU Int.* **2011**, *108*, 209–216. [CrossRef] [PubMed]
2. Cheng, L.; Bergstralh, E.J.; Cheville, J.C.; Slezak, J.; Corica, F.A.; Zincke, H.; Blute, M.L.; Bostwick, D.G. Cancer volume of lymph node metastasis predicts progression in prostate cancer. *Am. J. Surg. Pathol.* **1998**, *22*, 1491–1500. [CrossRef] [PubMed]
3. Cheng, L.; Zincke, H.; Blute, M.L.; Bergstralh, E.J.; Scherer, B.; Bostwick, D.G. Risk of prostate carcinoma death in patients with lymph node metastasis. *Cancer* **2001**, *91*, 66–73. [CrossRef]
4. Wawroschek, F.; Wagner, T.; Hamm, M.; Weckermann, D.; Vogt, H.; Märkl, B.; Gordijn, R.; Harzmann, R. The influence of serial sections, immunohistochemistry, and extension of pelvic lymph node dissection on the lymph node status in clinically localized prostate cancer. *Eur. Urol.* **2003**, *43*, 132–137. [CrossRef]
5. Choo, M.S.; Kim, M.; Ku, J.H.; Kwak, C.; Kim, H.H.; Jeong, C.W. Extended versus standard pelvic lymph node dissection in radical prostatectomy on oncological and functional outcomes: A systematic review and meta-analysis. *Ann. Surg. Oncol.* **2017**, *24*, 2047–2054. [CrossRef]
6. Seiler, R.; Studer, U.E.; Tschan, K.; Bader, P.; Burkhard, F.C. Removal of limited nodal disease in patients undergoing radical prostatectomy: Long-term results confirm a chance for cure. *J. Urol.* **2014**, *191*, 1280–1285. [CrossRef]
7. Winter, A.; Henke, R.P.; Wawroschek, F. Targeted salvage lymphadenectomy in patients treated with radical prostatectomy with biochemical recurrence: Complete biochemical response without adjuvant therapy in patients with low volume lymph node recurrence over a long-term follow-up. *BMC Urol.* **2015**, *15*, 1–8. [CrossRef]
8. Preisser, F.; van den Bergh, R.C.N.; Gandaglia, G.; Ost, P.; Surcel, C.I.; Sooriakumaran, P.; Montorsi, F.; Graefen, M.; van der Poel, H.; de la Taille, A.; et al. Effect of extended pelvic lymph node dissection on oncologic outcomes in patients with D'Amico intermediate and high risk prostate cancer treated with radical prostatectomy: A multi-institutional study. *J. Urol.* **2020**, *203*, 338–343. [CrossRef] [PubMed]
9. Bivalacqua, T.J.; Pierorazio, P.M.; Gorin, M.A.; Allaf, M.E.; Carter, H.B.; Walsh, P.C. Anatomic extent of pelvic lymph node dissection: Impact on long-term cancer-specific outcomes in men with positive lymph nodes at time of radical prostatectomy. *Urology* **2013**, *82*, 653–659. [CrossRef]
10. Briganti, A.; Blute, M.L.; Eastham, J.H.; Graefen, M.; Heidenreich, A.; Karnes, J.R.; Montorsi, F.; Studer, U.E. Pelvic lymph node dissection in prostate cancer. *Eur. Urol.* **2009**, *55*, 1251–1265. [CrossRef]
11. Hinev, A.I.; Anakievski, D.; Kolev, N.H.; Hadjiev, V.I. Validation of nomograms predicting lymph node involvement in patients with prostate cancer undergoing extended pelvic lymph node dissection. *Urol. Int.* **2014**, *92*, 300–305. [CrossRef]

12. Mottet, N.; Cornford, P.; van den Bergh, R.C.N.; Briers, E.; Expert Patient Advocate (European Prostate Cancer Coalition/Europa UOMO); De Santis, M.; Gillessen, S.; Grummet, J.; Henry, A.M.; van der Kwast, T.H.; et al. EAU-EANM-ESTRO-ESUR-ISUP-SIOG Guidelines on Prostate Cancer. Available online: https://uroweb.org/guideline/prostate-cancer/ (accessed on 6 September 2021).

13. Gandaglia, G.; Ploussard, G.; Valerio, M.; Mattei, A.; Fiori, C.; Fossati, N.; Stabile, A.; Beauval, J.-B.; Malavaud, B.; Roumiguié, M.; et al. A novel nomogram to identify candidates for extended pelvic lymph node dissection among patients with clinically localized prostate cancer diagnosed with magnetic resonance imaging-targeted and systematic biopsies. *Eur. Urol.* **2019**, *75*, 506–514. [CrossRef]

14. Briganti, A.; Chun, F.K.H.; Salonia, A.; Suardi, N.; Gallina, A.; Da Pozzo, L.F.; Roscigno, M.; Zanni, G.; Valiquette, L.; Rigatti, P.; et al. Complications and other surgical outcomes associated with extended pelvic lymphadenectomy in men with localized prostate cancer. *Eur. Urol.* **2006**, *50*, 1006–1013. [CrossRef]

15. Musch, M.; Klevecka, V.; Roggenbuck, U.; Kroepfl, D. Complications of pelvic lymphadenectomy in 1,380 patients undergoing radical retropubic prostatectomy between 1993 and 2006. *J. Urol.* **2008**, *179*, 923–929. [CrossRef]

16. Winter, A.; Vogt, C.; Weckermann, D.; Wawroschek, F. Complications of pelvic lymphadenectomy in clinically localised prostate cancer: Different techniques in comparison and dependency on the number of removed lymph nodes. *Aktuel. Urol.* **2011**, *12*, 179–183. [CrossRef] [PubMed]

17. Wawroschek, F.; Vogt, H.; Weckermann, D.; Wagner, T.; Harzmann, R. The sentinel lymph node concept in prostate cancer-first results of gamma probe-guided sentinel lymph node identification. *Eur. Urol.* **1999**, *36*, 595–600. [CrossRef] [PubMed]

18. Fossati, N.; Willemse, P.-P.M.; Van den Broeck, T.; van den Bergh, R.C.N.; Yuan, C.Y.; Briers, E.; Bellmunt, J.; Bolla, M.; Cornford, P.; De Santis, M.; et al. The benefits and harms of different extents of lymph node dissection during radical prostatectomy for prostate cancer: A systematic review. *Eur. Urol.* **2017**, *72*, 84–109. [CrossRef]

19. Wit, E.M.K.; Acar, C.; Grivas, N.; Yuan, C.; Horenblas, S.; Liedberg, F.; Valdes Olmos, R.A.; van Leeuwen, F.W.B.; van den Berg, N.S.; Winter, A.; et al. Sentinel node procedure in prostate cancer: A systematic review to assess diagnostic accuracy. *Eur. Urol.* **2017**, *71*, 596–605. [CrossRef]

20. Cabanas, R.M. An approach for the treatment of penile carcinoma. *Cancer* **1977**, *39*, 456–466. [CrossRef]

21. Gould, E.A.; Winship, T.; Philbin, P.H.; Kerr, H.H. Observations on a "sentinel node" in cancer of the parotid. *Cancer* **1960**, *13*, 77–78. [CrossRef]

22. Van der Poel, H.; Wit, E.; Acar, C.; van den Berg, N.S.; van Leeuwen, F.W.B.; Valdés Olmos, R.A.; Winter, A.; Wawroschek, F.; Liedberg, F.; Maclennan, S.; et al. Sentinel Node Prostate Cancer Consensus Panel Group members. Sentinel node biopsy for prostate cancer: Report from a consensus panel meeting. *BJU Int.* **2017**, *120*, 204–211. [CrossRef]

23. Winter, A.; Engels, S.; Reinhardt, L.; Wasylow, C.; Gerullis, H.; Wawroschek, F. Magnetic marking and intraoperative detection of primary draining lymph nodes in high-risk prostate cancer using superparamagnetic iron oxide nanoparticles: Additional diagnostic value. *Molecules* **2017**, *22*, 2192. [CrossRef] [PubMed]

24. Winter, A.; Kowald, T.; Paulo, T.S.; Goos, P.; Engels, S.; Gerullis, H.; Schiffmann, J.; Chavan, A.; Wawroschek, F. Magnetic resonance sentinel lymph node imaging and magnetometer-guided intraoperative detection in prostate cancer using superparamagnetic iron oxide nanoparticles. *Int. J. Nanomed.* **2018**, *13*, 6689–6698. [CrossRef]

25. Joniau, S.; Van den Bergh, L.; Lerut, E.; Deroose, C.M.; Haustermans, K.; Oyen, R.; Budiharto, T.; Ameye, F.; Bogaerts, K.; Van Poppel, H. Mapping of pelvic lymph node metastases in prostate cancer. *Eur. Urol.* **2013**, *63*, 450–458. [CrossRef]

26. Douek, M.; Klaase, J.; Monypenny, I.; Kothari, A.; Zechmeister, K.; Brown, D.; Wyld, L.; Drew, P.; Garmo, H.; Agbaje, O.; et al. Sentinel node biopsy using a magnetic tracer versus standard technique: The SentiMAG multicentre trial. *Ann. Surg. Oncol.* **2014**, *21*, 1237–1245. [CrossRef]

27. Winter, A.; Woenkhaus, J.; Wawroschek, F. A novel method for intraoperative sentinel lymph node detection in prostate cancer patients using superparamagnetic iron oxide nanoparticles and a handheld magnetometer: The initial clinical experience. *Ann. Surg. Oncol.* **2014**, *21*, 4390–4396. [CrossRef] [PubMed]

28. Thill, M.; Kurylcio, A.; Welter, R.; van Haasteren, V.; Grosse, B.; Berclaz, G.; Polkowski, W.; Hauser, N. The Central-European SentiMag study: Sentinel lymph node biopsy with superparamagnetic iron oxide (SPIO) vs. radioisotope. *Breast* **2014**, *23*, 175–179. [CrossRef]

29. Geißen, W.; Engels, S.; Aust, P.; Schiffmann, J.; Gerullis, H.; Wawroschek, F.; Winter, A. Diagnostic accuracy of magnetometer-guided sentinel lymphadenectomy after intraprostatic injection of superparamagnetic iron oxide nanoparticles in intermediate- and high-risk prostate cancer using the magnetic activity of sentinel nodes. *Front. Pharmacol.* **2019**, *10*, 1123. [CrossRef] [PubMed]

30. Winter, A.; Kneib, T.; Wasylow, C.; Reinhardt, L.; Henke, R.-P.; Engels, S.; Gerullis, H.; Wawroschek, F. Updated nomogram incorporating percentage of positive cores to predict probability of lymph node invasion in prostate cancer patients undergoing sentinel lymph node dissection. *J. Cancer* **2017**, *8*, 2692–2698. [CrossRef]

31. Wawroschek, F.; Vogt, H.; Weckermann, D.; Wagner, T.; Hamm, M.; Harzmann, R. Radioisotope guided pelvic lymph node dissection for prostate cancer. *J. Urol.* **2001**, *166*, 1715–1719. [CrossRef]

32. Winter, A.; Engels, S.; Süykers, M.-C.; Henke, R.-P.; Wawroschek, F. Radioisotope guided sentinel lymph node dissection in prostate cancer: Rate of lymph node involvment depending on preoperative tumor characteristics in more than 2100 patients. *SM J. Urol.* **2015**, *1*, 1002.

33. Weingärtner, K.; Ramaswamy, A.; Bittinger, A.; Gerharz, E.W.; Voge, D.; Riedmiller, H. Anatomical basis for pelvic lymphadenectomy in prostate cancer: Results of an autopsy study and implications for the clinic. *J. Urol.* **1996**, *156*, 1969–1971. [CrossRef]

34. R Core Team. *R: A Language and Environment for Statistical Computing*; R Foundation for Statistical Computing: Vienna, Austria, 2021. Available online: https://www.r-project.org/ (accessed on 4 June 2021).

35. Holl, G.; Dorn, R.; Wengenmair, H.; Weckermann, D.; Sciuk, J. Validation of sentinel lymph node dissection in prostate cancer: Experience in more than 2,000 patients. *Eur. J. Nucl. Med. Mol. Imaging* **2009**, *36*, 1377–1382. [CrossRef]

36. Wawroschek, F.; Vogt, H.; Wengenmair, H.; Weckermann, D.; Hamm, M.; Keil, M.; Graf, G.; Heidenreich, P.; Harzmann, R. Prostate lymphoscintigraphy and radio-guided surgery for sentinel lymph node identification in prostate cancer. *Urol. Int.* **2003**, *70*, 303–310. [CrossRef] [PubMed]

37. Abdollah, F.; Suardi, N.; Gallina, A.; Bianchi, M.; Tutolo, M.; Passoni, N.; Fossati, N.; Sun, M.; dell'Oglio, P.; Salonia, A.; et al. Extended pelvic lymph node dissection in prostate cancer: A 20-year audit in a single center. *Ann. Oncol.* **2013**, *24*, 1459–1466. [CrossRef]

38. Briganti, A.; Chun, F.K.H.; Salonia, A.; Gallina, A.; Farina, E.; Da Pozzo, L.F.; Rigatti, P.; Montorsi, F.; Karakiewicz, P.I. Validation of a nomogram predicting the probability of lymph node invasion based on the extent of pelvic lymphadenectomy in patients with clinically localized prostate cancer. *BJU Int.* **2006**, *98*, 788–793. [CrossRef] [PubMed]

39. Briganti, A.; Larcher, A.; Abdollah, F.; Capitanio, U.; Gallina, A.; Suardi, N.; Bianchi, M.; Sun, M.; Freschi, M.; Salonia, A.; et al. Updated nomogram predicting lymph node invasion in patients with prostate cancer undergoing extended pelvic lymph node dissection: The essential importance of percentage of positive cores. *Eur. Urol.* **2012**, *61*, 480–487. [CrossRef] [PubMed]

40. Godoy, G.; Chong, K.T.; Cronin, A.; Vickers, A.; Laudone, V.; Touijer, K.; Guillonneau, B.; Eastham, J.A.; Scardino, P.T.; Coleman, J.A. Extent of pelvic lymph node dissection and the impact of standard template dissection on nomogram prediction of lymph node involvement. *Eur. Urol.* **2011**, *60*, 195–201. [CrossRef]

41. Grivas, N.; Wit, E.; Tillier, C.; van Muilekom, E.; Pos, F.; Winter, A.; van der Poel, H. Validation and head-to-head comparison of three nomograms predicting probability of lymph node invasion of prostate cancer in patients undergoing extended and/or sentinel lymph node dissection. *Eur. J. Nucl. Med. Mol. Imaging* **2017**, *44*, 2213–2226. [CrossRef]

42. Anninga, B.; White, S.H.; Moncrieff, M.; Dziewulski, P.; Geh, J.L.C.; Klaase, J.; Garmo, H.; Castro, F.; Pinder, S.; Pankhurst, Q.A.; et al. Magnetic technique for sentinel lymph node biopsy in melanoma: The MELAMAG trial. *Ann. Surg. Oncol.* **2016**, *23*, 2070–2078. [CrossRef]

43. Baena Fustegueras, J.A.; González, F.H.; Calderó, S.G.; de la Fuente Juárez, M.C.; López, S.R.; Riu, F.R.; Petit, N.M.; Álvarez, P.M.; Torelló, A.L.; Matias-Guiu, X.; et al. Magnetic detection of sentinel lymph node in papillary thyroid carcinoma: The MAGIC-PAT study results. *Eur. J. Surg. Oncol.* **2019**, *45*, 1175–1181. [CrossRef]

44. Hernando, J.; Aguirre, P.; Aguilar-Salvatierra, A.; Leizaola-Cardesa, I.O.; Bidaguren, A.; Gómez-Moreno, G. Magnetic detection of sentinel nodes in oral squamous cell carcinoma by means of superparamagnetic iron oxide contrast. *J. Surg. Oncol.* **2020**, *121*, 244–248. [CrossRef]

45. Murakami, K.; Kotani, Y.; Suzuki, A.; Takaya, H.; Nakai, H.; Matsuki, M.; Sato, T.; Mandai, M.; Matsumura, N. Superparamagnetic iron oxide as a tracer for sentinel lymph node detection in uterine cancer: A pilot study. *Sci. Rep.* **2020**, *10*, 1–9. [CrossRef]

46. Mottet, N.; Bellmunt, J.; Bolla, M.; Briers, E.; Cumberbatch, M.G.; De Santis, M.; Fossati, N.; Gross, T.; Henry, A.M.; Joniau, S.; et al. EAU-ESTRO-SIOG guidelines on prostate cancer. Part 1: Screening, diagnosis, and local treatment with curative intent. *Eur. Urol.* **2017**, *71*, 618–629. [CrossRef]

47. Morgan-Parkes, J.H. Metastases: Mechanisms, pathways, and cascades. *Am. J. Roentgenol.* **1995**, *164*, 1075–1082. [CrossRef] [PubMed]

48. Weckermann, D.; Dorn, R.; Holl, G.; Wagner, T.; Harzmann, R. Limitations of radioguided surgery in high-risk prostate cancer. *Eur. Urol.* **2007**, *51*, 1549–1558. [CrossRef] [PubMed]

49. Kroon, B.K.; Horenblas, S.; Estourgie, S.H.; Lont, A.P.; Valdés Olmos, R.A.; Nieweg, O.E. How to avoid false-negative dynamic sentinel node procedures in penile carcinoma. *J. Urol.* **2004**, *171*, 2191–2194. [CrossRef]

50. Leijte, J.A.P.; van der Ploeg, I.M.C.; Valdés Olmos, R.A.; Nieweg, O.E.; Horenblas, S. Visualization of tumor blockage and rerouting of lymphatic drainage in penile cancer patients by use of SPECT/CT. *J. Nucl. Med.* **2009**, *50*, 364. [CrossRef]

51. Mattei, A.; Fuechsel, F.G.; Bhatta Dhar, N.; Warncke, S.H.; Thalmann, G.N.; Krause, T.; Studer, U.E. The template of the primary lymphatic landing sites of the prostate should be revisited: Results of a multimodality mapping study. *Eur. Urol.* **2008**, *53*, 118–125. [CrossRef] [PubMed]

Article

A Propensity Score Matched Analysis of Superparamagnetic Iron Oxide versus Radioisotope Sentinel Node Biopsy in Breast Cancer Patients after Neoadjuvant Chemotherapy

Zuzanna Pelc [1,*], Magdalena Skórzewska [1], Maria Kurylcio [1], Tomasz Nowikiewicz [2], Radosław Mlak [3], Katarzyna Sędłak [1], Katarzyna Gęca [1], Karol Rawicz-Pruszyński [1], Wojciech Zegarski [2], Wojciech P. Polkowski [1] and Andrzej Kurylcio [1]

[1] Department of Surgical Oncology, Medical University of Lublin, Radziwiłłowska 13 St., 20-080 Lublin, Poland; magdalena.skorzewska@umlub.pl (M.S.); m.kurylcio@gmail.com (M.K.); sedlak.katarz@gmail.com (K.S.); kasiaa.geca@gmail.com (K.G.); karolrawiczpruszynski@uml.edu.pl (K.R.-P.); wojciechpolkowski@umlub.pl (W.P.P.); andrzej.kurylcio@umlub.pl (A.K.)

[2] Department of Clinical Breast Cancer and Reconstructive Surgery, Oncology Center, Prof. Lukaszczyk Memorial Hospital, Romanowskiej 2 St., 85-796 Bydgoszcz, Poland; tomasz.nowikiewicz@cm.umk.pl (T.N.); zegarskiw@cm.umk.pl (W.Z.)

[3] Department of Human Physiology, Medical University of Lublin, Radziwiłłowska 11 St., 20-080 Lublin, Poland; radoslaw.mlak@gmail.com

* Correspondence: zuzanna.torun@gmail.com; Tel.: +48-81-53-18-126

Citation: Pelc, Z.; Skórzewska, M.; Kurylcio, M.; Nowikiewicz, T.; Mlak, R.; Sędłak, K.; Gęca, K.; Rawicz-Pruszyński, K.; Zegarski, W.; Polkowski, W.P.; et al. A Propensity Score Matched Analysis of Superparamagnetic Iron Oxide versus Radioisotope Sentinel Node Biopsy in Breast Cancer Patients after Neoadjuvant Chemotherapy. *Cancers* **2022**, *14*, 676. https://doi.org/10.3390/cancers14030676

Academic Editors: Moriaki Kusakabe and Akihiro Kuwahata

Received: 10 January 2022
Accepted: 27 January 2022
Published: 28 January 2022

Publisher's Note: MDPI stays neutral with regard to jurisdictional claims in published maps and institutional affiliations.

Simple Summary: This Propensity Score Matched Analysis aimed to assess the efficacy of superparamagnetic iron oxide (SPIO) and radioisotope sentinel lymph node biopsy (SNLB) in breast cancer (BC) patients after neoadjuvant chemotherapy (NAC). One hundred and twenty-four patients were eligible for final analysis. In the SPIO group, the median of retrieved sentinel lymph nodes (SLNs) was significantly higher than in the RI group. The SPIO method was associated with a significantly higher chance of retrieving at least three SLNs when compared to the RI method. SPIO-guided SLNB allows efficient retrieval and detection of SLNs in BC patients after NAC when compared to RI.

Abstract: The standard method for nodal staging in breast cancer (BC) patients after neoadjuvant chemotherapy (NAC) is sentinel lymph node biopsy (SLNB) with a radioisotope (RI) injection. However, SLNB after NAC results in high false-negative rates (FNR), and the RI method is restricted by nuclear medicine unit dependency. These limitations resulted in the development of the superparamagnetic iron oxide (SPIO) method, reducing FNR and presenting a comparable detection rate. This bi-institutional cohort comparison study aimed to assess the efficacy of SPIO and radioisotope SNLB in BC patients after NAC using Propensity Score Matching (PSM) analysis. The study group comprised 508 patients who underwent SLNB after NAC for ycT1-4N0M0 BC between 2013 and 2021 in two high volume centers. Data were retrieved from prospectively conducted databases. In the SPIO group, the median of retrieved sentinel lymph nodes (SLNs) was significantly higher than in the RI group (3 vs. 2; $p < 0.0001$). The SPIO method was associated with a significantly higher chance of retrieving at least three lymph nodes when compared to the RI method (71% vs. 11.3%; $p < 0.0001$). None of the analyzed demographic and clinical variables had a statistically significant influence on the efficacy of SLNs retrieval in the RI group, while in the SPIO group, patients with $\geq$ three harvested SLNs had lower weight and decreased BMI. Based on this PSM analysis, SPIO-guided SLNB allowed the efficient retrieval and detection of SLNs in BC patients after NAC compared to RI.

Keywords: breast cancer; neoadjuvant chemotherapy; sentinel lymph node biopsy; superparamagnetic iron oxide; SPIO; radioisotope

1. Introduction

Sentinel lymph node biopsy (SLNB) is an established procedure for clinically negative nodes (cN0) in early breast cancer (BC) [1,2]. Since the end of the 20th century, the introduction of the SLNB developed a minimally invasive staging procedure for BC patients [3]. However, the surgical management of the axilla has been a matter of debate. Neoadjuvant chemotherapy (NAC) combined with personalized, targeted therapy results in a high pathologic complete response in the primary tumor [2]. Moreover, SLNB performed after NAC decreases the axillary lymph node dissection (ALND) rate, reducing adverse effects such as seroma, wound infections, or haemorrhage [4,5]. One of the factors restricting proper axillary mapping after neoadjuvant therapy includes the alteration of the lymphatic drainage due to fibrosis and obstruction of lymphatic vessels or the apoptosis of tumor cells [6]. For this reason, SLNB after NAC results in false-negative rates (FNR) varying from 10 to 30%, as shown in the SENTINA trial [7]. The current standard for nodal staging in BC patients after NAC is a radioisotope (RI) SLNB. However, this method also contains disadvantages such as nuclear medicine unit dependency or radiation exposure [8]. The existing drawbacks resulted in new, non-radioactive methods of sentinel lymph node (SLN) identification. In a recent meta-analysis, fluorescence-guided SLNB using indocyanine green (ICG) occurred to present non-inferior IR to the current standard [9]. In the randomized study, Jung et al. compared SLNB with ICG versus dual technique for BC patients after NAC, proving ICG to be a feasible method, with no statistically significant higher IR than RI-alone [10]. Superparamagnetic iron oxide (SPIO) with a handheld magnetometer presents a comparable detection rate as the RI combined with blue dye, known as the dual technique [11]. Moreover, the SPIO tracer reduces the FNR, possibly due to its peculiar, one-dimensional nanostructure and various physicochemical properties dependent on high intrinsic anisotropy and surface activity [12–14]. However, Corso et al. suggest a significant discrepancy between the FNR and detection rate depends on the individual experience of the research center or incoherent structure of retrospective studies [15]. Numerous trials and meta-analyses revealed the noninferiority of the SPIO to the gold standard isotope technique [16–27]. However, these studies did not include BC patients after NAC. Recently, we have determined that SPIO is a feasible and oncologically safe method for identifying SN in BC patients after preoperative treatment [17]. Therefore, the present study aimed to compare the identification rate (IR) of SPIO-guided SLNB to RI in BC patients after receiving neoadjuvant treatment based on the Propensity Score Matching (PSM) analysis.

2. Materials and Methods

2.1. Study Design

This bi-institutional cohort comparison study was performed based on the data retrieved from prospectively conducted databases. The study group comprised 508 patients who underwent SLNB after NAC for non-recurrent, non-metastatic ycT1-4N0M0 BC between 2013 and 2021 in two high volumes centers. After PSM analysis, 124 patients were eligible for analysis. Institutional review board approval (Bioethical Committee of the Medical University of Lublin, Ethic Code: Ke-0221-34-2013) was granted. PSM was performed to eliminate selection bias of clinicopathological data of enrolled patients (Figure 1).

2.2. Neoadjuvant Chemotherapy

Each patient was qualified for neoadjuvant treatment based on the multidisciplinary team decision. NAC was administered in accordance with national guidelines, depending on the clinical stage of the disease and molecular subtype. Four cycles of conventional or dose-dense AC (doxorubicin 75 mg/m^2 with cyclophosphamide 750 mg/m^2) followed by 12 cycles of weekly paclitaxel (80 mg/m^2) or triweekly docetaxel (100 mg/m^2) were the preferred regimen. Patients with primary BC with human epidermal growth factor receptor, two (HER2) protein overexpression, and/or HER2 gene amplification (HER2–positive) additionally received anti-HER2 therapy. The evaluation of the pathological tumor response to the NAC was performed according to Pinder classification [23]: response—complete pathologic response or <10% residual tumor/tumor bed; no response—$\geq$10% residual tumor/tumor bed.

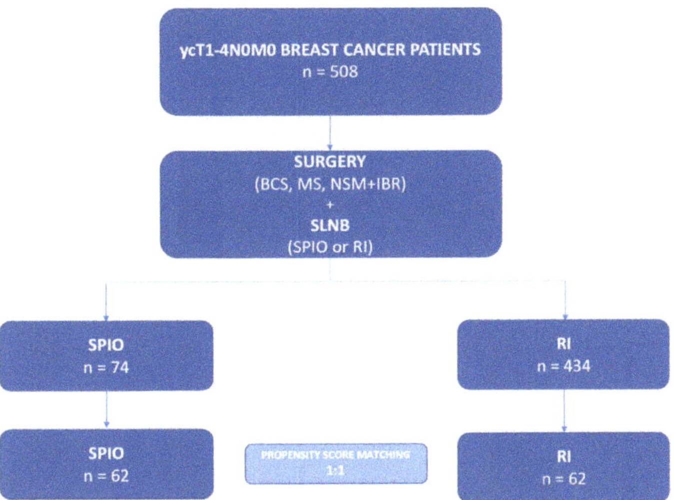

Figure 1. Flow chart of the Study Group. BCS—breast conserving surgery; MS—simple mastectomy, NSM—nipple-sparing mastectomy; IBR—immediate breast reconstruction; SLNB—sentinel lymph node biopsy; SPIO—superparamagnetic iron oxide; RI—radioisotope.

2.3. Sentinel Lymph Node Biopsy

In the RI method, the isotope 99mTc with an activity of 75–100 MBq was used and administered on an albumin carrier (Nanocol). About 2–3 h before the surgery, lymphoscintigraphy was performed. The radiotracer was administered as a periareolar intradermal injection into the lesion-specific breast quadrant. We used a manual gamma radiation detector (Crystal Prob GmBH, Berlin, Germany; Crystal Photonics GmBH, Berlin, Germany; NeoProbe, San Diego, California; Autosuture, Cornwalk, Conn) to perform intraoperative identification of areas of increased radiotracer capture within the axilla and measurements of radiation levels (hot spot).

A handheld magnetometer (SentiMag®, Sysmex Europe GmBH, Hamburg, Germany) was used in the SPIO method. This allowed for a non-radioactive detection and location of SLNs before their surgical retrieval. Sienna+® (Sysmex Europe GmBH, Hamburg, Germany) and Magtrace® (Endomagnetics Limited, Cambridge, UK) were used. SPIO was injected deeply into the subareolar interstitial tissue, followed by SentiMag® probe measurements as previously described [17].

All visualized SLNs were removed regardless of the method until the background signal was less than 10% of its highest value during SLNB. The method was assessed as efficient when retrieving at least three SLNs, confirmed in detailed histopathological reports. These assumptions are in accordance with the 2019 St. Galen Consensus Conference [18]. For the study purposes, the following definitions were adopted: SLN retrieval—intraoperative assessment and removal of at least three SLNs for further pathological analysis; SLN evaluation—the objective histopathological verification of separately submitted SLNs specimens.

2.4. Statistical Analysis

Data were analyzed using MedCalc Statistical Software version 20.009 (MedCalc Software, Ostend, Belgium). Due to the two-institutional character of the study, Propensity Score Matching (PSM) was used, enabling matching patients by their clinicopathological features before further analysis. Initially, the SPIO group counted 74 patients, whereas the RI group had 434 patients. After PSM, considering age, ypTN, and biological subtype, each group consisted of 62 patients. Due to the non-normal distribution of continuous data (assessed using the D'Agostino-Pearson test), non-parametric tests were used (Mann U–Whitney test) to compare demographic and clinical variables between SPIO and RI groups). The chi-square test was used to compare the distribution of individual variants of categorical variables in both studied groups. The odds ratio (OR) test was used to assess the chance of achieving the desired efficacy of a given method of SLNs detection/retrieval (evaluated by surgeon or pathologist) (SPIO vs. RI). In all cases, p-value < 0.05 was considered statistically significant. The results for which p had values were >0.05, and <0.06 was considered a trend towards significance.

3. Results

3.1. Characteristics and Comparison of the Study Groups

3.1.1. RI

The RI group consisted of 62 women with a median age of 52. The median BMI was 25.72. The overweight or obese women dominated (56.5%). The dominant clinical features were as follows: ypT0 (56.5%) and ypN0 (91.9%). The most common biological types were B1 (33.9%), TN (29%), B2 (22.6%), HER2 + (11.3%), and A (3.2%), respectively. Most patients had pathological tumor responses to NAC (69.4%). Almost half of the patients underwent breast conserving surgery (BCS) (48.4%). ALND was performed in all patients with lymph nodes metastases (9.7%).

3.1.2. SPIO

The SPIO group consisted of 62 women with a median age of 53.5 years. The median BMI was 25.97. Overweight or obese patients dominated (61.3%). ypT0 (56.5%), and ypN0 (91.9%) were dominant clinical features. The distribution of biological types in this group was identical to that in the RI group. Fifty-eight point one percent (58.1%) of the patients presented pathological tumor responses to NAC. Fifty-one point six percent (51.6%) of the patients underwent BCS. ALND was performed in all cases with lymph nodes metastases (8.1%). Detailed characteristics and comparisons of the study groups are presented in Table 1.

Table 1. Characteristics and comparison of the study groups.

Variable	RI	SPIO	p
	(n = 62)	(n = 62)	
Age (years) Median (interquartile range)	52 (44–61)	53.5 (43–62)	0.7910
Weight (kg) Median (interquartile range)	70 (62–75)	68 (58–77)	0.7413
BMI Median (interquartile range)	25.72 (23.31–29.33)	25.97 (21.60–28.63)	0.6601
BMI			
Underweight	2 (3.2%)	1 (1.6%)	
Healthy body weight	25 (40.3%)	23 (37.1%)	
Overweight	21 (33.9%)	27 (43.5%)	0.3933
Obese (Grade I)	10 (16.1%)	8 (12.9%)	
Obese (Grade II)	-	2 (3.2%)	
Obese (Grade III)	4 (6.5%)	1 (1.6%)	
Tumor highest diameter (mm)	19.5 (15–30)	25 (15–30)	0.4044

Table 1. *Cont.*

Variable	RI (*n* = 62)	SPIO (*n* = 62)	*p*
ypT			
0	36 (58.1%)	35 (56.5%)	
1	14 (22.6%)	15 (24.2%)	0.9760
2	12 (19.4%)	12 (19.4%)	
ypN			
Negative	57 (91.9%)	57 (91.9%)	0.7415
Positive	5 (8.1%)	5 (8.1%)	
ypTN			
T0N0	36 (58.1%)	36 (58.1%)	
T1N0	11 (17.7%)	11 (17.7%)	
T1N1	3 (4.8%)	3 (4.8%)	1.0000
T2N0	10 (16.1%)	10 (16.1%)	
T2N1	2 (3.2%)	2 (3.2%)	
Biological subtypes of cancer			
A	2 (3.2%)	2 (3.2%)	
B1	21 (33.9%)	21 (33.9%)	
B2	14 (22.6%)	14 (22.6%)	1.0000
HER2+	7 (11.3%)	7 (11.3%)	
TN	18 (29%)	18 (29%)	
NAC	62 (100%)	62 (100%)	1.0000
Response to NAC			
No response (small, medium)	19 (30.6%)	26 (41.9%)	0.2625
Response (high, complete)	43 (69.4%)	36 (58.1%)	
Type of surgery			
BCS	32 (51.6)	30 (48.4%)	
MRM	3 (4.8%)	1 (1.6%)	0.1419
MS	23 (37.1%)	19 (30.6%)	
NSM+IBR	4 (6.5%)	12 (19.4%)	
Site			
Left	34 (54.8%)	33 (53.2%)	1.0000
Right	28 (45.2%)	29 (46.8%)	
Margin			
R0	62 (100%)	58 (93.5%)	0.1273
R1	-	4 (6.5%)	
Lymphadenectomy			
No	56 (90.3%)	57 (91.9%)	1.0000
Yes	6 (9.7%)	5 (8.1%)	
ycSNB (retrieved)			
Median (interquartile range)	2 (2–2)	3 (2–4)	<0.0001 *
ypSNB (evaluated)			
Median (interquartile range)	3 (2–3)	4 (3–5)	0.0005 *
ypSN			
Negative	57 (91.9%)	57 (91.9%)	0.7415
Positive	5 (8.1%)	5 (8.1%)	

BMI—body mass index; RI—radioisotope; SPIO—superparamagnetic iron oxide; ypT—post neoadjuvant therapy T stage; ypN—post neoadjuvant therapy N stage; ypTN—post neoadjuvant therapy T and N stage; A—luminal A; B1– luminal B HER2 negative; B2—luminal B HER2 positive; HER2+—human epithelial receptor-positive; TN—triple-negative; NAC—neoadjuvant chemotherapy; BCS—breast conserving surgery; MRM—modified radical mastectomy; MS—simple mastectomy; NSM—nipple-sparing mastectomy; IBR—immediate breast reconstruction; R0—radical microscopic margin; R1—non-radical microscopic margin; ycSNB—post neoadjuvant therapy clinical sentinel lymph node biopsy number; ypSNB—post neoadjuvant therapy pathological sentinel lymph node biopsy number; ypSN—post neoadjuvant therapy pathological sentinel lymph node number; *—statistically significant result.

3.2. Comparison of SLNs Detection Efficacy Depending on the Method Implemented (RI vs. SPIO)

In the SPIO group, the median of retrieved SLNs was significantly higher than in the RI group (3 vs. 2, $p < 0.0001$). Similarly, the median of evaluated SLNs in the SPIO group was significantly higher than in the RI group (four vs. three, $p = 0.0005$). The SPIO method was associated with a significant, over 19-fold higher chance of retrieving at least three SLNs when compared to the RI method (71% vs. 11.3%; OR = 19.21; 95% CI: 7.36–50.10; $p < 0.0001$). Moreover, the SPIO method was associated with a significant, over 3-fold higher chance of evaluation of at least three SLNs when compared to the RI method (71.4% vs. 51.6%; OR = 3.21; 95% CI: 1.48–6.98; $p = 0.0032$). Detailed comparisons of the effectiveness of SLNs retrieval in both groups are shown in Table 2.

Table 2. Comparison of LN's retrieval efficacy in RI and SPIO methods.

Variable	RI	SPIO	OR [95%CI]
	(*n* = 62)	(*n* = 62)	*p*
SLN retrieval			
<3 retrieved SLNs	55 (88.7%)	18 (29%)	19.21 [7.36–50.10]
≥3 retrieved SLNs	7 (11.3%)	44 (71%)	<0.0001 *
SLN evaluation			
<3 evaluated SLNs	30 (48.4%)	14 (22.6%)	3.21 [1.48–6.98]
≥3 evaluated SLNs	32 (51.6%)	48 (77.4%)	0.0032 *
Efficacy of positive SLNs detection			
Negative	57 (91.9%)	57 (91.9%)	1.00 [0.30–3.28]
Positive	5 (8.1%)	5 (8.1%)	1.0000
IR of SLNs retrieval			
Undetected SLNs	-	-	0.33 [0.01–8.21]
Detected SLNs	62 (100%)	62 (100%)	0.4974
IR of SLNs evaluation			
Undetected SLNs	-	-	0.19 [0.01–4.42]
Detected SLNs	62 (100%)	62 (100%)	0.2924

SLNs—sentinel lymph nodes; RI—radioisotope; SPIO—superparamagnetic iron oxide; IR—identification rate; OR—odds ratio; *—statistically significant result.

In the SPIO group, patients with SLN harvest ≥ 3 had lower weight (median: 66 vs. 77 kg; $p = 0.0280$) and lower BMI (25.55 vs. 28.26 kg/m^2; $p = 0.0323$). None of the analyzed demographic and clinical variables had a statistically significant influence on the efficacy of SLNs retrieval in the RI group. Detailed data on the influence of selected demographic and clinical variables on the efficacy of SLNs retrieval in the RI and SPIO groups are presented in Table 3.

Table 3. Influence of selected demographic and clinical variables on the efficacy of SLNs detection/retrieval (based on surgeon evaluation) using the RI or SPIO method.

Variable	RI (*n* = 62)		$p^{\,a}$ or OR (95%CI)	SPIO (*n* = 62)		$p^{\,a}$ or OR (95%CI)
	SLNs Retrieval Efficacy			SLNs Retrieval Efficacy		
	<3 Retrieved SLNs	≥3 Retrieved SLNs	$p^{\,b}$	$p^{\,b}$	≥3 Retrieved SLNs	$p^{\,b}$
Age (years) Median (interquartile range)	53 (44–62)	50 (44–58)	0.4763	52 (43–62)	55 (43–62)	0.6194
Weight (kg) Median (interquartile range)	69 (60–75)	70 (67–85)	0.4227	77 (63–85)	66 (57–74)	0.0280 *
BMI Median (interquartile range)	25.51 (23.09–28.20)	29.33 (25.19–31.93)	0.1391	28.26 (23.23–30.84)	25.55 (20.98–27.51)	0.0323 *

Table 3. *Cont.*

Variable	RI (*n* = 62)		*p* [a] or OR (95%CI)	SPIO (*n* = 62)		*p* [a] or OR (95%CI)
	SLNs Retrieval Efficacy			SLNs Retrieval Efficacy		
	<3 Retrieved SLNs	≥3 Retrieved SLNs	*p* [b]	*p* [b]	≥3 Retrieved SLNs	*p* [b]
BMI Underweight or Healthy body weight	25 (92.6%)	2 (7.4%)	2.08 [0.40–9.20]	5 (20.8%)	19 (79.2%)	0.51 [0.15–1.67]
Overweight or Obese (classes 1–3)	30 (85.7%)	5 (14.3%)	0.4039	13 (34.2%)	25 (65.8%)	0.2626
Tumor highest diameter (mm)	20.5 (15–30)	18.5 (12–27.5)	1	25 (20–35)	20 (13.5–28.7)	0.0889
ypT 0	33 (91.7%)	3 (8.3%)	2.00 [0.41–9.82]	10 (28.6%)	25 (71.4%)	0.95 (0.31–2.87)
1 or 2	22 (84.6%)	4 (15.4%)	0.3932	8 (29.6%)	19 (70.4%)	0.9275
ypN Negative	52 (91.2%)	5 (8.8%)	6.93 [0.93–51.79]	18 (31.6%)	39 (68.4%)	5.15 [0.27–98.7]
Positive	3 (60%)	2 (40%)	0.0591 ˆ	-	5 (100%)	0.2756
Biological subtype of cancer A, B1, B2, HER2+	39 (88.6%)	5 (11.4%)	0.97 (0.17–5.55)	13 (29.5%)	31 (70.5%)	1.09 (0.32–3.68) 0.8893
TN	16 (88.9%)	2 (11.1%)	0.9772	5 (27.8%)	13 (72.2%)	
Biological subtype of cancer A, B1, B2, TN	49 (89.1%)	6 (10.9%)	1.36 (0.14–13.31)	17 (30.9%)	38 (69,1%)	2.68 (0.30–24.05) 0.3775
HER2+	6 (85.7%)	1 (14.3%)	0.791	1 (14.3%)	6 (85.7%)	
Response to NAC No response	16 (84.2%)	3 (15.8%)	1.83 (0.37–9.11)	8 (30.8%)	18 (69.2%)	1.16 (0.38–3.50)
Response	39 (90.7%)	4 (9.3%)	0.4616	10 (27.8%)	26 (72.2%)	0.798

RI—radioisotope; SPIO—superparamagnetic iron oxide; OR—odds ratio; *p* [a]—chi-square test result; *p* [b]—odds ratio test result; SLNs—sentinel lymph nodes; BMI—body mass index; ypT—post neoadjuvant therapy pathological T stage; ypN—post neoadjuvant therapy pathological N stage; A—luminal A; B1—luminal B human epithelial receptor-negative; B2—luminal B human epithelial receptor-positive; HER2+—human epithelial receptor-positive; TN—triple-negative; NAC—neoadjuvant chemotherapy *—statistically significant result; ˆ—a trend into statistically significant result.

None of the analyzed demographic and clinical variables had a statistically significant influence on the efficacy of SLNs evaluation in the RI and SPIO groups. Detailed data on the influence of selected demographic and clinical variables on the efficacy of SLNs evaluation in the RI and SPIO groups are shown in Table S1.

4. Discussion

To the best of our knowledge, apart from the recent IMAGINE study, this is the first research that compared RI versus SPIO for SLNB after NAC in BC patients [28]. In both studies, the SPIO method was compared to the RI-alone.

Our results indicate that retrieved and evaluated SLNs' median was significantly higher in the SPIO group. Higher levels of detected SLNs in the magnetic tracer group were observed in the present and IMAGINE studies. However, the median number of detected SLNs in the latter study ranged from 1.3 to 1.4. In order to perform a proper evaluation of SLNB after systemic treatment and decrease the risk of FNR, we aimed to assess at least three SLNs [20,21,29,30]. As shown in ACOSOG Z1071 trial, SLN IR and FNR were approximate between SLNB before and after NAC in patients with cN0 [29]. Notably, the sensitivity of the assessment increased with the retrieval of two or more SLNs, supporting the recommendations to perform SLNB in the post-neoadjuvant setting.

Since AMAROS' trial results, the gold standard for SLNB in BC patients remains the RI technique, which presents a comparable detection rate to the dual technique [12,19,31]. Numerous studies confirmed the noninferiority of SPIO to RI in the upfront surgery setting [16,17,24,25,32]. Since 2014, when Rubio et al. presented promising data on the outcome of SLNB after NAC using a dual tracer (SPIO-TC99) at the ASCO annual meeting [33], the results of the SENTINAC-01 clinical trial are awaited [34]. The study primary and secondary outcomes measures FNR and detection rate, respectively. After neoadjuvant treatment, BC patients are randomized into three groups, with SLNB guided by dual technique, combined RI + SPIO or SPIO alone. Our previous study established SPIO-guided SLNB after NAC as a safe and feasible method [17].

Surprisingly, Aksoy et al. indicated that SLNB after NAC did not influence overall survival or disease-free survival, underlining the necessity of adequate NAC response assessment [35]. However, the evaluation of SLNs after NAC provides more precise predictions of patients' response to systemic therapy and residual disease severity [36], indicating that SLNB remains an essential tool for evaluating systemic treatment [36]. Jatoi et al. concluded that SLNB after NAC presented disadvantages such as decreased IR and increased FNR due to the lymphatic drainage alteration [37]. However, in the SPIO group of the present study, IR and FNR were 98.4% and 0%, respectively. These results support the hypothesis that SPIO seems to be an optimal tracer for SLNB after NAC in BC patients due to high SLN retrieval number and low FNR compared with the conventional methods [22]. This conclusion was confirmed by Mok et al., suggesting SPIO-guided SLBN improves the clinical value of SLNB with similar accuracy but avoids irradiation or risks of allergy to blue dye [8,23]. Furthermore, SLNBs using the Resovist magnetic nanoparticles and a new handheld, lightweight magnetic probe in BC patients were considered equivalent to the conventional RI method, as shown in a recent multicenter Japanese trial [24].

There are several limitations associated with the use of magnetic-guided surgery, including limited depth of detection and restricted use of metal instruments [38]. Moreover, since the magnetometer must seek out a small tracer collection point, the detection of SLNs by SPIO may be limited. In contrast, the RI probe can detect the radiation beam directly from SLNs [39]. Peek et al. investigated injection characteristics of the magnetic tracer on the rat model, proving that iron uptake appears to be proportional to the injection dose before reaching a plateau level [40]. This outcome suggests that each lymph node presents maximum magnetic particles load depending on the size and location of the SLN. In order to increase the iron uptake, the injection preferably should be performed one day before the SLNB, rather than directly prior to the surgery.

The RI method is also associated with specific limitations such as short half-life limiting the timeframe of SLNB and dependency on the availability of nuclear medicine units or disturbed SLN detection. Additionally, the magnetic tracer is related to no radiation exposure, easier implementation, longer retention in the SLNs, and more convenient workflow than the dual technique [30]. Although SPIO-guided SLNB seems to be an equivalent method to RI, the standardization of the axillary nodal management in BC patients after NAC is warranted.

The presented study has some limitations. Despite PSM analysis, the associated selection bias may exist. The databases were conducted since 2013 when the guidelines did not specify the necessity of retrieval and detection of at least three SLNs during SLNB.

5. Conclusions

In patients with BC after NAC, SLNB using magnetic technique allows high IR of SLNs and may result in more efficient retrieval and detection of at least three SLNs compared to RI. Considering the increasing role of preoperative chemotherapy and noninferiority of SPIO to the current standard, further studies establishing ferromagnetic assessment of SLNB after NAC are indicated.

Supplementary Materials: The following supporting information can be downloaded at https://www.mdpi.com/article/10.3390/cancers14030676/s1; Table S1: Influence of selected demographic and clinical variables on the efficacy of LNs evaluation in the RI and SPIO groups.

Author Contributions: Conceptualization, Z.P., M.S., M.K., K.R.-P., and T.N.; methodology, R.M., K.S., and K.G.; software, R.M., K.S., K.G., and K.R.-P.; validation, Z.P., M.S., M.K., and T.N.; formal analysis, R.M., K.S., and K.R.-P.; investigation, Z.P., M.S., A.K., and K.R.-P.; data curation, Z.P., M.S., M.K., K.R.-P., and K.G.; writing—original draft preparation, Z.P., K.R.-P., M.S., and M.K.; writing—review and editing, A.K., W.Z., and W.P.P.; supervision, A.K., W.Z., and W.P.P. All authors have read and agreed to the published version of the manuscript.

Funding: This research received no external funding.

Institutional Review Board Statement: The study was conducted in accordance with the Declaration of Helsinki and approved by Bioethical Committee of the Medical University of Lublin (Ethic Code: Ke-0221-34-2013).

Informed Consent Statement: Informed consent was obtained from all subjects involved in the study.

Data Availability Statement: The data presented in this study are available on request from the corresponding author.

Conflicts of Interest: The authors declare no conflict of interest.

References

1. Banys-Paluchowski, M.; Gasparri, M.L.; de Boniface, J.; Gentilini, O.; Stickeler, E.; Hartmann, S.; Thill, M.; Rubio, I.T.; Di Micco, R.; Bonci, E.A.; et al. Surgical management of the axilla in clinically node-positive breast cancer patients converting to clinical node negativity through neoadjuvant chemotherapy: Current status, knowledge gaps, and rationale for the EUBREAST-03 AXSANA study. *Cancers* **2021**, *13*, 1565. [CrossRef] [PubMed]
2. Curigliano, G.; Burstein, H.J.; Winer, E.P.; Gnant, M.; Dubsky, P.; Loibl, S.; Colleoni, M.; Regan, M.M.; Piccart-Gebhart, M.; Senn, H.J.; et al. De-escalating and escalating treatments for early-stage breast cancer: The St. Gallen International Expert Consensus Conference on the Primary Therapy of Early Breast Cancer 2017. *Ann. Oncol.* **2017**, *28*, 1700–1712. [CrossRef] [PubMed]
3. Cirocchi, R.; Amabile, M.I.; De Luca, A.; Frusone, F.; Tripodi, D.; Gentile, P.; Tabola, R.; Pironi, D.; Forte, F.; Monti, M.; et al. New classifications of axillary lymph nodes and their anatomical-clinical correlations in breast surgery. *World J. Surg. Oncol.* **2021**, *19*, 93. [CrossRef] [PubMed]
4. King, T.A.; Morrow, M. Surgical issues in patients with breast cancer receiving neoadjuvant chemotherapy. *Nat. Rev. Clin. Oncol.* **2015**, *12*, 335–343. [CrossRef] [PubMed]
5. De Luca, A.; Tripodi, D.; Frusone, F.; Leonardi, B.; Cerbelli, B.; Botticelli, A.; Vergine, M.; D'Andrea, V.; Pironi, D.; Sorrenti, S.; et al. Retrospective evaluation of the effectiveness of a synthetic glue and a fibrin-based sealant for the prevention of seroma following axillary dissection in breast cancer patients. *Front. Oncol.* **2020**, *10*, 1061. [CrossRef]
6. Classe, J.-M.; Bordes, V.; Campion, L.; Mignotte, H.; Dravet, F.; Leveque, J.; Sagan, C.; Dupre, P.F.; Body, G.; Giard, S. Sentinel lymph node biopsy after neoadjuvant chemotherapy for advanced breast cancer: Results of ganglion sentinelle et chimiothérapie neoadjuvante, a French prospective multicentric study. *J. Clin. Oncol.* **2009**, *27*, 726–732. [CrossRef] [PubMed]
7. Kuehn, T.; Bauerfeind, I.; Fehm, T.; Fleige, B.; Hausschild, M.; Helms, G.; Lebeau, A.; Liedtke, C.; von Minckwitz, G.; Nekljudova, V.; et al. Sentinel-lymph-node biopsy in patients with breast cancer before and after neoadjuvant chemotherapy (SENTINA): A prospective, multicentre cohort study. *Lancet Oncol.* **2013**, *14*, 609–618. [CrossRef]
8. Ahmed, M.; Purushotham, A.D.; Douek, M. Novel techniques for sentinel lymph node biopsy in breast cancer: A systematic review. *Lancet Oncol.* **2014**, *15*, e351–e362. [CrossRef]
9. Kedrzycki, M.S.; Leiloglou, M.; Ashrafian, H.; Jiwa, N.; Thiruchelvam, P.T.R.; Elson, D.S.; Leff, D.R. Meta-analysis comparing fluorescence imaging with radioisotope and blue dye-guided sentinel node identification for breast cancer surgery. *Ann. Surg. Oncol.* **2021**, *28*, 3738–3748. [CrossRef]
10. Jung, S.-Y.; Han, J.H.; Park, S.J.; Lee, E.-G.; Kwak, J.; Kim, S.H.; Lee, M.H.; Lee, E.S.; Kang, H.-S.; Lee, K.S.; et al. The sentinel lymph node biopsy using indocyanine green fluorescence plus radioisotope method compared with the radioisotope-only method for breast cancer patients after neoadjuvant chemotherapy: A prospective, randomized, open-label, single-center phase 2 trial. *Ann. Surg. Oncol.* **2019**, *26*, 2409–2416. [CrossRef]
11. Warnberg, F.; Stigberg, E.; Obondo, C.; Olofsson, H.; Abdsaleh, S.; Warnberg, M.; Karakatsanis, A. Long-term outcome after retro-areolar versus peri-tumoral injection of superparamagnetic iron oxide nanoparticles (SPIO) for sentinel lymph node detection in breast cancer surgery. *Ann. Surg. Oncol.* **2019**, *26*, 1247–1253. [CrossRef] [PubMed]
12. Thill, M.; Kurylcio, A.; Welter, R.; van Haasteren, V.; Grosse, B.; Berclaz, G.; Polkowski, W.; Hauser, N. The Central-European SentiMag study: Sentinel lymph node biopsy with superparamagnetic iron oxide (SPIO) vs. radioisotope. *Breast* **2014**, *23*, 175–179. [CrossRef] [PubMed]
13. Burlizzi, S.; Giacobbe, F.; Ranieri, E.; Stasolla, S.; Villanucci, A.; D'Amuri, A.; Niccoli, A. Supermagnetic iron oxide tracer in association with radioisotope, for sentinel node biopsy in patients with complete axillary response, after neoadjuvant chemoterapy: A single center, prospective study. *J. Surg. Res.* **2021**, *4*, 465–472.
14. Wu, W.; Wu, Z.; Yu, T.; Jiang, C.; Kim, W.-S. Recent progress on magnetic iron oxide nanoparticles: Synthesis, surface functional strategies and biomedical applications. *Sci. Technol. Adv. Mater.* **2015**, *16*, 023501. [CrossRef] [PubMed]
15. Corso, G.; Scalzi, A.M.D.; Vicini, E.; Morigi, C.; Veronesi, P.; Galimberti, V. Sentinel lymph node biopsy management after neoadjuvant treatment for breast cancer care. *Future Oncol.* **2018**, *14*, 1423–1426. [CrossRef]

16. Alvarado, M.D.; Mittendorf, E.A.; Teshome, M.; Thompson, A.M.; Bold, R.J.; Gittleman, M.A.; Beitsch, P.D.; Blair, S.L.; Kivilaid, K.; Harmer, Q.J.; et al. SentimagIC: A non-inferiority trial comparing superparamagnetic iron oxide versus technetium-99m and blue dye in the detection of axillary sentinel nodes in patients with early-stage breast cancer. *Ann. Surg. Oncol.* **2019**, *26*, 3510–3516. [CrossRef]
17. Kurylcio, A.; Pelc, Z.; Skorzewska, M.; Rawicz-Pruszynski, K.; Mlak, R.; Geca, K.; Sedlak, K.; Kurylcio, P.; Malecka-Massalska, T.; Polkowski, W. Superparamagnetic iron oxide for identifying sentinel lymph node in breast cancer after neoadjuvant chemotherapy: Feasibility study. *J. Clin. Med.* **2021**, *10*, 3149. [CrossRef]
18. Balic, M.; Thomssen, C.; Wurstlein, R.; Gnant, M.; Harbeck, N. St. Gallen/Vienna 2019: A brief summary of the consensus discussion on the optimal primary breast cancer treatment. *Breast Care* **2019**, *14*, 103–110. [CrossRef]
19. Straver, M.E.; Meijnen, P.; van Tienhoven, G.; van de Velde, C.J.; Mansel, R.E.; Bogaerts, J.; Duez, N.; Cataliotti, L.; Klinkenbijl, J.H.; Westenberg, H.A.; et al. Sentinel node identification rate and nodal involvement in the EORTC 10981-22023 AMAROS trial. *Ann. Surg. Oncol.* **2010**, *17*, 1854–1861. [CrossRef]
20. Cardoso, F.; Kyriakides, S.; Ohno, S.; Penault-Llorca, F.; Poortmans, P.; Rubio, I.T.; Zackrisson, S.; Senkus, E.; ESMO Guidelines Commettee. Early breast cancer: ESMO Clinical Practice Guidelines for diagnosis, treatment and follow-updagger. *Ann. Oncol.* **2019**, *30*, 1194–1220. [CrossRef]
21. Gradishar, W.J.; Anderson, B.O.; Abraham, J.; Aft, R.; Agnese, D.; Allison, K.H.; Blair, S.L.; Burstein, H.J.; Dang, C.; Elias, A.D.; et al. Breast cancer, version 3.2020, NCCN clinical practice guidelines in oncology. *J. Natl. Compr. Cancer Netw.* **2020**, *18*, 452–478. [CrossRef] [PubMed]
22. Qiu, S.Q.; Zhang, G.J.; Jansen, L.; de Vries, J.; Schroder, C.P.; de Vries, E.G.E.; van Dam, G.M. Evolution in sentinel lymph node biopsy in breast cancer. *Crit. Rev. Oncol. Hematol.* **2018**, *123*, 83–94. [CrossRef] [PubMed]
23. Mok, C.W.; Tan, S.M.; Zheng, Q.; Shi, L. Network meta-analysis of novel and conventional sentinel lymph node biopsy techniques in breast cancer. *BJS Open* **2019**, *3*, 445–452. [CrossRef] [PubMed]
24. Taruno, K.; Kurita, T.; Kuwahata, A.; Yanagihara, K.; Enokido, K.; Katayose, Y.; Nakamura, S.; Takei, H.; Sekino, M.; Kusakabe, M. Multicenter clinical trial on sentinel lymph node biopsy using superparamagnetic iron oxide nanoparticles and a novel handheld magnetic probe. *J. Surg. Oncol.* **2019**, *120*, 1391–1396. [CrossRef] [PubMed]
25. Houpeau, J.L.; Chauvet, M.P.; Guillemin, F.; Bendavid-Athias, C.; Charitansky, H.; Kramar, A.; Giard, S. Sentinel lymph node identification using superparamagnetic iron oxide particles versus radioisotope: The French Sentimag feasibility trial. *J. Surg. Oncol.* **2016**, *113*, 501–507. [CrossRef]
26. Douek, M.; Klaase, J.; Monypenny, I.; Kothari, A.; Zechmeister, K.; Brown, D.; Wyld, L.; Drew, P.; Garmo, H.; Agbaje, O.; et al. Sentinel node biopsy using a magnetic tracer versus standard technique: The SentiMAG Multicentre Trial. *Ann. Surg. Oncol.* **2014**, *21*, 1237–1245. [CrossRef]
27. Ghilli, M.; Carretta, E.; Di Filippo, F.; Battaglia, C.; Fustaino, L.; Galanou, I.; Di Filippo, S.; Rucci, P.; Fantini, M.P.; Roncella, M. The superparamagnetic iron oxide tracer: A valid alternative in sentinel node biopsy for breast cancer treatment. *Eur. J. Cancer Care* **2017**, *26*, e12385. [CrossRef]
28. Giménez-Climent, J.; Marín-Hernández, C.; Fuster-Diana, C.A.; Torró-Richart, J.A.; Navarro-Cecilia, J. Sentinel lymph node biopsy in breast cancer after neoadjuvant therapy using a magnetic tracer versus standard technique: A multicentre comparative non-inferiority study (IMAGINE-II). *Int. J. Surg. Open* **2021**, *35*, 100404. [CrossRef]
29. Boughey, J.C.; Suman, V.J.; Mittendorf, E.A.; Ahrendt, G.M.; Wilke, L.G.; Taback, B.; Leitch, A.M.; Kuerer, H.M.; Bowling, M.; Flippo-Morton, T.S.; et al. Sentinel lymph node surgery after neoadjuvant chemotherapy in patients with node-positive breast cancer: The ACOSOG Z1071 (Alliance) clinical trial. *JAMA* **2013**, *310*, 1455–1461. [CrossRef]
30. Jassem, J.; Krzakowski, M. Breast cancer. *Oncol. Clin. Pract.* **2018**, *14*, 171–215. [CrossRef]
31. Mamounas, E.P.; Brown, A.; Anderson, S.; Smith, R.; Julian, T.; Miller, B.; Bear, H.D.; Caldwell, C.B.; Walker, A.P.; Mikkelson, W.M.; et al. Sentinel node biopsy after neoadjuvant chemotherapy in breast cancer: Results from National Surgical Adjuvant Breast and Bowel Project Protocol B-27. *J. Clin. Oncol.* **2005**, *23*, 2694–2702. [CrossRef] [PubMed]
32. Rubio, I.T.; Diaz-Botero, S.; Esgueva, A.; Rodriguez, R.; Cortadellas, T.; Cordoba, O.; Espinosa-Bravo, M. The superparamagnetic iron oxide is equivalent to the Tc99 radiotracer method for identifying the sentinel lymph node in breast cancer. *Eur. J. Surg. Oncol.* **2015**, *41*, 46–51. [CrossRef] [PubMed]
33. Franceschini, G.; Di Leone, A.; Sanchez, A.M.; D'Archi, S.; Terribile, D.; Magno, S.; Scardina, L.; Masetti, R. Update on sentinel lymph node biopsy after neoadjuvant chemotherapy in breast cancer patient. *Ann. Ital. Chir.* **2020**, *91*, 465–468. [PubMed]
34. ClinicalTrials.gov. Available online: https://clinicaltrials.gov/ct2/show/NCT02249208 (accessed on 10 January 2022).
35. Aksoy, S.O.; Sevinc, A.; Ünal, M.; Balci, P.; Görkem, İ.B.; Durak, M.G.; Ozer, O.; Bekiş, R.; Emir, B. Management of the axilla with sentinel lymph node biopsy after neoadjuvant chemotherapy for breast cancer: A single-center study. *Medicine* **2020**, *99*, e23538. [CrossRef] [PubMed]
36. Erdahl, L.M.; Boughey, J.C. Use of sentinel lymph node biopsy to select patients for local-regional therapy after neoadjuvant chemotherapy. *Curr. Breast Cancer Rep.* **2014**, *6*, 10–16. [CrossRef] [PubMed]
37. Jatoi, I.; Benson, J.R.; Toi, M. De-escalation of axillary surgery in early breast cancer. *Lancet Oncol.* **2016**, *17*, e430–e441. [CrossRef]

38. Kurita, T.; Taruno, K.; Nakamura, S.; Takei, H.; Enokido, K.; Kuwayama, T.; Kanada, Y.; Akashi-Tanaka, S.; Matsuyanagi, M.; Hankyo, M.; et al. Magnetically guided localization using a Guiding-Marker System® and a handheld magnetic probe for nonpalpable breast lesions: A multicenter feasibility study in Japan. *Cancers* **2021**, *13*, 2923. [CrossRef] [PubMed]
39. Makita, M.; Manabe, E.; Kurita, T.; Takei, H.; Nakamura, S.; Kuwahata, A.; Sekino, M.; Kusakabe, M.; Ohashi, Y. Moving a neodymium magnet promotes the migration of a magnetic tracer and increases the monitoring counts on the skin surface of sentinel lymph nodes in breast cancer. *BMC Med. Imaging* **2020**, *20*, 58. [CrossRef] [PubMed]
40. Peek, M.C.L.; Saeki, K.; Ohashi, K.; Chikaki, S.; Baker, R.; Nakagawa, T.; Kusakabe, M.; Douek, M.; Sekino, M. Optimization of SPIO injection for sentinel lymph node dissection in a rat model. *Cancers* **2021**, *13*, 5031. [CrossRef]

MDPI

St. Alban-Anlage 66

4052 Basel

Switzerland

Tel. +41 61 683 77 34

Fax +41 61 302 89 18

www.mdpi.com

Cancers Editorial Office

E-mail: cancers@mdpi.com

www.mdpi.com/journal/cancers